Cancer Vaccines as Immunotherapy of Cancer

AF784210

Cancer Vaccines as Immunotherapy of Cancer

Edited by

Luigi Buonaguro
Innovative Immunological Models Unit,
National Cancer Institute "Pascale", Naples, Italy

Sjoerd Van Der Burg
Experimental Cancer Immunology and Therapy Institute,
Oncode Institute, Department of Medical Oncology,
Leiden University Medical Center, Leiden, The Netherlands

ELSEVIER

ACADEMIC PRESS
An imprint of Elsevier

Academic Press is an imprint of Elsevier
125 London Wall, London EC2Y 5AS, United Kingdom
525 B Street, Suite 1650, San Diego, CA 92101, United States
50 Hampshire Street, 5th Floor, Cambridge, MA 02139, United States
The Boulevard, Langford Lane, Kidlington, Oxford OX5 1GB, United Kingdom

Copyright © 2022 Elsevier Inc. All rights reserved.

No part of this publication may be reproduced or transmitted in any form or by any means, electronic or mechanical, including photocopying, recording, or any information storage and retrieval system, without permission in writing from the publisher. Details on how to seek permission, further information about the Publisher's permissions policies and our arrangements with organizations such as the Copyright Clearance Center and the Copyright Licensing Agency, can be found at our website: www.elsevier.com/permissions.

This book and the individual contributions contained in it are protected under copyright by the Publisher (other than as may be noted herein).

Notices
Knowledge and best practice in this field are constantly changing. As new research and experience broaden our understanding, changes in research methods, professional practices, or medical treatment may become necessary.

Practitioners and researchers must always rely on their own experience and knowledge in evaluating and using any information, methods, compounds, or experiments described herein. In using such information or methods they should be mindful of their own safety and the safety of others, including parties for whom they have a professional responsibility.

To the fullest extent of the law, neither the Publisher nor the authors, contributors, or editors, assume any liability for any injury and/or damage to persons or property as a matter of products liability, negligence or otherwise, or from any use or operation of any methods, products, instructions, or ideas contained in the material herein.

ISBN: 978-0-12-823901-8

For Information on all Academic Press publications
visit our website at https://www.elsevier.com/books-and-journals

Publisher: Stacy Masucci
Acquisitions Editor: Rafael E. Teixeira
Editorial Project Manager: Samantha Allard
Production Project Manager: Kiruthika Govindaraju
Cover Designer: Miles Hitchen

Typeset by MPS Limited, Chennai, India

Working together
to grow libraries in
developing countries
www.elsevier.com • www.bookaid.org

Contents

List of contributors

Juraj Adamik Parker Institute for Cancer Immunotherapy, University of California San Francisco, Microbiology and Immunology, San Francisco, CA, United States

Luigi Aurisicchio Evvivax s.r.l., Rome, Italy; Takis s.r.l., Rome, Italy; NeoMatrix s. r.l., Rome, Italy

John C. Bell Cancer Therapeutics Program, Ottawa Hospital Research Institute, Ottawa, ON, Canada

Marie-Claude Bourgeois-Daigneault "Centre Hospitalier de l'Université de Montréal" CRCHUM, Montreal, QC, Canada; "Institut du cancer de Montréal", Montreal, QC, Canada; Department of Microbiology, Infectious Diseases and Immunology, Faculty of Medicine, University of Montreal, Montreal, QC, Canada

Lisa H. Butterfield Parker Institute for Cancer Immunotherapy, University of California San Francisco, Microbiology and Immunology, San Francisco, CA, United States

Vincenzo Cerullo Laboratory of Immunovirotherapy, Drug Research Program, Faculty of Pharmacy, University of Helsinki, Helsinki, Finland; TRIMM, Translational Immunology Research Program, University of Helsinki, Helsinki, Finland; iCAN Digital Precision Cancer Medicine Flagship, University of Helsinki, Helsinki, Finland; Department of Molecular Medicine and Medical Biotechnology and CEINGE, Naples University Federico II, Naples, Italy

Cheryl L.-L. Chiang Department of Oncology, Lausanne University Hospital (CHUV), Ludwig Institute for Cancer Research, University of Lausanne, Lausanne, Switzerland

Gennaro Ciliberto Scientific Direction, IRCCS Regina Elena National Cancer Institute, Rome, Italy

Antonella Conforti Evvivax s.r.l., Rome, Italy; Takis s.r.l., Rome, Italy

Olivera J. Finn Department of Immunology, University of Pittsburgh School of Medicine, Pittsburgh, PA, United States

Maria Cristina Galli National Centre for the Control and Evaluation of Medicines, Italian National Institute of Health, Rome, Italy

Sine Reker Hadrup Section for Experimental and Translational Immunology, Department of Health Technology, The Technical University of Denmark, Kgs Lyngby, Denmark

Duane H. Hamilton Laboratory of Tumor Immunology and Biology, Center for Cancer Research, National Cancer Institute, National Institutes of Health, Bethesda, MD, United States

Jenci Hawthorne Department of Obstetrics, Gynecology, and Women's Health, University of Louisville School of Medicine, Louisville, KY, United States

Hidetomo Himuro Kanagawa Cancer Center Research Institute, Kanagawa, Japan

Camille Jacqueline Department of Immunology, University of Pittsburgh School of Medicine, Pittsburgh, PA, United States

Caroline Jochems Laboratory of Tumor Immunology and Biology, Center for Cancer Research, National Cancer Institute, National Institutes of Health, Bethesda, MD, United States

Mohammad Kadivar Section for Experimental and Translational Immunology, Department of Health Technology, The Technical University of Denmark, Kgs Lyngby, Denmark

Lana E. Kandalaft Ludwig Institute for Cancer Research, University of Lausanne, Lausanne, Switzerland; Center of Experimental Therapeutics, Department of Oncology, Lausanne University Hospital (CHUV), Lausanne, Switzerland

Cornelis J.M. Melief ISA Pharmaceuticals, Oegstgeest, The Netherlands

Paul V. Munson Parker Institute for Cancer Immunotherapy, University of California San Francisco, Microbiology and Immunology, San Francisco, CA, United States

Filomena Nappi National Centre for the Control and Evaluation of Medicines, Italian National Institute of Health, Rome, Italy

Fabio Palombo Takis s.r.l., Rome, Italy; NeoMatrix s.r.l., Rome, Italy

Hans-Georg Rammensee Institute for Cell Biology, Department of Immunology, University of Tübingen, Tübingen, Germany; Cluster of Excellence iFIT (EXC2180) "Image-Guided and Functionally Instructed Tumor Therapies," University of Tübingen, Tübingen, Germany; German Cancer Consortium (DKTK) and German Cancer Research Center (DKFZ), Partner Site Tübingen, Tübingen, Germany

Tetsuro Sasada Kanagawa Cancer Center Research Institute, Kanagawa, Japan

Jeffrey Schlom Laboratory of Tumor Immunology and Biology, Center for Cancer Research, National Cancer Institute, National Institutes of Health, Bethesda, MD, United States

Haval Shirwan Department of Child Health, Molecular Microbiology and Immunology, School of Medicine, University of Missouri, Columbia, MO, United States

Maria Tagliamonte Laboratory of Innovative Immunological Models, National Cancer Institute "Fond. Pascale" - IRCCS, Naples, Italy

Esmé T.I. van der Gracht ISA Pharmaceuticals, Oegstgeest, The Netherlands

Anna-Sophia Wiekmeijer ISA Pharmaceuticals, Oegstgeest, The Netherlands

Erkko Ylösmäki Laboratory of Immunovirotherapy, Drug Research Program, Faculty of Pharmacy, University of Helsinki, Helsinki, Finland; TRIMM, Translational Immunology Research Program, University of Helsinki, Helsinki, Finland

Esma Yolcu Department of Child Health, Molecular Microbiology and Immunology, School of Medicine, University of Missouri, Columbia, MO, United States

About the editors

Luigi Buonaguro, MD, is Director of the Innovative Immunological Models Unity at National Cancer Institute "Pascale" in Italy. His primary interest is the identification of novel target tumor antigens, their optimization for improving antigenicity and immunogenicity, as well as a combination with immunoregulatory drugs. In particular, Dr. Buonaguro is coordinating the development of a novel therapeutic cancer vaccine for liver cancer. Currently, he is focusing on the impact of the antimicrobial memory T-cell immunity in cancer development and progression.

Sjoerd Van Der Burg, PhD, is Head of the Experimental Cancer Immunology and Therapy program, Senior Investigator at the Oncode Institute, Department of Medical Oncology, Leiden University Medical Center, The Netherlands. His primary interest is the exploration of key factors in host-tumor interactions that determine successes and failures in the immunotherapy of cancer. Dr. Van Der Burg co-developed the synthetic long peptide vaccine concept, in particular for HPV-induced cancers, and currently is developing a vaccine to stimulate the immune response to TAP-impaired tumors.

Luigi Buonaguro[1] and Sjoerd Van Der Burg[2]
[1]Innovative Immunological Models Unit, National Cancer Institute "Pascale", Naples, Italy, [2]Experimental Cancer Immunology and Therapy Institute, Oncode Institute, Department of Medical Oncology, Leiden University Medical Center, Leiden, The Netherlands

Preface

Immunotherapy has established itself as one of the main pillars of cancer therapy. In the last years, immunotherapy is dominated by checkpoint inhibitors and adoptive cell therapy approaches. However, many patients do not benefit from checkpoint therapy as they are only effective when a preexistent tumor-reactive T-cell response is present. In addition, adoptive cell therapy for solid tumors is hampered by problems to sustain the levels of activated T cells in vivo. In this framework, the cancer vaccine field is redeveloping itself to fill this void. Indeed, a vaccine is the only strategy able to amplify and/or reactivate the pool of Type 1 cytokine-oriented tumor-reactive T cells needed for tumor control. Undeniable the development of therapeutic cancer vaccines has come a long way. Several different scientific achievements on the identification of the first human tumor antigen, the selection of peptides by HLA, and how this could be used for stimulation of T cells in vivo, spurred the rational design of vaccines against cancer in the past. Unfortunately, a series of clinical trials showing limited efficacy have severely reduced the enthusiasm for this form of immunotherapy. Now therapeutic cancer vaccines are back. New insights in the regulation of activated T cells, the need to activate both CD4 and CD8 T cells, and the impact of the hostile tumor micro and macro environments drive the development of strong vaccine platforms which can be used in combination treatment strategies to optimally exploit their anticancer potential. New trials have shown the first signs of success and light the path forward. This book covers a number of important aspects in the development of clinically useful vaccines, including the different vaccine platforms, the tumor antigens that can be targeted, the adjuvants that can be used, and potential combination strategies. The insights and views of the renowned experts in this field turn it into a reference book for the clinical oncologists, scientists, and students for the coming years.

Chapter 1

Oncolytic viruses for antigen delivery

Erkko Ylösmäki[1,2], Vincenzo Cerullo[1,2,3,4], John C. Bell[5] and Marie-Claude Bourgeois-Daigneault[6,7,8]

[1]Laboratory of Immunovirotherapy, Drug Research Program, Faculty of Pharmacy, University of Helsinki, Helsinki, Finland, [2]TRIMM, Translational Immunology Research Program, University of Helsinki, Helsinki, Finland, [3]iCAN Digital Precision Cancer Medicine Flagship, University of Helsinki, Helsinki, Finland, [4]Department of Molecular Medicine and Medical Biotechnology and CEINGE, Naples University Federico II, Naples, Italy, [5]Cancer Therapeutics Program, Ottawa Hospital Research Institute, Ottawa, ON, Canada, [6]"Centre Hospitalier de l'Université de Montréal" CRCHUM, Montreal, QC, Canada, [7]"Institut du cancer de Montréal", Montreal, QC, Canada, [8]Department of Microbiology, Infectious Diseases and Immunology, Faculty of Medicine, University of Montreal, Montreal, QC, Canada

Introduction

Cancer immunotherapies have led to dramatic improvements in patient outcomes and have revolutionized cancer treatment. With the objective of redirecting a patient's own immune system against cancer, immunotherapeutic approaches exploit the natural ability of the immune system to eliminate foreign pathogens or, in the case of cancer, altered self. The various immunotherapeutic treatment strategies aim at inducing tumor-specific adaptive immune cells, such as T cells, as well as alleviating immune suppression. Various strategies are being explored for the treatment of different cancers. As such, anticancer vaccines aim at immunizing a patient against specific antigens that are found in cancer[1]; chimeric antigen receptor-T cells are engineered T cells that are redirected toward cancer cells by the expression of a surface receptor that gets activated upon binding to a molecule that is found specifically at the surface of cancer cells[2]; adoptive cell therapy is an approach by which antitumor immune cells are amplified from blood or tumor samples and reinfused to the patient in large numbers[3]; bispecific T-cell engagers are engineered molecules that bridge killer T cells and tumor cells together to allow for tumor cell killing[4]; and oncolytic viruses (OVs) are viruses that have the ability of specifically killing tumor cells while also

Cancer Vaccines as Immunotherapy of Cancer. DOI: https://doi.org/10.1016/B978-0-12-823901-8.00018-2
© 2022 Elsevier Inc. All rights reserved.

inducing antitumor immunity.[5–8] All of these strategies provide an army of immune cells with the capacity to readily and specifically attack cancer cells. However, to avoid immune attacks and persist, malignant tumor cells evolve to have an immune-refractory microenvironment. Mechanisms of dampening antitumor immunity include immune suppression by regulatory T cells, myeloid-derived suppressor cells, and antiinflammatory M2 macrophages, as well as the surface expression of immune checkpoint inhibitory ligands which, when they interact with their corresponding receptors at the surface of immune cells, inhibit their function. Treatment modalities, such as immune checkpoint inhibitors (anti-PD-L1, -PD1, and -CTLA-4 blocking antibodies), have been developed to prevent these interactions, release the full power of antitumor immunity, and allow for better immune-mediated tumor control.[9] With an effective tumor-specific adaptive immune response in place, immune cells can directly eliminate tumor cells and cancer reemergence is kept in check, therefore minimizing relapse. In this chapter, we review the evidence as well as the importance of OV-induced antitumor immunity and discuss different strategies that have been developed to improve this response (see Fig. 1.1 for various vaccination strategies used with OVs).

Antitumor immunity induced by OVs

Various mechanisms of action have been described as being important for OV-induced antitumor immunity. OV infection leads to the local production of various chemokines by tumor cells as well as tumor-infiltrating immune cells, which in turn recruit more immune cells to the tumor microenvironment.[10,11] Many OVs have also been shown to induce immunogenic cell death, an immune-stimulating type of death in which danger signals are released in the tumor microenvironment to allow for immune activation.[8,12–15] OV infection also triggers the production of proinflammatory cytokines including antiviral interferons (IFNs) by infected tumor cells as well as local immune cells, which activate nearby innate and adaptive immune cells.[16] Finally, tumor cell killing by OVs leads to the release of tumor antigen in the context of a highly immune-stimulating environment containing danger signals, such as pathogen- and danger-associated molecular patterns that confer the costimulatory signals required for the activation of antigen-presenting cells and the initiation of adaptive immune responses.[17] These various mechanisms by which OVs promote antitumor immunity have been extensively reviewed recently[18–20] and will therefore not be the focus of this chapter.

The importance of the immune system for the full efficacy of OV therapy is now well recognized. Key early findings include the observation that OVs were more effective in immune-competent compared to immune-deficient mice engrafted with the same tumor cell lines. Most of the evidence of OV-induced antitumor immunity arise from experiments conducted in animal

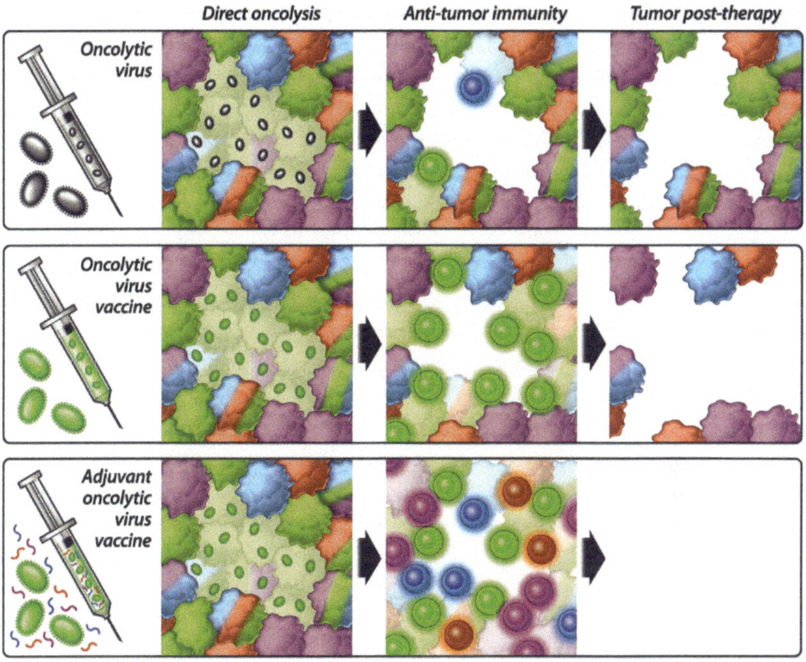

FIGURE 1.1 Illustration of different vaccination approaches using oncolytic viruses. Tumor cells are shown in different colors to illustrate cancer heterogeneity. Most of the cells share the green antigen, but some cells lost expression. The top panel represents the use of an oncolytic virus for cancer therapy. The virus directly infects and kills cancer cells while inducing a moderate antitumor immune response. Some cancer cells remain after treatment. The middle panel illustrates the use of oncolytic virus vaccines (oncolytic viruses that are modified to encode a cancer antigen: the green antigen). Treatment with oncolytic virus vaccines leads to direct oncolysis, as well as an important response directed against the antigen of choice. In this case, the cancer cells that lost expression of the targeted antigen persist after treatment. The bottom panel represents the use of oncolytic viruses as vaccination adjuvants. In this case, an oncolytic virus vaccine is mixed with antigenic peptides corresponding to multiple cancer-specific mutations. The virus directly destroys cancer cells and the vaccine induces a powerful immune response that targets both the shared cancer antigen and additional cancer-specific mutations. In this last scenario, cancer cells are completely eliminated by the treatment.

cancer models or cell culture systems. In sharp contrast with this abundance of data in animals, in-human data on this topic are rather limited. Nevertheless, one study by Andtbacka et al. has found in-patient evidence of OV-induced antitumor immunity.[21] In this study, the authors administered the oncolytic herpes simplex virus (HSV) Talimogene Laherparepvec (TVec) intratumorally to patients with melanoma. While only a fraction of the visible tumors were injected, the authors observed that 15% of the noninjected measurable lesions reduced in size after treatment. Furthermore, many patients developed vitiligo, an autoimmune condition in which patches of skin lose their pigmentation as a result of an immune reaction against melanocytes. Importantly, many patients with melanoma who undergo immunotherapy develop vitiligo,[22] which is a sign of immune-activity against melanoma cells, and it has been described as a consequence that is predictive of successful outcome.[23] TVec is the commercial name of a HSV-1 virus with both copies of the virulence factor ICP34.5 deleted, which confers specificity for cancer cells, and the gene coding for the granulocyte-macrophage colony-stimulating factor (GM-CSF) inserted into the viral genome.[24] Proficient infection thus leads to the local expression of GM-CSF, a cytokine that promotes the proliferation and maturation of many immune cell types. To date, TVec remains the only OV approved by the FDA for the treatment of patients with cancer in the United States and is also approved in Europe. While there is, to our knowledge, no other study that directly demonstrates the induction of antitumor immunity following OV therapy, other groups have reported evidence of improved immune activation in OV-treated patients. As such, a phase I trial conducted in patients with advanced cancer who received oncolytic Reovirus (Reo) treatment showed an increase in both CD4 and CD8 T cells and NK cells with the OV.[25] Another example is the oncolytic Measles virus (MV), which increased intratumoral T-cell infiltration as well as the production of proinflammatory cytokines in patients with T-cell lymphoma.[26] A myriad of clinical studies testing different OVs in all kinds of patient cohorts are currently underway (see Table 1.1) and data obtained from these trials should soon provide more information on the importance of the antitumor immune response in patients with cancer.

Targeting tumor antigens for cancer treatment

One strategy to induce antitumor immunity is to directly vaccinate against tumor antigens and all kinds of vaccine platforms are currently being explored for cancer treatment. As such, vaccination with tumor antigen-coding RNA, which allows for the production and presentation of tumor antigens by antigen-presenting cells,[27–29] blood-purified dendritic cells loaded with tumor cell lysates,[30] tumor cell lysate-loaded proinflammatory matrix polymers[31] or immune-stimulating vaccine adjuvants,[32] such as the cytokines interleukin (IL)-2,[33] aluminum salts (alum),[34] oil-in-water emulsion,[35] TLR-3-simulating polyI:C (viral RNA mimic), or TLR-5-stimulating bacterial flagellin[36] mixed with either synthetic antigenic peptides or recombinant proteins, are all being explored as possible vaccine formulations.

For an anticancer immune response to be effective, it has to be directed against elements that are found specifically in malignant cells: tumor antigens. Otherwise, the immune response would not only eliminate cancer cells but would also target healthy tissues and causes damaging autoimmune reactions. The tumor antigens can either be shared among several patients with cancer and types [tumor-associated antigen (TAA)] or specific to a given cancer [tumor-specific antigen (TSA) or neoantigen].[37]

TABLE 1.1 Currently ongoing clinical trials with OVs.

OV	Transgene	In combination with	Indication	Clinical phase	Identifier
TVec (HSV-1)	GM-CSF	Atezolizumab	Early breast cancer	Exploratory study	NCT03802604
TVec	GM-CSF	Pembrolizumab	Metastatic and/or locally advanced sarcoma	Phase II	NCT03069378
TVec	GM-CSF	Pembrolizumab	Melanoma	Phase II	NCT04068181
TVec	GM-CSF	Atezolizumab	Triple negative breast cancer and colorectal cancer with liver metastases	Phase Ib	NCT03256344
TVec	GM-CSF	Pembrolizumab	Melanoma	Phase II	NCT02965716
TVec	GM-CSF	Nivolumab	Sarcoma	Phase II	NCT03886311
TVec	GM-CSF	Pembrolizumab	Melanoma	Phase II	NTC03842943
TVec	GM-CSF	Pembrolizumab	Liver tumors	Phase Ib/II	NCT02509507
Pexa-Vec; JX-594 (VV)	GM-CSF	Nivolumab	Hepatocellular carcinoma	Phase I/IIa	NCT03071094
Pexa-Vec; JX-594	GM-CSF	Ipilimumab	Metastatic/advanced solid tumors	Phase I	NCT02977156
Pexa-Vec; JX-594	GM-CSF	Durvalumab and/or tremelilumab	Colorectal cancer	Phase I/II	NCT03206073
DNX-2401 (Ad)	None	Pembrolizumab	Brain cancers	Phase II	NCT02798406
ONCOS-102 (Ad)	GM-CSF	Pemetrexed/cisplatin (carboplatin)	Malignant pleural mesothelioma	Phase II	NCT02879669

(*Continued*)

TABLE 1.1 (Continued)

OV	Transgene	In combination with	Indication	Clinical phase	Identifier
ONCOS-102	GM-CSF	Pembrolizumab	Melanoma	Phase I	NCT03003676
ONCOS-102	GM-CSF	Durvalumab	Advanced peritoneal malignancies	Phase I/II	NCT02963831
Ad-MAGE A3 (Ad) MG1-MAGEA3 (MRB)	Melanoma-associated antigen 3	Pembrolizumab	Melanoma or cutaneous squamous cell carcinoma	Phase Ib	NCT03773744
Ad-MAGEA3 MG1-MAGEA3	Melanoma-associated antigen 3	Pembrolizumab	Nonsmall-cell lung cancer	Phase I/II	NCT02879760
TBio-6517 (VV)	Flt3 ligand, anti-CTLA-4 antibody, and IL-12	Pembrolizumab	Triple negative breast cancer or microsatellite stable colorectal cancer	Phase I/IIa	NCT04301011
ASP9801 (VV)	IL-7 and IL-12	–	Metastatic/advanced solid tumors	Phase I	NCT03954067
LOAd703 (Ad)	CD40L and 4–1BBL	Gemcitabine and nab-paclitaxel +/-atezolizumab	Pancreatic cancer	Phase I/IIa	NCT02705196
OH2 (HSV-2)	GM-CSF	–	Pancreatic cancer	Phase Ib/II	NCT04637698
TILT-123 (Ad)	TNFα and IL-2	–	Advanced solid tumors	Phase I	NCT04695327
TILT-123	TNFα and IL-2	Adoptive T-cell therapy	Metastatic melanoma	Phase I	NCT04217473

HSV-1, Herpes simplex virus 1; *VV*, vaccinia virus; *Ad*, adenovirus; *MRB*, Maraba virus; *HSV-2*, herpes simplex virus 2; *GM-CSF*, granulocyte-macrophage colony-stimulating factor; *Flt3 ligand*, FMS-related receptor tyrosine kinase 3 ligand; *CTLA-4*, cytotoxic T-lymphocyte-associated protein 4; *IL*, interleukin; *CD40L*, CD40 antigen ligand; *4–1BBL*, receptor 4–1BB ligand; *TNFα*, tumor necrosis factor-alpha.

When an immune response is directed against a single antigen, it results in a selective pressure that favors the evolution of escape variants, especially if the given antigen is not essential for the survival of the cells. Escape variants lacking the expression of antigens against which immune cells are directed emerge through a process known as cancer immunoediting.[38] Given that they lack expression of the target antigen, such variants are resistant to the mounted antitumor immune response and this escape from immune recognition allows for tumor outgrowth. To avoid such scenario, a vaccine can be designed to target multiple antigens, and TSAs are well-suited targets for such an approach. Indeed, every cancer has a collection of mutations that are different from self-proteins and are therefore only found in cancer cells. The collection of cancer and patient-unique mutations is called the tumor mutanome and can vary even between lesions found in a same patient or cancer cells from the same tumor mass.[39] Given that every cancer has a mutanome, vaccines targeting patient-specific TSAs could theoretically be used to treat all patients with cancer.

OVs confer neoantigen immunity

The various OVs have different abilities at blocking or stimulating inflammation and inducing antitumor immune responses. For example, the matrix protein of vesicular stomatitis virus (VSV) has the capacity to block messenger RNA export from the nucleus, which prevents the production and secretion of antiviral IFNs and other cytokines by infected cells and therefore limits local inflammation.[40] While this mechanism is beneficial for virus persistence, it would certainly impair the optimal induction of antitumor immunity. An oncolytic mutant of VSV for which a methionine of the matrix protein is deleted (Δ51) no longer possesses this function and therefore allows for IFN production. This not only confers specificity for tumor cells with defects in the IFN pathway but also leads to more local inflammation in the tumor bed, which benefits the therapy.[41] Another example is VV, which encodes many viral antiinflammatory proteins, such as B18R, a soluble type I IFN receptor that is secreted by infected cells to prevent the autocrine and paracrine antiviral effects of IFNs.[42] While B18R allows for the virus to replicate more, it prevents the local activation of immune cells. Notably, the immune-suppressive effects of B18R can even be harnessed to shelter the replication of other viruses, such as VSV in the context of VV and VSV coinfection or by encoding the antiinflammatory protein directly into VSV.[43] Also, HSV encodes antiviral proteins, such as ICP34.5, ICP0, and US3, which inhibit TLR stimulation and IFN production. To properly stimulate an immune response, the oncolytic mutants of the virus are often deleted for these genes and also encode immune-activating cytokines.[44]

In a recent study, we compared Ad, VSV, MRB, Reo, and HSV in their capacities at inducing long-term protection in a tumor-rechallenge model of breast cancer where tumors are treated with the different viruses intratumorally and secondary untreated tumors are seeded upon surgical resection of the primary tumors. We found the different OVs to have different abilities at controlling the primary versus the secondary tumors. For instance, Reo, VSV, Ad, and MRB could efficiently control the primary tumor upon direct injection while HSV had no effect. However, HSV, VSV, Ad, and MRB could all control secondary tumor growth without direct treatment while previous Reo administration had no effect in this context, suggesting that it may not induce potent antitumor immune responses despite being able to control the primary tumor.[45] More work is required to compare the immune-stimulating properties of the different OV candidates and identify viruses that are optimal at both directly killing cancers and inducing immune-mediated tumor control.

Many studies have demonstrated that OV treatment induces immunity against cancer neoantigens. In most cases, this response was improved by the insertion of immune-stimulating transgenes in the viral genomes or by combination with immune checkpoint blockade. For example, an early study demonstrated that the expression of IL-15 by oncolytic VSV greatly improved the anti-AH1 (a CT26 colon carcinoma antigen) immune response induced by the virus.[46] Another study published by our group showed that splenocytes from 4T1 breast tumor-bearing mice treated with oncolytic VSV encoding IFNγ could more efficiently recognize and react against 4T1 cells ex vivo compared to splenocytes from mice treated with the parental virus.[47] Interestingly, a VSV variant encoding IFNβ could induce anti-AH1 immunity in CT26 tumor-bearing mice, as well as anti-gp100, a B16F10 antigen, immunity in B16F10 melanoma tumor-bearing mice.[48] Many other cytokine encoding OVs have been shown to have superior therapeutic benefits compared to the parental viruses. A few examples are: a Newcastle disease virus (NDV) encoding IL-7,[49] a HSV-1 encoding IL-4,[50] VSV,[51] Ad,[52] and HSV-1[53] encoding IL-12, VSV encoding cluster of differentiation 40 ligand (CD40L),[54] and MV encoding IFNβ.[55] These different viruses were shown to induce more cytotoxic T cells and fewer regulatory T cells, increase the activity of NK cells, and allow for the generation of memory antitumor CD8 T cells. In another study, oncolytic Maraba virus (MRB) administered intratumorally was shown to induce anti-4T1 breast tumor CD8 T cells and this antitumor immune response was dependent on type I IFNs.[56] Finally, an oncolytic HSV variant targeting HER2 via an affibody inserted into the ICP10 gene could efficiently induce an immune response against MHCII neoepitopes using the CT26-HER2 cancer model.[57] Taken together, these studies demonstrate that OVs, especially OVs engineered to have enhanced immune-stimulating abilities, can induce antitumor immunity even in the absence of active vaccination.

OVs in combination therapies

The combination of OVs and immune checkpoint inhibitors is also a promising approach to augment antitumor responses induced by OVs. The group of

James Allison was the first to publish a very important study that investigated the combination in 2014. Using the B16F10 tumor models, the authors showed that tumor-infiltrating lymphocytes from oncolytic NDV-treated mice were reactive against B16F10 cell lysates and that the immune cells produced more IFNγ if the animals also received anti-CTLA-4 antibodies.[58] Several other studies have since followed to demonstrate similar findings with various OVs.[59] For instance, an oncolytic MV engineered to encode antibodies against PD-L1 or CTLA-4 had improved therapeutic efficacy compared to the parental virus when used in immune competent mice.[60] Also, the combination of VSV-IFNβ with immune checkpoint blockade (either anti-PD-L1 or anti-CTLA-4) improved the magnitude of the antitumor immune response compared to the virus treatment only.[48] Similarly, the intratumoral injection of the oncolytic HSV TVec could induce antitumor immunity in two different murine cancer models (A20 and CT26) and the percentage of cells specific to the tumor antigens was further increased by the combination with anti-CTLA-4 antibodies.[61] In another study, the combination of oncolytic MRB virus with both anti-PD1 and anti-CTLA-4 antibodies dramatically improved treatment efficacy.[56] Moreover, oncolytic Reo combined with an anti-PD1 could control B16F10 tumors by activating both NK cells and CD8 T cells.[62] Similarly, Vaccinia virus (VV) combined with anti-CTLA-4 blocking antibodies could control Renca murine renal tumors.[63] Also, while oncolytic VV expressing GM-CSF could trigger a modest but detectable antitumor immune response, a variant encoding GM-CSF as well as an inhibitor of PD-L1 administered intratumorally to MC38 colon carcinoma tumor-bearing mice could efficiently induce an immune response against five MC38 tumor neoantigens.[64] Furthermore, the authors found that tumor-specific IFNγ-secreting memory CD8 T cells were detected in all animals that received the VV-GM-CSF-PD-L1 inhibitor virus. Another study by Groeneveldt et al. found that treatment of immune-cold tumors with a replication competent Reovirus could efficiently attract T cells to B16F10 melanoma and KPC3 pancreatic cancers in mice, which enhanced the efficacy of bispecific T-cell engagers.[65] Taken together, these preclinical studies show that the combination of OVs and immune checkpoint blockade is a very promising approach for cancer treatment and many OV-immune checkpoint inhibitor combinations are currently being tested in the clinic (Table 1.1).

OV-based anticancer vaccines

One way to ensure the induction of a powerful antitumor immune response against a chosen antigen is to use OVs as vaccination platforms.[66] Indeed, while OV infection in the tumor niche activates antitumor immunity, it is not possible to control which tumor antigen is targeted by this response and the success of such passive vaccination relies, at least in part, on the virus successfully reaching and infecting the tumor cells. Therefore tumor antigens that are poorly expressed or that are found only in a small fraction of cancer cells can become the main target of the antitumor immune response and result in incomplete tumor control. Given the intrinsic heterogeneity in viral sensitivities observed in both mouse and human cancers, it is extremely difficult to predict which tumor will be sensitive to infection by a given virus and the most resistant cancers might not benefit from neither viral oncolysis nor OV-induced immune responses. By using OVs as anticancer vaccines, the virus-resistant cancers can still benefit from a powerful antitumor immune response while the virus-sensitive ones are destroyed by both the vaccine and the virus.

Given that most viruses can be engineered to express transgenes, an OV can be modified to encode a tumor antigen that will then be expressed upon proficient infection. The production of that tumor antigen in the proinflammatory context of an infected tumor bed leads to the induction of antigen-specific immune cells. This creates an anticancer vaccine. When used as vaccination platforms, OVs, unlike other vaccine adjuvants, confer the added benefit of directly killing cancer cells, thanks to their oncolytic abilities (Fig. 1.1, middle panel). Since viruses and other pathogen-based vaccines platforms are themselves recognized by the immune system and induce pathogen-specific immune responses, multidose vaccination regimens benefit from using different viral platforms for each dose. That way the immune response is boosted toward the vaccinated antigen rather that the vaccine platform.[67] Using this strategy, an early study by Bridle et al.[68] showed that oncolytic VSV modified to encode the B16F10 antigen dopachrome tautomerase (DCT) could efficiently induce an antigen-specific immune response and that this immunity could be further boosted by using another virus, in this case a nonreplicative Ad, modified to encode the same tumor antigen. This approach, called the heterologous OV prime-boost vaccine, could protect more than 60% of B16F10 subcutaneous tumor-bearing mice when the vaccine was administered on the day of tumor challenge (prime) and 1 week later (boost).[68] In a follow-up study, using Ad and VSV as priming and boosting agents, respectively, the authors found that when VSV encoding a tumor antigen was used as a boosting vector, the vaccine induced a greater amount of antigen-specific CD8 T cells and that these immune cells were more effective compared to Ad-primed animals only and the characterization of these cells revealed that they were of the central memory phenotype.[69] In another similar study, the same group demonstrated that oncolytic Maraba virus encoding DCT could also boost a powerful antitumor immune response when used after Ad priming. The authors found that the vaccine could cure more than 20% of animals bearing established B16F10 lung tumors and 20% of B16F10 brain tumors, 2 very aggressive models in which untreated mice usually succumb to the disease in less than 30 and 20 days, respectively.[70] Importantly, this heterologous OV prime-boost vaccine is currently undergoing clinical testing for the treatment of MAGE-A3-expressing cancers

(NCT02879760) and another clinical study is investigating the combination of anti-E6/E7 Ad-MRB prime-boost vaccine with immune checkpoint blockade (NCT03618953). The results of these clinical studies have not been published to date.

Many other heterologous prime-boost strategies using antigen-encoding OVs have also been explored, for instance, Ad boosting following DNA priming,[71] Ad boosting after transduced dendritic cell priming,[72] and Ad priming with VV boosting.[73] Also, a vaccine targeting the Ova antigen in which MRB-Ova was used as a boosting vector after MIS416Vax (a microparticle-based adjuvant mixed with Ova protein) priming was also found to be effective against ID8 ovarian cancer in mice.[74] Taken together, these studies show therapeutic efficacy and reveal that multidose anticancer vaccines using OVs are promising treatment approaches. Future work aimed at determining the optimal combination of priming and boosting vaccine platforms, as well as optimization of critical parameters, such as route of administration and the amount of time between the prime and the boost, will certainly guide the clinical implementation of this already promising approach.

OVs as platforms for personalized anticancer vaccines

Few strategies that use OVs to actively vaccinate against chosen cancer neoantigens have been explored. The Cerullo group developed one of the first approaches that allowed for an OV vaccine to be easily adapted to an individual patient without requiring any genetic modification of the virus. This personalized anticancer vaccine platform is based on the coating of OVs with specifically designed tumor epitope peptides. The vaccine platform has so far been tested with oncolytic Ad,[75–79] VV and HSV-1.[78] Intratumoral administration of Ad coated with modified tumor epitope peptides (PeptiCRAd) was shown to increase tumor-specific T-cell responses, enhance tumor growth control, and induce systemic anticancer immunity in various murine models of melanoma[75,77,79] and in 4T1 murine models of triple negative breast cancer.[76] Similarly, intratumoral administration of VV and HSV-1 viruses coated with modified tumor epitope peptides (PeptiENV) was shown to increase intratumoral peptide-specific T-cell responses in B16. OVA and B16F10 murine melanoma models.[78] By engineering the oncolytic Ad used in the PeptiCRAd platform to encode two immunostimulatory molecules, CD40L and tumor necrosis factor receptor superfamily member 4 ligand (OX40L), tumor growth control and antitumor immune responses were shown to be enhanced compared to PeptiCRAd using an unarmed parental Ad.[79] In addition, in murine models of B16.OVA and B16.F10.9/K1 melanoma and 4T1 triple negative breast cancer, the Cerullo group demonstrated that PeptiCRAd-induced antitumor effects could be augmented with immune checkpoint inhibitors.[76,77,79] Furthermore, they showed that preexisting pathogen-specific immunity derived from vaccinations against common diseases, such as tetanus, diphtheria, or pertussis, could be re-engaged by the PeptiCRAd platform to support the induction of more robust tumor-specific $CD8^+$ T-cell responses.[77] The Cerullo group has also developed a bioinformatics-based approach to identify tumor antigens and neoantigens based on their similarity to pathogen antigens, to exploit molecular mimicry and cross-reactive T cells in anticancer vaccine development.[80] By using this approach, they have developed a streamlined pipeline for the generation of personalized anticancer vaccines that includes the isolation and selection of the most immunogenic neoantigen epitopes displayed on the tumor cells, and the generation of personalized PeptiCRAd anticancer vaccines based on the newly identified neoantigens.[81] Recently, the Cerullo group has expanded their peptide-based anticancer vaccine platforms to include the use of live-attenuated viral vaccines against common pathogens, such as a vaccine against measles, mumps, and rubella (MMR vaccine) in the PeptiENV anticancer vaccine platform. They showed that using the MMR vaccine in the PeptiENV platform induced robust tumor-specific T-cell responses and achieved efficient tumor growth control.[82] The group also developed a novel personalized anticancer vaccine platform based on Bacillus Calmette-Guérin, one of the earliest immunotherapies still in use.[83] This novel peptide-based personalized anticancer vaccine platform (PeptiBAC) was used in combination with the PeptiCRAd platform in a heterologous prime-boost vaccination regime and was shown to enhance the tumor-specific T-cell responses elicited by the PeptiCRAd platform.

Another study tested the combination of intratumoral oncolytic Ad treatment with DNA vaccination. Using the B16F1 murine melanoma model, the authors showed that the DNA vaccine could effectively induce neoantigen-specific T cells and that the combination with Ad injection allowed for more cancer-specific CD8 T cells to be produced, as well as for the infiltration of the tumor by immune cells.[84] The results show that the vaccine conferred therapeutic benefits.

Our group has recently developed a personalized heterologous OV prime-boost approach (Fig. 1.1, bottom panel).[85] Using synthetic 27-amino-acid-long mutanome peptides coadministered with different OVs, we found that we were able to induce poly-epitope neoantigen-specific immunity in two different murine cancer models: B16F10 and CT26. Our results show that OVs can be mixed together with synthetic antigenic peptides to create personalized vaccines and that their vaccine adjuvant activities are comparable to that of the widely used adjuvant polyI:C. This approach has been successfully tested with oncolytic MRB, oncolytic VSV, VV, and Ad, suggesting that it may be amenable to other OV platforms as well. When used in heterologous prime-boost regimens, the vaccine induced important amounts of memory CD8 T cells that persisted to high levels in the blood of the animals for at least 35 days. Furthermore, we found that antigen-encoding viruses could be used as vaccine adjuvants together with neoantigen peptides to create effective personalized vaccines and immunize against both a TAA and multiple TSAs at the same time. Most of the work was performed using Ad and MRB as priming and boosting vaccine adjuvants, respectively.

To our knowledge, only one report investigated the use of neoantigen OV vaccines in patients with cancer. In a compassionate use study, two heavily treated patients with cancer (one metastatic colorectal cancer and one neuroendocrine carcinoma of the pancreas) received injections of oncolytic Ads that were engineered to encode 50 mutanome epitopes specific to each cancer and both patients still had a stable disease 12 months after treatment.[86]

Conclusions

OVs have an intrinsic capability to modulate the tumor microenvironment toward a less immunosuppressive phenotype and induce anticancer immunity. These characteristics mean that OVs are well-suited to combination with other cancer immunotherapies, such as therapies targeted against immune checkpoint proteins. Multiple clinical trials are ongoing to assess the combination of OVs and immune checkpoint inhibitors. Recent advances in enhancing neoantigen-specific anticancer immunity using OVs, together with progress in identifying tumor neoantigens, are moving the OV field toward personalized cancer immunotherapy. With many exciting strategies being developed to trigger cancer neoantigen-specific immunity in patients and the promising responses observed in clinical trials using OVs, we are likely to see more innovative and successful OV-based treatments in the future. Data emerging from clinical trials that explore a range of OV vaccination techniques are producing promising results and will be key for developing a better understanding of the determinants of favorable responses.

References

1. Apostolopoulos V. Cancer vaccines: research and applications. *Cancers (Basel)*. 2019;11. Available from: https://doi.org/10.3390/cancers11081041.
2. Mohanty R, Chowdhury CR, Arega S, Sen P, Ganguly P, Ganguly N. CAR T cell therapy: a new era for cancer treatment (review). *Oncol Rep*. 2019;42:2183–2195. Available from: https://doi.org/10.3892/or.2019.7335.
3. Yang JC, Rosenberg SA. Adoptive T-cell therapy for cancer. *Adv Immunol*. 2016;130:279–294. Available from: https://doi.org/10.1016/bs.ai.2015.12.006.
4. Goebeler M-E, Bargou RC. T cell-engaging therapies—BiTEs and beyond. *Nat Rev Clin Oncol*. 2020;17:418–434. Available from: https://doi.org/10.1038/s41571-020-0347-5.
5. Diaz RM, Galivo F, Kottke T, Wongthida P, Qiao J, Thompson J, et al. Oncolytic immunovirotherapy for melanoma using vesicular stomatitis virus. *Cancer Res*. 2007;67:2840–2848. Available from: https://doi.org/10.1158/0008-5472.CAN-06-3974.
6. Prestwich RJ, Ilett EJ, Errington F, Diaz RM, Steele LP, Kottke T, et al. Immune-mediated antitumor activity of reovirus is required for therapy and is independent of direct viral oncolysis and replication. *Clin Cancer Res*. 2009;15:4374–4381. Available from: https://doi.org/10.1158/1078-0432.CCR-09-0334.

7. Sobol PT, Boudreau JE, Stephenson K, Wan Y, Lichty BD, Mossman KL. Adaptive antiviral immunity is a determinant of the therapeutic success of oncolytic virotherapy. *Mol Ther.* 2011;19:335–344. Available from: https://doi.org/10.1038/mt.2010.264.
8. Workenhe ST, Simmons G, Pol JG, Lichty BD, Halford WP, Mossman KL. Immunogenic HSV-mediated oncolysis shapes the antitumor immune response and contributes to therapeutic efficacy. *Mol Ther.* 2014;22:123–131. Available from: https://doi.org/10.1038/mt.2013.238.
9. Postow MA, Callahan MK, Wolchok JD. Immune checkpoint blockade in cancer therapy. *J Clin Oncol.* 2015;33:1974–1982. Available from: https://doi.org/10.1200/JCO.2014.59.4358.
10. Pol JG, Workenhe ST, Konda P, Gujar S, Kroemer G. Cytokines in oncolytic virotherapy. *Cytokine Growth Factor Rev.* 2020;56:4–27. Available from: https://doi.org/10.1016/J.CYTOGFR.2020.10.007.
11. Morales-Molina A, Rodríguez-Milla MÁ, Gimenez-Sanchez A, Perisé-Barrios AJ, García-Castro J. Cellular virotherapy increases tumor-infiltrating lymphocytes (TIL) and decreases their PD-1 + subsets in mouse immunocompetent models. *Cancers (Basel).* 2020;12:1920. Available from: https://doi.org/10.3390/cancers12071920.
12. Workenhe ST, Mossman KL. Oncolytic virotherapy and immunogenic cancer cell death: sharpening the sword for improved cancer treatment strategies. *Mol Ther.* 2014;22:251–256. Available from: https://doi.org/10.1038/mt.2013.220.
13. Liikanen I, Ahtiainen L, Hirvinen MLM, Bramante S, Cerullo V, Nokisalmi P, et al. Oncolytic adenovirus with temozolomide induces autophagy and antitumor immune responses in cancer patients. *Mol Ther.* 2013;21:1212–1223. Available from: https://doi.org/10.1038/mt.2013.51.
14. Donnelly OG, Errington-Mais F, Steele L, Hadac E, Jennings V, Scott K, et al. Measles virus causes immunogenic cell death in human melanoma. *Gene Ther.* 2013;20:7–15. Available from: https://doi.org/10.1038/gt.2011.205.
15. Miyamoto S, Inoue H, Nakamura T, Yamada M, Sakamoto C, Urata Y, et al. Coxsackievirus B3 is an oncolytic virus with immunostimulatory properties that is active against lung adenocarcinoma. *Cancer Res.* 2012;72:2609–2621. Available from: https://doi.org/10.1158/0008-5472.CAN-11-3185.
16. Geoffroy K, Bourgeois-Daigneault M-C. The pros and cons of interferons for oncolytic virotherapy. *Cytokine Growth Factor Rev.* 2020;56:49–58. Available from: https://doi.org/10.1016/j.cytogfr.2020.07.002.
17. Woller N, GÃ¼rlevik E, Ureche C-I, Schumacher A, KÃ¼hnel F. Oncolytic viruses as anticancer vaccines. *Front Oncol.* 2014;4:188. Available from: https://doi.org/10.3389/fonc.2014.00188.
18. Breitbach CJ, Lichty BD, Bell JC. Oncolytic viruses: therapeutics with an identity crisis. *EBioMedicine.* 2016;9:31–36. Available from: https://doi.org/10.1016/j.ebiom.2016.06.046.
19. Lichty BD, Breitbach CJ, Stojdl DF, Bell JC. Going viral with cancer immunotherapy. *Nat Rev Cancer.* 2014;14:559–567. Available from: https://doi.org/10.1038/nrc3770.
20. Lemos de Matos A, Franco LS, McFadden G. Oncolytic viruses and the immune system: the dynamic duo. *Mol Ther Methods Clin Dev.* 2020;17:349–358. Available from: https://doi.org/10.1016/j.omtm.2020.01.001.
21. Andtbacka RHI, Kaufman HL, Collichio F, Amatruda T, Senzer N, Chesney J, et al. Talimogene laherparepvec improves durable response rate in patients with advanced melanoma. *J Clin Oncol.* 2015;33:2780–2788. Available from: https://doi.org/10.1200/JCO.2014.58.3377.
22. Failla CM, Carbone ML, Fortes C, Pagnanelli G, D'Atri S. Melanoma and vitiligo: in good company. *Int J Mol Sci.* 2019;20. Available from: https://doi.org/10.3390/ijms20225731.

23. Quaglino P, Marenco F, Osella-Abate S, Cappello N, Ortoncelli M, Salomone B, et al. Vitiligo is an independent favourable prognostic factor in stage III and IV metastatic melanoma patients: results from a single-institution hospital-based observational cohort study. *Ann Oncol J Eur Soc Med Oncol*. 2010;21:409−414. Available from: https://doi.org/10.1093/annonc/mdp325.
24. Conry RM, Westbrook B, McKee S, Norwood TG. Talimogene laherparepvec: first in class oncolytic virotherapy. *Hum Vaccin Immunother*. 2018;14:839−846. Available from: https://doi.org/10.1080/21645515.2017.1412896.
25. White CL, Twigger KR, Vidal L, De Bono JS, Coffey M, Heinemann L, et al. Characterization of the adaptive and innate immune response to intravenous oncolytic reovirus (Dearing type 3) during a phase I clinical trial. *Gene Ther*. 2008;15:911−920. Available from: https://doi.org/10.1038/gt.2008.21.
26. Heinzerling L, Künzi V, Oberholzer PA, Kündig T, Naim H, Dummer R. Oncolytic measles virus in cutaneous T-cell lymphomas mounts antitumor immune responses in vivo and targets interferon-resistant tumor cells. *Blood*. 2005;106:2287−2294. Available from: https://doi.org/10.1182/blood-2004-11-4558.
27. Castle JC, Kreiter S, Diekmann J, Löwer M, van de Roemer N, de Graaf J, et al. Exploiting the mutanome for tumor vaccination. *Cancer Res*. 2012;72:1081−1091. Available from: https://doi.org/10.1158/0008-5472.CAN-11-3722.
28. Boisguérin V, Castle JC, Loewer M, Diekmann J, Mueller F, Britten CM, et al. Translation of genomics-guided RNA-based personalised cancer vaccines: towards the bedside. *Br J Cancer*. 2014;111:1469−1475. Available from: https://doi.org/10.1038/bjc.2013.820.
29. Türeci Ö, Vormehr M, Diken M, Kreiter S, Huber C, Sahin U. Targeting the heterogeneity of cancer with individualized neoepitope vaccines. *Clin Cancer Res*. 2016;22:1885−1896. Available from: https://doi.org/10.1158/1078-0432.CCR-15-1509.
30. DeMatos P, Abdel-Wahab Z, Vervaert C, Hester D, Seigler H. Pulsing of dendritic cells with cell lysates from either B16 melanoma or MCA-106 fibrosarcoma yields equally effective vaccines against B16 tumors in mice. *J Surg Oncol*. 1998;68:79−91. https://doi.org/10.1002/(sici)1096-9098(199806)68:2<79::aid-jso3>3.0.co;2-h.
31. Ali OA, Verbeke C, Johnson C, Sands RW, Lewin SA, White D, et al. Identification of immune factors regulating antitumor immunity using polymeric vaccines with multiple adjuvants. *Cancer Res*. 2014;74:1670−1681. Available from: https://doi.org/10.1158/0008-5472.CAN-13-0777.
32. Vogel FR. Improving vaccine performance with adjuvants. *Clin Infect Dis*. 2000;30(Suppl. 3):S266−S270. Available from: https://doi.org/10.1086/313883.
33. Schwartzentruber DJ, Lawson DH, Richards JM, Conry RM, Miller DM, Treisman J, et al. gp100 peptide vaccine and interleukin-2 in patients with advanced melanoma. *N Engl J Med*. 2011;364:2119−2127. Available from: https://doi.org/10.1056/NEJMoa1012863.
34. Marrack P, McKee AS, Munks MW. Towards an understanding of the adjuvant action of aluminium. *Nat Rev Immunol*. 2009;9:287−293. Available from: https://doi.org/10.1038/nri2510.
35. Mantile F, Trovato M, Santoni A, Barba P, Ottonello S, De Berardinis P, et al. Alum and squalene-oil-in-water emulsion enhance the titer and avidity of anti-Aβ antibodies induced by multimeric protein antigen (1−11)E2, preserving the Igg1-skewed isotype distribution. *PLoS One*. 2014;9:e101474. Available from: https://doi.org/10.1371/journal.pone.0101474.
36. Coffman RL, Sher A, Seder RA. Vaccine adjuvants: putting innate immunity to work. *Immunity*. 2010;33:492−503. Available from: https://doi.org/10.1016/j.immuni.2010.10.002.
37. Vigneron N. Human tumor antigens and cancer immunotherapy. *Biomed Res Int*. 2015;2015:948501. Available from: https://doi.org/10.1155/2015/948501.

38. Dunn GP, Bruce AT, Ikeda H, Old LJ, Schreiber RD. Cancer immunoediting: from immunosurveillance to tumor escape. *Nat Immunol.* 2002;3:991–998. Available from: https://doi.org/10.1038/ni1102-991.
39. Xue R, Li R, Guo H, Guo L, Su Z, Ni X, et al. Variable intra-tumor genomic heterogeneity of multiple lesions in patients with hepatocellular carcinoma. *Gastroenterology.* 2016;150:998–1008. Available from: https://doi.org/10.1053/j.gastro.2015.12.033.
40. Stojdl DF, Lichty B, Knowles S, Marius R, Atkins H, Sonenberg N, et al. Exploiting tumor-specific defects in the interferon pathway with a previously unknown oncolytic virus. *Nat Med.* 2000;6:821–825. Available from: https://doi.org/10.1038/77558.
41. Stojdl DF, Lichty BD, enOever BR t, Paterson JM, Power AT, Knowles S, et al. VSV strains with defects in their ability to shutdown innate immunity are potent systemic anti-cancer agents. *Cancer Cell.* 2003;4:263–275. Available from: http://www.sciencedirect.com/science/article/pii/S1535610803002411. Accessed September 26, 2013.
42. Symons JA, Alcami A, Smith Sir William GL. *Vaccinia Virus Encodes a Soluble Type I Interferon Receptor of Novel Structure and Broad Species Specificity*; 1995. https://www.cell.com/cell/pdf/0092-8674(95)90076-4.pdf?_returnURL = https%3A%2F%2Flinkinghub.elsevier.com%2Fretrieve%2Fpii%2F0092867495900764%3Fshowall%3Dtrue. Accessed May 20, 2020.
43. Le Boeuf F, Diallo J-S, McCart JA, Thorne S, Falls T, Stanford M, et al. Synergistic interaction between oncolytic viruses augments tumor killing. *Mol Ther.* 2010;18:888–895. Available from: https://doi.org/10.1038/mt.2010.44.
44. Ma W, He H, Wang H. Oncolytic herpes simplex virus and immunotherapy. *BMC Immunol.* 2018;19:40. Available from: https://doi.org/10.1186/s12865-018-0281-9.
45. Martin NT, Roy DG, Workenhe ST, van den Wollenberg DJM, Hoeben RC, Mossman KL, et al. Pre-surgical neoadjuvant oncolytic virotherapy confers protection against rechallenge in a murine model of breast cancer. *Sci Rep.* 2019;9:1865. Available from: https://doi.org/10.1038/s41598-018-38385-7.
46. Stephenson KB, Barra NG, Davies E, Ashkar AA, Lichty BD. Expressing human interleukin-15 from oncolytic vesicular stomatitis virus improves survival in a murine metastatic colon adenocarcinoma model through the enhancement of anti-tumor immunity. *Cancer Gene Ther.* 2012;19:238–246. Available from: https://doi.org/10.1038/cgt.2011.81.
47. Bourgeois-Daigneault M-C, Roy DG, Falls T, Twumasi-Boateng K, St-Germain LE, Marguerie M, et al. Oncolytic vesicular stomatitis virus expressing interferon-γ has enhanced therapeutic activity. *Mol Ther Oncolytics.* 2016;3:16001. Available from: https://doi.org/10.1038/mto.2016.1.
48. Durham NM, Mulgrew K, McGlinchey K, Monks NR, Ji H, Herbst R, et al. Oncolytic VSV primes differential responses to immuno-oncology therapy. *Mol Ther.* 2017;25:1917–1932. Available from: https://doi.org/10.1016/j.ymthe.2017.05.006.
49. Yu X, Hu B, Liu X, Liu H, Zhao L, Mei Y, et al. Immune response expressing IL-7 promotes antitumor recombinant new castle disease virus autologous tumor vaccine modified with. 2014. https://doi.org/10.4049/jimmunol.1400004.
50. Terada K, Wakimoto H, Tyminski E, Chiocca EA, Saeki Y. Development of a rapid method to generate multiple oncolytic HSV vectors and their in vivo evaluation using syngeneic mouse tumor models. *Gene Ther.* 2006;13:705–714. Available from: https://doi.org/10.1038/sj.gt.3302717.
51. Shin EJ, Wanna GB, Choi B, Aguila D, Ebert O, Genden EM, et al. Interleukin-12 expression enhances vesicular stomatitis virus oncolytic therapy in murine squamous cell carcinoma. *Laryngoscope.* 2007;117:210–214. Available from: https://doi.org/10.1097/01.mlg.0000246194.66295.d8.

52. Choi I-K, Lee J-S, Zhang S-N, Park J, Sonn CH, Lee K-M, et al. Oncolytic adenovirus co-expressing IL-12 and IL-18 improves tumor-specific immunity via differentiation of T cells expressing IL-12Rβ2 or IL-18Rα. *Gene Ther.* 2011;18:898–909. Available from: https://doi.org/10.1038/gt.2011.37.
53. Derubertis BG, Stiles BM, Bhargava A, Gusani NJ, Hezel M, D'Angelica M, et al. Cytokine-secreting herpes viral mutants effectively treat tumor in a murine metastatic colorectal liver model by oncolytic and T-cell-dependent mechanisms. *Cancer Gene Ther.* 2007;14:590–597. Available from: https://doi.org/10.1038/sj.cgt.7701053.
54. Galivo F, Diaz RM, Thanarajasingam U, Jevremovic D, Wongthida P, Thompson J, et al. Interference of CD40L-mediated tumor immunotherapy by oncolytic vesicular stomatitis virus. *Hum Gene Ther.* 2010;21:439–450. Available from: https://doi.org/10.1089/hum.2009.143.
55. Li H, Peng K-W, Dingli D, Kratzke RA, Russell SJ. Oncolytic measles viruses encoding interferon beta and the thyroidal sodium iodide symporter gene for mesothelioma virotherapy. *Cancer Gene Ther.* 2010;17:550–558. Available from: https://doi.org/10.1038/cgt.2010.10.
56. Bourgeois-Daigneault M-C, Roy DG, Aitken AS, El Sayes N, Martin NT, Varette O, et al. Neoadjuvant oncolytic virotherapy before surgery sensitizes triple-negative breast cancer to immune checkpoint therapy. *Sci Transl Med.* 2018;10. Available from: https://doi.org/10.1126/scitranslmed.aao1641. eaao1641.
57. Fu X, Tao L, Wu W, Zhang X. Arming HSV-based oncolytic viruses with the ability to redirect the host's innate antiviral immunity to attack tumor cells. *Mol Ther Oncolytics.* 2020;19:33–46. Available from: https://doi.org/10.1016/j.omto.2020.09.002.
58. Zamarin D, Holmgaard RB, Subudhi SK, Park JS, Mansour M, Palese P, et al. Localized oncolytic virotherapy overcomes systemic tumor resistance to immune checkpoint blockade immunotherapy. *Sci Transl Med.* 2014;6:226ra32. Available from: https://doi.org/10.1126/scitranslmed.3008095.
59. Sivanandam V, LaRocca CJ, Chen NG, Fong Y, Warner SG. Oncolytic viruses and immune checkpoint inhibition: the best of both worlds. *Mol Ther Oncolytics.* 2019;13:93–106. Available from: https://doi.org/10.1016/J.OMTO.2019.04.003.
60. Engeland CE, Grossardt C, Veinalde R, Bossow S, Lutz D, Kaufmann JK, et al. CTLA-4 and PD-L1 checkpoint blockade enhances oncolytic measles virus therapy. *Mol Ther.* 2014;22:1949–1959. Available from: https://doi.org/10.1038/mt.2014.160.
61. Moesta AK, Cooke K, Piasecki J, Mitchell P, Rottman JB, Fitzgerald K, et al. Local delivery of OncoVEXmGM-CSF generates systemic antitumor immune responses enhanced by cytotoxic T-lymphocyte-associated protein blockade. *Clin Cancer Res.* 2017;23:6190–6202. Available from: https://doi.org/10.1158/1078-0432.CCR-17-0681.
62. Rajani K, Parrish C, Kottke T, Thompson J, Zaidi S, Ilett L, et al. Combination therapy with reovirus and anti-PD-1 blockade controls tumor growth through innate and adaptive immune responses. *Mol Ther.* 2016;24:166–174. Available from: https://doi.org/10.1038/mt.2015.156.
63. Rojas JJ, Sampath P, Hou W, Thorne SH. Defining effective combinations of immune checkpoint blockade and oncolytic virotherapy. *Clin Cancer Res.* 2015;21:5543–5551. Available from: https://doi.org/10.1158/1078-0432.CCR-14-2009.
64. Wang G, Kang X, Chen KS, Jehng T, Jones L, Chen J, et al. An engineered oncolytic virus expressing PD-L1 inhibitors activates tumor neoantigen-specific T cell responses. *Nat Commun.* 2020;11:1395. Available from: https://doi.org/10.1038/s41467-020-15229-5.
65. Groeneveldt C, Kinderman P, van den Wollenberg DJM, van den Oever RL, Middelburg J, Mustafa DAM, et al. Preconditioning of the tumor microenvironment with oncolytic reovirus converts CD3-bispecific antibody treatment into effective immunotherapy. *J Immunother Cancer.* 2020;8. Available from: https://doi.org/10.1136/jitc-2020-001191. e001191.

66. Aitken A, Roy D, Bourgeois-Daigneault M-C. Taking a stab at cancer; oncolytic virus-mediated anti-cancer vaccination strategies. *Biomedicines*. 2017;5:3. Available from: https://doi.org/10.3390/biomedicines5010003.
67. Irvine KR, Chamberlain RS, Shulman EP, Surman DR, Rosenberg SA, Restifo NP. Enhancing efficacy of recombinant anticancer vaccines with prime/boost regimens that use two different vectors. *J Natl Cancer Inst*. 1997;89:1595–1601. Available from: https://doi.org/10.1093/jnci/89.21.1595.
68. Bridle BW, Boudreau JE, Lichty BD, Brunellière J, Stephenson K, Koshy S, et al. Vesicular stomatitis virus as a novel cancer vaccine vector to prime antitumor immunity amenable to rapid boosting with adenovirus. *Mol Ther*. 2009;17:1814–1821. Available from: https://doi.org/10.1038/mt.2009.154.
69. Bridle BW, Clouthier D, Zhang L, Pol J, Chen L, Lichty BD, et al. Oncolytic vesicular stomatitis virus quantitatively and qualitatively improves primary CD8(+) T-cell responses to anticancer vaccines. *Oncoimmunology*. 2013;2:e26013. Available from: https://doi.org/10.4161/onci.26013.
70. Pol JG, Zhang L, Bridle BW, Stephenson KB, Rességuier J, Hanson S, et al. Maraba virus as a potent oncolytic vaccine vector. *Mol Ther*. 2014;22:420–429. Available from: https://doi.org/10.1038/mt.2013.249.
71. Aurisicchio L, Mennuni C, Giannetti P, Calvaruso F, Nuzzo M, Cipriani B, et al. Immunogenicity and safety of a DNA prime/adenovirus boost vaccine against rhesus CEA in nonhuman primates. *Int J Cancer*. 2007;120:2290–2300. Available from: https://doi.org/10.1002/ijc.22555.
72. Ribas A, Butterfield LH, McBride WH, Jilani SM, Bui LA, Vollmer CM, et al. Genetic immunization for the melanoma antigen MART-1/Melan-A using recombinant adenovirus-transduced murine dendritic cells. *Cancer Res*. 1997;57:2865–2869. Available from: http://www.ncbi.nlm.nih.gov/pubmed/9230191. Accessed May 27, 2021.
73. Zhang L, Bridle BW, Chen L, Pol J, Spaner D, Boudreau JE, et al. Delivery of viral-vectored vaccines by B cells represents a novel strategy to accelerate CD8(+) T-cell recall responses. *Blood*. 2013;121:2432–2439. Available from: https://doi.org/10.1182/blood-2012-06-438481.
74. McGray AJR, Huang R-Y, Battaglia S, Eppolito C, Miliotto A, Stephenson KB, et al. Oncolytic Maraba virus armed with tumor antigen boosts vaccine priming and reveals diverse therapeutic response patterns when combined with checkpoint blockade in ovarian cancer. *J Immunother Cancer*. 2019;7:189. Available from: https://doi.org/10.1186/s40425-019-0641-x.
75. Capasso C, Hirvinen M, Garofalo M, Romaniuk D, Kuryk L, Sarvela T, et al. Oncolytic adenoviruses coated with MHC-I tumor epitopes for a new oncolytic vaccine platform. *J Immunother Cancer*. 2015;3:P333. Available from: https://doi.org/10.1186/2051-1426-3-S2-P333.
76. Feola S, Capasso C, Fusciello M, Martins B, Tähtinen S, Medeot M, et al. Oncolytic vaccines increase the response to PD-L1 blockade in immunogenic and poorly immunogenic tumors. *Oncoimmunology*. 2018;7:e1457596. Available from: https://doi.org/10.1080/2162402X.2018.1457596.
77. Tähtinen S, Feola S, Capasso C, Laustio N, Groeneveldt C, Ylösmäki EO, et al. Exploiting pre-existing immunity to enhance oncolytic cancer immunotherapy. *Cancer Res*. 2020;80. Available from: https://doi.org/10.1158/0008-5472.CAN-19-2062. canres.2062.2019.
78. Ylösmäki E, Malorzo C, Capasso C, Honkasalo O, Fusciello M, Martins B, et al. Personalized cancer vaccine platform for clinically relevant oncolytic enveloped viruses. *Mol Ther*. 2018;26:2315–2325. Available from: https://doi.org/10.1016/j.ymthe.2018.06.008.

79. Ylösmäki E, Ylösmäki L, Fusciello M, Martins B, Ahokas P, Cojoc H, et al. Characterization of a novel OX40 ligand and CD40 ligand-expressing oncolytic adenovirus used in the PeptiCRAd cancer vaccine platform. *Mol Ther Oncolytics*. 2021;20:459–469. Available from: https://doi.org/10.1016/j.omto.2021.02.006.
80. Chiaro J, Kasanen HHE, Whalley T, Capasso C, Grönholm M, Feola S, et al. Viral molecular mimicry influences the antitumor immune response in murine and human melanoma. *MedRxiv*. 2020;. Available from: https://doi.org/10.1101/2020.09.09.20191171.
81. Feola S, Chiaro J, Martins B, Russo S, Fusciello M, Ylösmäki E, et al. A novel immunopeptidomic-based pipeline for the generation of personalized oncolytic cancer vaccines. *BioRxiv*. 2021;. Available from: https://doi.org/10.1101/2021.06.08.447483.
82. Ylösmäki E, Fusciello M, Uoti A, Feola S, Martins B, Aalto K, et al. A personalized anticancer vaccine for melanoma based on an approved vaccine against measles, mumps, and rubella. *BioRxiv*. 2021;. Available from: https://doi.org/10.1101/2021.06.07.447462.
83. Ylösmäki E, Martins B, Fusciello M, Feola S, Hamdan F, Chiaro J, et al. Novel personalized cancer vaccine platform based on Bacillus Calmette-Guèrin. *BioRxiv*. 2021;. Available from: https://doi.org/10.1101/2021.03.05.434062.
84. Lopes A, Feola S, Ligot S, Fusciello M, Vandermeulen G, Préat V, et al. Oncolytic adenovirus drives specific immune response generated by a poly-epitope pDNA vaccine encoding melanoma neoantigens into the tumor site. *J Immunother Cancer*. 2019;7:174. Available from: https://doi.org/10.1186/s40425-019-0644-7.
85. Roy DG, Geoffroy K, Marguerie M, Khan ST, Martin NT, Kmiecik J, et al. Adjuvant oncolytic virotherapy for personalized anti-cancer vaccination. *Nat Commun*. 2021;12:2626. Available from: https://doi.org/10.1038/s41467-021-22929-z.
86. Bouvet M, Reid TR, Larson C, Oronsky B, Carter C, Morris JC. Extended treatment with MY-NEOVAX, personalized neoantigen-enhanced oncolytic viruses, for two end-stage cancer patients. *Oxf Med Case Rep*. 2019;2019:461–463. Available from: https://doi.org/10.1093/omcr/omz105.

Chapter 2

Tumor lysates cancer vaccine

Cheryl L.-L. Chiang[1] and Lana E. Kandalaft[2,3]
[1]*Department of Oncology, Lausanne University Hospital (CHUV), Ludwig Institute for Cancer Research, University of Lausanne, Lausanne, Switzerland,* [2]*Ludwig Institute for Cancer Research, University of Lausanne, Lausanne, Switzerland,* [3]*Center of Experimental Therapeutics, Department of Oncology, Lausanne University Hospital (CHUV), Lausanne, Switzerland*

Introduction

Cancer is an imperative health issue with 19.3 million new cases detected and almost 10 million deaths in 2020 (GLOBOCAN 2020 database; http://gco.iarc.fr/). In the United States and the United Kingdom, 1 in 2 adults has an estimated lifetime risk of developing cancer.[1,2] Conventional cancer treatments, such as chemotherapy and radiotherapy, are effective in certain subsets of patients, but they can produce undesirable long-term side effects, toxicity, and drug resistance.[3,4] The emerging field of cancer immunotherapy introduces a new paradigm for cancer treatment based on the concept of harnessing the human immune system against cancer. Seminal work from Schreiber et al. provided key evidence that the immune system played a role in conferring protection against cancers.[5,6] The notion of cancer immunoediting (i.e., editing, equilibrium, and escape) further underscores the complex interplay between cancer immunity and control.[7] Currently in clinics, several types of cancer immunotherapy are being actively pursued. Among these, the use of cancer vaccines to elicit and/or boost patient's antitumor immune response is an attractive approach. Here, we discuss using whole-tumor cell lysate (WTL) as a form of cancer vaccination and increasing its overall efficacy with different immunomodulating adjuvants and therapies (Fig. 2.1). In Table 2.1, we summarized the active and completed cancer clinical trials that utilized WTL as a source of antigens in the past 11 years with the keywords cancer and tumor lysate at http://www.clinicaltrials.gov.

Personalized cancer treatment with whole-tumor lysate vaccine

Major advancements in the field of molecular biology and genomics have assisted in the identification of more than 100 distinct types of human

Cancer Vaccines as Immunotherapy of Cancer. DOI: https://doi.org/10.1016/B978-0-12-823901-8.00001-7
© 2022 Elsevier Inc. All rights reserved.

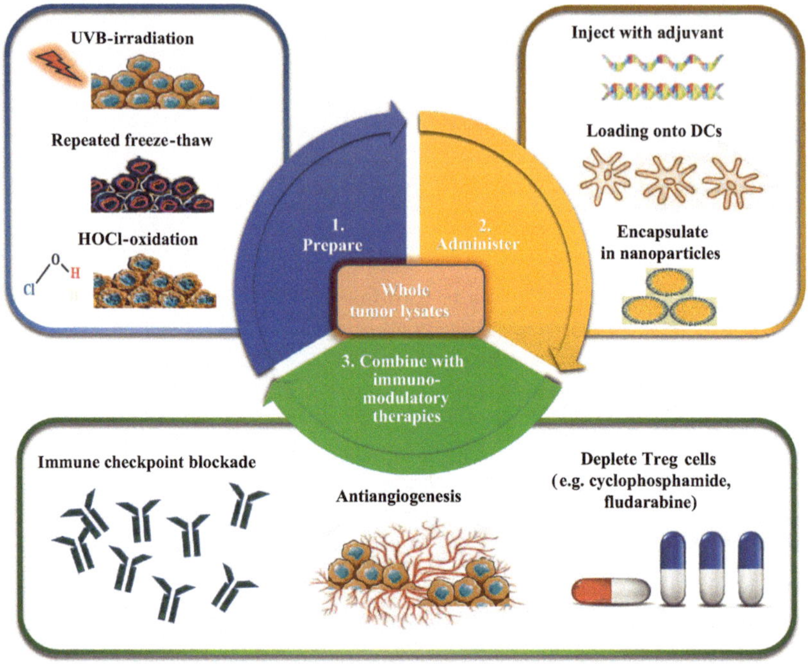

FIGURE 2.1 Methods of preparing whole-tumor lysate cancer vaccines, increasing their immunogenicity with adjuvants and other immunomodulatory therapies. Tumor cells could be obtained from patients during surgery and treated with different methods (ultraviolet B, repeated freeze–thaw or hypochlorous acid oxidation) to inactivate tumor cells and to produce whole-tumor lysate (*WTL*) cancer vaccine. The WTL could be administered directly with adjuvants to increase their immunogenicity, or loaded on to dendritic cells or nanoparticles for administration. To further enhance the overall efficacy of the WTL vaccine, different immunomodulatory therapies (e.g., immune checkpoint blockade, depleting T regulatory cells with chemotherapy or anti-angiogenesis) could be apply.

cancers so far. With increasing knowledge gained from studying these cancers, it is becoming clear that a personalized therapeutic approach based on the patient's unique cancer characteristics could effectively improve the overall efficacy of the cancer treatment.[8] In terms of cancer vaccination, personalized WTL cancer vaccine created from patient's own tumor cells would contain a vast array of characterized and uncharacterized antigenic T-cell epitopes for eliciting a broad-based immunity. Such polyclonal $CD4^+$ and $CD8^+$ antitumor T-cell responses, raised against shared tumor-associated antigens (TAAs) and/or tumor neoantigens,[9,10] would be useful in preventing tumor escape variants through the loss of a single or multiple antigens in the tumor cell. Long-term immunological memory against the tumor could also be induced via the cognate help provided by $CD4^+$ T helper cells to $CD8^+$ T cells.[11] Furthermore, a *meta*-analysis of various cancer clinical trials of

TABLE 2.1 Active and complete cancer clinical trials that utilize whole-tumor cell lysate (WTL) vaccine from year 2010 to 2021 (http://www.clinicaltrials.gov).

Identifier	Phases	Status (reference if available)	Tumor type	Interventions
Source of WTL: autologous				
NCT01946373	I	Recruiting	Melanoma	DC vaccine, T cells, cyclophosphamide, Fludarabine, IL-2
NCT01973322	I	Recruiting	Melanoma	DC vaccine, IFN-α, radiotherapy
NCT04201873	I	Recruiting	Recurrent glioblastoma	DC vaccine, Pembrolizumab, poly-ICLC
NCT03671720	I	Recruiting	Metastatic cancer	DC vaccine
NCT03879512	I	Recruiting	Glioblastoma	DC vaccine, metronomic cyclophosphamide, checkpoint blockade (nivolumab and ipilimumab)
NCT03395587	II	Recruiting	Glioblastoma	DC vaccine, standard chemotherapy
NCT01204684	I	Active, not recruiting	Glioblastoma, glioma, anaplastic astrocytoma, and anaplastic astro-oligodendroglioma	DC vaccination, 0.2% resiquimod, poly-ICLC
NCT02678741	I	Active, not recruiting	Melanoma	DC vaccine
NCT01753089	I/II	Active, not recruiting	Melanoma	DC vaccine
NCT03803397	I/II	Not yet recruiting	Melanoma	DC vaccine

(Continued)

TABLE 2.1 (Continued)

Identifier	Phases	Status (reference if available)	Tumor type	Interventions
NCT01803152	II	Active, not recruiting	Soft tissue and bone sarcomas	DC vaccine, Gemcitabine, Imiquimod
NCT01808820	II	Active, not recruiting	Glioblastoma Multiforme, high grade or metastic glioma, and anaplastic astrocytoma	DC vaccine, Imiquimod
NCT02301611	II	Active, not recruiting	Melanoma	DC vaccine
NCT01671592	I	Completed	Colorectal carcinoma	DC vaccine
NCT01132014	I	Completed (9, 10, 155)	Ovarian cancer	DC vaccine (OCDC; oxidized autologous whole-tumor cell lysate)
NCT01312389	I/II	Completed	Ovarian cancer, fallopian tube cancer and primary peritoneal cancer	Oxidized whole-tumor lysate (OC-L), Montanide ISA 51 VG, poly(I:C12U)
NCT03114631	I/II	Completed (165)	Pancreatic neoplasms	DC vaccine pulsed with WTL or with mucin (MUC)-1/Wilms' tumor (WT)-1 synthetic peptides
NCT02496520	I/II	Completed	Sarcoma and central nervous system tumor	DC vaccine, standard chemotherapy, standard radiation therapy
NCT01213407	II	Completed (166)	Glioblastoma multiforme	DC vaccine, Temozolomide, radiotherapy
NCT01635283	II	Completed	Adult diffuse astrocytoma, mixed glioma, oligodendroglioma and oligoastrocytoma	DC vaccine

NCT01998542	II	Completed (167)	Head and neck squamous cell carcinoma	Chaperone-rich tumor cell lysate (CRCL)
NCT02409524	II	Completed (168)	Hepatocellular carcinoma	CRCL, mismatched allogeneic T helper 1 memory cells (AlloStim) conjugated with CD3/CD28-coated microbeads
Source of WTL: allogenic				
NCT03827967	I	Recruiting	Colon cancer	Colon cancer cell line lysated loaded onto yeast cell wall particles
NCT02549833	I	Recruiting	Oligodendrogliom, astrocytoma, glioma, and oligoastrocytoma	GBM6-AD tumor cell line lysate vaccine, poly-ICLC
NCT01883518	I/II	Enrolling by invitation	Sarcoma	DC vaccine
NCT04212377	II	Recruiting	Endometrial cancer	DC vaccine
NCT03610360	II/III	Recruiting	Mesothelioma	DC vaccine
NCT01957956	I	Active, not recruiting	Glioblastoma and gliosarcoma	DC vaccine, Temozolomide
NCT02010606	I	Active, not recruiting	Glioblastoma, glioblastoma multiforme, glioma, and astrocytoma	DC vaccine, standard temozolomide chemotherapy, field radiation therapy, bevacizumab
NCT01678352	I	Completed	Glioma	GBM6-AD tumor cell line lysate vaccine, Imiquimod
NCT03086564	I/II	Completed	Hepatocellular carcinoma	DC vaccine, HepG2 cell line lysate, hepatitis B virus-specific antigen peptides, transarterial chemoembolization

IFN, Interferon; *IL*, interleukin; *WTL*, whole tumor lysate.

about 1800 patients demonstrated that whole-tumor vaccines induced a higher objective response (8.1%) in the patients when compared to using defined tumor antigens (3.6%).[12]

Patient's tumor could be harvested during debulking surgery for WTL preparation. Malignancies, such as ovarian, breast, and lung carcinomas and glioblastomas, could benefit from WTL therapy as sufficient tumor materials could be obtained. In cases where autologous tumor materials are limited, allogeneic tumor cell lines expressing shared TAAs could serve as an alternative. For example, the allogenic melanoma tumor cell lysate Melacine (Corixa Corp.) has shown some antitumor activities when used in combined with the adjuvant DETOX [monophosphoryl lipid A (MPLA) and purified mycobacterial cell wall skeleton].[13] Accumulating evidence now demonstrated that targeting tumor neoantigen could elicit a stronger and longer-lasting antitumor immunity. Unlike TAAs, neoantigens are private non-self-mutated tumor antigens derived directly from transformation processes (driver mutations) or as byproducts from increasing genetic instability (passenger mutations).[14] As neoantigens are not subjected to thymic selection and central tolerance, such as TAAs, high-avidity T cells targeting neoantigens are highly likely to exist. Wölfel et al. was the first to show that a nonsynonymous mutation in cyclin-dependent kinase 4 was a neoantigen for cytotoxic T lymphocytes (CTLs) in melanoma.[15] Furthermore, a melanoma patient who received adoptively transferred tumor-infiltrating lymphocytes (TILs) containing neoantigen-specific T cells experienced complete tumor regressions.[16] Verdegaal et al. successfully treated two patients with stage IV melanoma with autologous tumor-reactive T cells that were generated by coculturing peripheral blood mononuclear cells and autologous resected melanoma.[17] The expanded T cells were mainly $CD8^+$ (80%–70%) and showed specificity against tumor neoantigens, as well as prolonging progression-free survival of up to 9 years in one of the patients.[17] Another study found that rare long-term pancreatic cancer survivors (median survival of 10.5 years) had the highest number of neoantigens and the most abundant $CD8^+$ TILs.[18] Similarly, Linnemann et al. detected intratumoral neoantigen-specific $CD4^+$ T cells in four out of the five stage IV melanoma patients who were treated with adoptive T-cell transfer and showed clinical responses.[19] Hence, a broad-based neoantigen-specific $CD8^+$ and $CD4^+$ T-cell responses could be important for suppressing tumors. These findings strongly supported the rationale of using autologous WTL as a source of tumor neoantigens that are unique to the patient for cancer vaccination. As WTL is essentially self and has poor immunostimulatory potential, different preparation methods have been explored to increase its immunogenicity.

Preparation of whole-tumor cell lysate vaccine

In clinics, WTL is commonly prepared using ultraviolet B (UVB) ray-irradiation and/or repeated freeze–thaw cycles to inactivate and completely fragment the tumor cells. Studies have shown that UVB irradiation-induced apoptotic cell death and the exposure of "find me" and "eat me" signals on the tumor cells in the early phase of apoptosis to facilitate uptake by phagocytes.[20] The engulfment of the apoptotic tumor cells could be achieved via three major steps—first, phagocytes, such as dendritic cells (DCs) and macrophages, are attracted by "find me" signals including fractalkine,[21] lysophosphatidylcholine,[22] adenosine triphosphate and uridine 5′ triphosphate,[23] and sphingosine 1-phosphate,[24] released by the apoptotic cells. Second, the apoptotic cells also displayed major "eat me" signals, such as phosphatyidylserine, that are normally asymmetrically distributed and restricted to the inner leaflet of the plasma membrane[25,26] and calreticulin which is a Ca2 + -binding protein in the endoplasmic reticulum on their cell surface for specific recognition and engulfment by the phagocytes.[27] Following the ingestion of the apoptotic cells by the phagocytes, the third step is to process and degrade them through a series of phagosome maturation steps.[28,29] In the late-stage apoptosis, the tumor cells could also release other immunogenic cellular components, such as high mobility group box 1 (HMGB-1)[30] and pentraxin-3,[31,32] to stimulate DC maturation and activation.

Direct freezing and thawing of tumor cells, on the other hand, induces primary necrotic cell death that is characterized by the release of "danger signals" or damage-associated molecular patterns (DAMPs) from the dying cells. Such DAMPs, including ribonucleic acid (RNA), deoxyribonucleic acid (DNA), uric acid, cellular membrane, heat shock proteins (HSP) 70 and 90, mitochondria, and HMGB-1, have shown the ability to induce partial maturation in DCs.[33] It has been shown that HMGB-1[34] and HSP70[35] could interact with the scavenger receptor-A and Toll-like receptor (TLR) 4 on DCs. It is also shown that monocytes could secrete proinflammatory cytokines/chemokines, including tumor necrosis factor (TNF)-α, interleukin (IL)-1α, IL-1β, IL-6, IL-8, macrophage inflammatory protein (MIP)-1α, and MIP-1β, after coculturing with purified HMGB-1.[36] Similarly, HSP70, HSP90, and GP96 could induce the secretions of TNF-α, IL-1β, IL-6, and IL-12 in human monocytes and DCs.[37–39] Uric acid, which is the natural end product of the purine metabolic pathway of degraded cellular RNAs and DNAs, has shown the ability to activate $CD8^+$ T-cell responses in mice when coinjected with an antigen.[34] Furthermore, uric acid could enhance $CD4^+$ T-cell responses toward particulate antigens[40,41] and exert tumor suppression.[42]

The use of hypochlorous acid (HOCl), a strong oxidizing agent, has been investigated for the rapid induction of primary necrosis in tumor cells as well as enhancing their immunogenicity in a DC-based immunotherapy.[43,44] When compared to heat-killed or hydrochloric acid (HCl) treatment that did not induce oxidation, tumor cells treated with HOCl-oxidization successfully primed T cells that were antigen and tissue-type specific.[43] Following large-scale optimization of this approach, we conducted an ovarian cancer (OC) clinical trial to evaluation its safety and effectiveness.[9] In four out of five patients, we were able to elicit polyclonal $CD8^+$ and $CD4^+$ antigen-specific T-cell responses against known ovarian TAAs following HOCl-oxidized WTL vaccination. Two subjects showed prolonged progression-free survival intervals of 36 and 43 months that were longer than subjects in historical control groups treated with available second- or third-line chemotherapy.[9] Furthermore, we observed an amplification of T-cell responses against mutated neoantigens with increased avidity in certain patients as well as priming of T cells against previously unrecognized neoepitopes.[10] The effect of HOCl oxidation could be explained by the following proposed mechanisms. First, HOCl oxidation could rapidly induce necrotic tumor cell death to help release the different types of DAMPs as discussed in primary necrosis with repeated freeze–thaw cycles. Second, HOCl oxidation could lead to modification of numerous biomolecules in the tumor cells including surface proteins, lipids, and glycoproteins. It has been demonstrated that the scavenger receptor lectin-like oxidized low-density lipoprotein receptor (LOX-1) on murine DCs showed increased binding to HOCl-oxidized ovalbumin (OVA).[45] This suggested that HOCl could play a role in targeting oxidized antigens to scavenger receptors on the surface of antigen-presenting cells (APCs).[45] Third, HOCl oxidation could assist in inactivating immunosuppressive cytokines, such as transforming growth factor (TGF)-β and IL-10, that are released by live tumor cells.[46,47]

Methods of administrating whole-tumor cell lysate vaccine

Direct injection with Toll-like receptor agonist as adjuvant

A straightforward method of administrating WTL vaccine is with direct injection via different routes, such as intradermally, subcutaneously, or intravenously. The main aim of direct injection method is to deliver the WTL to professional APCs in vivo to promote WTL uptake, processing, and presentation to elicit antigen-specific T-cell responses. Among the professional APCs in the immune system, DCs are the most efficient in priming and activating naive T cells to become antigen-specific. DCs are also highly phagocytic and are strong producers of IL-12, an important T-cell-priming cytokine, upon appropriate stimulation.[9,48] Thus targeting DCs in vivo could be an attractive approach. As discussed earlier, WTL could be prepared in different ways to enhance their immunogenicity and uptake by phagocytes. Specific adjuvants could be coadministered with WTL to further enhance the latter's immunostimulatory capacity. Indeed, TLR agonists have shown to be potent adjuvants for activating DCs in vivo in preclinical and clinical studies. Depending on the desired immune responses, different TLR agonists could be used in combination to target more than one DC subpopulations.

Two major populations of DCs, that is, lymphoid-derived plasmacytoid DCs (pDCs) or myeloid-derived conventional DCs (cDCs), are generated from common $CD34^+$ bone marrow hematopoietic progenitor cells following different developmental pathways. pDCs reside in different lymphoid organs, including thymus, spleen, lymph nodes, mucosal-associated tissues, and liver in normal conditions. They could also migrate to the skin in response to infection and pathological conditions.[49] As pDCs express TLR7 and 9, they could be targeted in vivo with TLR7 and 9 agonists. The United States Food and Drug Administration (FDA) has approved the synthetic TLR7 agonist, Imiquimod (R-837), for use as a topical treatment for superficial basal cell carcinomas.[50,51] It has demonstrated the ability to induce strong infiltrations of pDCs and T and natural killer (NK) cells at the site of application. The activated pDCs would produce high levels of interferon (IFN)-α to facilitate cross-presentation of antigens to T cells, activating $CD4^+$, $CD8^+$ T, and NK cell responses, as well as inducing antibody responses.[52–54] Another synthetic TLR7 agonist, 852A (3M-001), was able to induce stable diseases when administered subcutaneously to patients with recurrent ovarian, cervical, and breast cancers,[55] as well as intravenously in patients with melanoma.[56] Lastly, Resiquimod (R-848; S-28463) could interact with TLR7 and 8 on pDCs to induce T helper (Th) 1-polarizing cytokine secretions, including IFN-α, IL-6, -8, and -12, and TNF-α.[57] Melanoma patients who received a topical application of Resiquimod and subcutaneous injection of full-length NY-ESO-1 protein emulsified in Montanide (an immune modulator[58]) developed therapeutic humoral responses against NY-ESO-1 antigen.[59]

Synthetic CpG-oligodeoxynucleotide (CpG ODN) ligands could be used to target pDCs via their TLR9. Different types of CpG ODNs have been developed (classes A, B, C, and P) to mimic the immunostimulatory activity of bacterial CpG[60,61] for inducing pDC maturation and production of Th 1-polarizing cytokines.[62] Melanoma patients receiving subcutaneous injections of CpG ODN 7909, MART-1 synthetic peptide and incomplete Freund's adjuvant demonstrated increased MART-1-specific $CD8^+$ effector T cells in the peripheral blood.[63] Similarly, patients treated with CpG ODN 7909 and NY-ESO-1 peptides/proteins emulsified in Montanide showed increased numbers of circulating NY-ESO-1-specific $CD8^+$ T cells.[64–66] Given these promising results, it could be envisaged that WTL could be emulsified in Montanide with TLR7 or 9 (or both) agonist for subcutaneous injection. Alternatively, TLR7 agonist could be given as a topical treatment (Imiquimod and Resiquimod have shown effectiveness in this format) together with subcutaneous injection of WTL (alone or with TLR9 agonists) emulsified in Montanide.

Myeloid-derived cDCs, that is, Langerhan cells (LCs), dermal DCs, and blood DCs, express different sets of TLRs as compared to pDCs. Both LCs and dermal DCs reside in the human skin where they act as the first line of defense against invading pathogens. LCs are characterized by TLR1, 2, 3, 6, and 10 expression, while dermal DCs express TLR2, 4, 5, 6, 8, and 10.[67] These two DC subsets also regulate immune responses differently; LCs have demonstrated the ability to elicit antigen-specific $CD8^+$ T cells expressing cytolytic molecules, such as granzymes and perforin,[68] and preferentially produced IL-15 to promote naive $CD8^+$ T-cell proliferation and memory T-cell development.[69] On the other hand, dermal DCs are involved in regulating B-cell differentiation and antibody class switching through IL-12-mediated activation of $CD4^+$ Th cells.[70,71] Blood DCs could be further classified into two subsets based on their BDCA-1^+ or BDCA-3^+ expression. BDCA-1^+ DC population, which constitutes the majority of blood DCs, express TLR1 and 6. BRAC-3^+ DCs, which are present in a lower frequency, express TLR3 and 8. It was shown that BRAC-3^+ DCs secreted IL-12 upon TLR3 stimulation and efficiently cross-presented antigens to $CD8^+$ T cells.[72,73]

Poly I:C, a synthetic double-strand RNA that binds TLR3, could be used to target LCs and BDCA-3^+ blood DCs based on their TLR3 expression. The Poly-ICLC (Hiltonol), which is RNAse-resistant and stabilized with poly-lysine, has demonstrated the ability to increase type I and II IFN secretions from DCs.[74] When given intramuscularly to glioma patients, 20 out of 38 of the treated patients experienced tumor regression or stabilization.[75] Similarly, 13 out of 45 patients with advanced malignancies experienced stable disease after treatment with Poly-ICLC topically or subcutaneously in combination with Resiquimod and CDX-1401 (monoclonal anti-DEC-205 fused to full-length NY-ESO-1).[76] Another Poly I:C formulation, Poly(I:C12U) [Ampligen], has been developed to markedly reduce systemic toxicity as compared to unmodified Poly I:C.[77] It could stimulate IL-12 production in ex vivo generated DCs.[78] Currently, it is being evaluated in OC (NCT03734692, NCT03599453) breast (NCT03403634), prostate (NCT02432378), colorectal (NCT03734692), and pancreatic (NCT03899987) cancers in combination with IFN-α2b (Intron A), cyclooxygenase 2 inhibitor (Celecoxib), and/or chemotherapy (http://www.clinicaltrials.gov). Hence, WTL and Poly-ICLC could potentially be emulsified in Montanide and administer via intramuscular or subcutaneous route.

MPLA, a dual TLR2 and 4 agonist, could be used to target dermal DCs via their TLRs. It is a derivative of the lipid A fraction of lipopolysaccharide (LPS) found in the outer membrane of the Gram-negative *Salmonella minnesota*. It retains the ability to activate DCs via its immunologically active lipid A portion but without the physiological toxicity associated with the parental LPS.[79–81] Compared to LPS, MPLA has shown to elicit a less pronounced inflammatory cytokine response following TLR4 stimulation through the Toll/IL-1 receptor domain-containing adaptor-inducing IFN-β (TRIF) signaling adaptor.[80] MPLA has demonstrated the ability to activate DCs to increase their phagocytic capacity and antigen presentation, as well inducing them to secrete a cascade of proinflammatory cytokines including IL-1, IL-2, granulocyte-macrophage colony-stimulating factor (GM-CSF), and TNF-α.[82]

In addition, MPLA is able to stimulate Th1 cells to produce IL-2 and IFN-γ to augment antigen-specific $CD8^+$ T-cell response and inducing complement-fixing antibodies.[83,84] MPLA is the first and only TLR agonist to be licensed for use as an adjuvant in different human vaccines, for example, in Cervarix, which is a prophylactic vaccine against human papillomavirus (HPV) types 16 and 18, and in Melacine, which is an allogenic melanoma cell lysate vaccine.[13] MPLA could potentially be coadministered with WTL via intradermal or intramuscular route to activate dermal DCs.

The small molecule TLR8 agonist, VTX-2337 (Motolimod), could potentially be used to target blood DCs via their TLR8. It has demonstrated the ability to elicit TNF-α and IL-12 productions from DCs, as well as inducing IFN-γ production from NK cells to augmented antibody-dependent cell-mediated cytotoxicity by rituximab and trastuzumab.[85] Subcutaneous injection of VTX-2337 in patients with advanced solid tumors and lymphoma led to increased plasma granulocyte-colony-stimulating factor, monocyte chemoattractant protein-1, TNF-α, and MIP-1β.[86] Eight out of the 38 treated patients experienced stable diseases.[86] In a study, 7 out of 13 patients with head and neck squamous cell carcinoma treated with VTX-2337 and cetuximab showed partial responses or stable diseases.[87] Thus WTL and VTX-2337 could potentially be coinjected subcutaneously to target DCs.

Whole-tumor cell lysate pulsed on ex vivo generated dendritic cells

DC-based immunotherapy with ex vivo generated myeloid-derived DCs has been widely investigated in clinics.[88] The main advantage of using this DC subpopulation is that a large number of circulating monocytes as the starting material could be obtained via apheresis from the patients. They could easily be generated ex vivo from blood monocytes with recombinant differentiation cytokines GM-CSF and IL-4. As discussed earlier, different DC subpopulations exist in different tissue compartments and they could distinctly regulate different types of immune responses. For example, pDCs could be generated ex vivo from blood in the presence of IL-3 and Fms-like tyrosine kinase 3 ligand (Flt3L).[89] When administered to melanoma patients, the pDCs loaded with TAA peptide antigens were able to elicit tumor-specific $CD8^+$ and $CD4^+$ T-cell responses.[90] The efficacy of ex vivo cultured autologous LCs and monocyte-derived DCs had been compared in advanced melanoma patients. These two DC populations were pulsed with tyrosinase-related protein-2 peptide prior to administering to the patients. It was shown that the peptide-pulsed LCs produced significantly more IL-15 than monocyte-derived DCs, and they were more efficient than monocyte-derived DCs in activating antigen-specific T cells with a cytolytic profile.[91] So far, few studies have investigated the use of ex vivo generated pDCs, LCs, and other DC subpopulations in cancer immunotherapy. Hence, more studies are warranted.

To date, DC-based immunotherapy is effective only in a small percentage of cancer patients.[92] Thus there is a strong need to optimize the preparation of ex vivo DCs. Upon uptake of tumor antigens, DCs need to be properly activated to become potent APCs for efficient antigen presentation and stimulation of tumor-specific T cells.[93,94] For this purpose, different maturation stimuli have been investigated. The "gold standard" cytokine cocktail comprised of IL-1β, IL-6, TNF-α, and prostaglandin E2 (PGE2) is widely used for activating myeloid-derived DCs.[95,96] It had been demonstrated that DCs activated with this cytokine combination elicited $CD4^+$ and $CD8^+$ T cells that produced IFN-γ and no IL-4 and IL-10.[96] However, a study showed that PEG2 could mediate the expression of indoleamine 2,3-dioxygenase 1 (IDO1) in DCs and that in turn impaired their ability to induce IFN-γ producing $CD8^+$ and $CD4^+$ T cells.[97] IDO1-expressing DCs had also been shown to induce $Foxp3^+$ T regulatory (Treg) cells and suppress effector T-cell functions.[97,98]

We developed alternative stimuli comprised of a clinical grade LPS (*Escherichia coli* O:113) and recombinant IFN-γ to activate myeloid-derived DCs that were pulsed with HOCl-oxidized WTL.[9] We demonstrated that DCs treated with this stimuli combination significantly upregulated surface expressions of costimulatory molecules (CD80, CD86, and CD40), antigen-presentation molecules [major histocompatibility (MHC) class I and II], as well as the migration marker CCR7. These activated DCs also produced high levels of IL-12p70 and IFN-inducible protein 10 (CXCL10) that were essential for priming antitumor T cells. Importantly, these DCs could elicit polyclonal $CD8^+$ and $CD4^+$ T-cell responses against known ovarian TAAs and tumor neoantigens in the treated patient.[9,10] In a previous study, myeloid-derived DCs stimulated with either MPLA or CD40 agonistic antibodies demonstrated upregulations of surface CD83, CD86, and MHC class I and II.[44] These HOCl-oxidized WTL-pulsed DCs also successfully elicited $CD8^+$ T cells that recognized TAAs expressed by the WTL (HER-2/neu and MUC1). $CD4^+$ class II-restricted T-cell responses, as well as responses against autologous tumor cells isolated from the patient's ascites, were induced.[44]

TLR3 agonist Poly(I:C12U) has shown the ability to activate myeloid-derived DCs in ex vivo culture by increasing their surface expressions of CD83, CCR7, CD86, CD40, and MHC class I and II.[61] It could also elicit significantly stronger IL-12 secretion and lower IL-10 from the DCs when compared to the unmodified Poly I:C.[78] Poly(I:C12U) used in combination with IL-1β, IFN-α, IFN-γ, and TNF-α had been demonstrated to effectively activate α-type-1-polarized DCs (αDC1) in a recurrent high grade glioma clinical study.[99] Myeloid-derived DCs pulsed with WT1, HER-2/neu, MAGE-A3, and MAGE-A1 synthetic peptides and keyhole limpet hemocyanin and then activated with the above stimuli cocktail. Following intradermal injection of these antigen-pulsed αDC1s, six out of nine treated glioma patients demonstrated responses to the immunized antigens. One patient also experienced stable disease after the treatment.[99] When compared to DCs activated with the "gold standard" cytokine cocktail (IL-1β, IL-6, TNF-α, and PGE2), such αDC1s demonstrated a high migratory responses to the CCR7 ligand, producing higher levels of IL-12p70, and elicited 40-fold higher numbers of long-lived CTLs against melanoma TAAs (MART-1, gp100, and tyrosinase) in another study.[100]

Whole-tumor cell lysate encapsulated in nanoparticles

Nanoparticles are small particles between 1 and 100 nm in size. They could be sourced from naturally occurring biomaterials or engineered to possess unique physical and chemical properties for specific purposes. Based on these characteristics, nanoparticles are well suited for diverse use including in medical and imaging applications.[101–103] In cancer immunotherapy, different types of nanoparticles have been evaluated.[103,104] Particularly, FDA has approved a variety of polymer-based nanoparticles including poly(lactide-o-glycolic) acid (PLGA), polyethylene glycol, and chitosan for clinical use based on their biocompatible, nontoxic, and biodegradable features.[105] Gold and lipid-based (liposomes) nanoparticle have also been approved for therapeutic purposes.[105]

Solbrig et al. demonstrated that defined protein antigens, such as glycoprotein 100 and OVA, and WTL–protein complex mixtures could be efficiently encapsulated by PLGA nanoparticles.[106] When compared to free soluble antigens, these encapsulated antigens retained their antigenicity and were more superior in stimulating antigen-specific T-cell responses in ex vivo cell culture systems and in controlling in vivo tumor growth.[106] In a study, a liposome-based nanoparticle delivery system was being investigated to deliver WTL in vivo in the B16.F10 melanoma tumor model.[107] Controlled release of WTL from the liposomes in vivo was achieved by using near-infrared irradiation. Furthermore, antigen-specific $CD8^+$ CTLs were elicited in the mice following treatment with the liposome-WTL preparation. When combined with liposomes that had encapsulated doxorubicin (a chemotherapeutic drug), significantly greater efficacy was observed in vivo.[107] Another study evaluated the use of chitosan nanoparticles (CTS-NPs) to encapsulate WTL.[108] These CTS-NPs were also engineered to display surface mannose moieties for specific targeting to DCs and shown to induce DC maturation following ex vivo coculture. In the B16 tumor model, mice that were immunized subcutaneously with CTS-NPs loaded with WTL and rechallenged with B16 tumor cell inoculation demonstrated delayed tumor growth. On further analysis, it was found that the injected CTS-NPs loaded with WTL were readily taken up by endogenous DCs in the draining lymph node. A systemic effect was induced, as demonstrated by increased sera IFN-γ and IL-4 levels, as well as CTL responses against the tumor.[108]

Different adjuvants including TLR agonists could be encapsulated with WTL in the nanoparticle in order to increase the latter's overall immunostimulatory capacity. In a study, calcium phosphate nanoparticles that had encapsulated murine Class B CpG ODN 1826 and tumor-associated antigen peptides were evaluated in a colorectal tumor model.[109] The treated mice demonstrated an induction of CTL response as well as delayed tumor growth, and these responses could be further enhanced by combining with immune checkpoint inhibitor against programmed death-ligand 1 (PD-L1).[109] In another study, PLGA particles encapsulating Pam3CSK4 (TLR2 agonist) and Poly I:C and further coating with an agonistic αCD40-mAb to target DCs were evaluated in the B16-OVA tumor model.[110] Mice treated with this PLGA nanoparticle preparation were able to suppress tumor growth. Moreover, 50% of the mice survived significantly longer at day 40 posttumor inoculation.[110] Hence, these studies demonstrated the feasibility of using nanoparticles to target WTL to DCs in vivo.

Combinatorial therapeutic approaches to increase whole-tumor cell lysate vaccine efficacy

Immune checkpoint blockade

Immune checkpoint blockade (ICB) therapy has revolutionized the management of cancer and demonstrated survival benefits as a monotherapy in patients. T-cell inhibitory mechanisms, such as programmed cell death protein-1 (PD-1), PD-L1, and cytotoxic T-lymphocyte antigen-4 (CTLA-4), could hinder the effectiveness of cancer immunotherapies that aim to break self-tolerance. Several immune checkpoint inhibitors (i.e., anti-PD-1 monoclonal antibodies—nivolumab and pembrolizumab; anti-PD-L1 monoclonal antibodies—atezolizumab and durvalumab; and anti-CTLA-4 mononclonal antibody—ipilimumab) have been approved by FDA as a frontline treatment in different cancers, including metastatic melanoma, renal cell carcinoma, non-small-cell lung cancer (NSCLC), and bladder cancer.[111]

ICB therapy has been evaluated in combination with DC-based immunotherapy with some success. In a phase II clinical trial, advanced melanoma patients treated with autologous monocyte-derived DCs electroporated with synthetic mRNA (TriMixDC-MEL) and ipilimumab.[112] A total of 39 patients were treated, and 7 patients showed complete responses and 1 patient had a partial tumor response (38% overall response rate). Grade 3 and 4 adverse events were observed in 36% of the patients.[112] In a phase I advanced melanoma trial, patients were treated with monocyte-derived DCs pulsed with MART-1_{26-35} peptide and a dose escalation of the CTLA-4-blocking antibody tremelimumab (CP-675,206).[113] The combinatorial treatment was well tolerated and side effects were limited to grade 2 hypophysitis and grade 3 diarrhea in 2 out of the 3 patients receiving monthly tremelimumab at10 mg/kg. Of the 16 treated patients, 2 patients demonstrated partial responses and another 2 patients showed complete responses.[113] Similarly, encouraging results were observed in preclinical studies. In a glioma tumor model, mice treated with two doses of bone marrow-derived DCs pulsed with glioma WTL in combination with monoclonal PD-1 blocking antibodies demonstrated prolonged overall survival when compared to mice treated with single therapy.[114] It was further demonstrated that $CD8^+$ T cells were essential for tumor suppression, and the TILs showed increased expression of integrin homing and immunologic memory markers.[114] In a murine lung tumor model, 80% of the mice treated with bone marrow-derived DCs loaded with WTL in combination with anti-PD-1 blocking antibody showed prolonged survival.[115] Both $CD4^+$ and $CD8^+$ T cells were found to be necessary for tumor control. In the tumor microenvironment, increased granzyme B and perforin expressions, as well as increased number of TILs, were detected.[115] Based on these encouraging results, more studies are warranted to evaluate the combinatorial regime of DC-WTL cancer vaccine and ICB therapy.

Depleting Treg cells

Naturally occurring Treg cells play an important role in maintaining peripheral self-tolerance and immune homeostasis.[116,117] However, in the context of cancers, peripheral Treg cells could be induced or converted from conventional $CD4^+$ T cells (Teffs) via different stimulatory signals.[118,119] Such Treg cells demonstrated the ability to produce immunosuppressive cytokines, such as TGF-β and IL-10 and IL-35, to impair effector T-cell functions.[120–123] Treg cells could also induce the downregulation of costimulatory molecules CD80 and CD86 on DCs, impairing the latters' maturation and ability to activate effector T cells.[124,125] Furthermore, studies showed that Treg cells could stimulate the expression of immunosuppressive IDO and inhibit the production of TNF-α and IL-6 in DCs via the engagement of CTLA-4 on Tregs to cognate receptors on DCs.[126,127] Thus depleting Treg cells prior to cancer immunotherapy could strengthen the activation of tumor-reactive T cells and help increase the overall effectiveness of immunotherapies.

Cyclophosphamide (CY), a commonly used chemotherapeutic drug, has shown the ability to deplete Treg cells when used at low metronomic doses. Treg cells were targeted by CY due to their high rate of proliferation, and their depletion led to improved activation of antitumor T cells.[128–130] A study indicated that CY could specifically deplete a distinct population of cycling $Ki\text{-}67^{hi}ICOS^{hi}TNFR2^{hi}$ Treg cells that were enriched in the tumor and shown to be maximally suppressive.[131–134] In malignant pleural mesothelioma, the use of CY in combination with a DC vaccine pulsed with autologous WTL was shown to be safe and induced radiographic disease control in 8 of the 10 treated patients.[135] Seven patients survived for at least 24 months following treatment and 2 patients remained alive after 50 and 66 months. The percentages of Treg cells in the peripheral blood of the patients were found to be reduced following 7 days of low-dose CY treatment.[135] In a phase II study, recurrent glioblastoma (GBM) patients were treated with a vaccine (ERC1671; Gliovac) that comprised of allogeneic/autologous whole-tumor lysate and inactivated tumor cells derived from the patient and three GBM donors in combination with CY, GM-CSF, and bevacizumab.[136] Patients receiving this combination showed a prolonged median OS when compared with patient receiving placebo and bevacizumab (12 vs 7.5 months).[136] Recently, we demonstrated that a DC vaccine pulsed with HOCl-oxidized autologous WTL given in combination with bevacizumab and CY led to a prolonged 2-year OS in OC patients when compared to a historical group that received CY plus bevacizumab (100% vs 44%).[10] Eight out of 10 patients who received the DC-CY-bevacizumab combination showed an immune response when compared to patients who received DC vaccine only, or DC vaccine plus bevacizumab (3 out of 12 patients).[10] Another chemotherapeutic drug, fludarabine, had shown the ability to deplete Treg cells.[137] Hence, CY and fludarabine could potentially be used in combination to target Treg cells.

Antiangiogenesis

Angiogenesis, the process of new blood vessel formation, is important in both physiological (such as embryonic development, reproduction, and wound healing) and pathological conditions (such as cancer and psoriasis).[138] It is a multistep process that is tightly governed by different angiogenic stimulators and inhibitors. In cancer, however, there is a sustained tumor angiogenesis to provide nutrients and oxygen for rapid tumor growth, invasion, and metastasis. Vascular endothelial growth factor (VEGF) is shown to be a key angiogenic stimulator secreted by tumor cells to promote an "angiogenic switch," whereby the balance of the angiogenic factors is tipped toward a proangiogenic outcome.[139] The tumor vasculatures are structurally and functionally abnormal; the blood vessels are often leaky, hemorrhagic with high interstitial pressure. They are also hypoxic, leading to the stimulation of more VEGF production. Hence, targeting VEGF could help to control tumor angiogenesis, normalize vasculature, and facilitate chemotherapeutic drug delivery to the tumor.[140,141] Bevacizumab (Avastin), a humanized monoclonal antibody that binds to VEGF-A,[142,143] has demonstrated the ability to prolong the overall survival of metastatic colon cancer patients when use in combination with fluorouracil-based chemotherapy.[144,145] It has also been approved by FDA for treating NSCLC,[146] OC,[145] and cervical carcinoma.[147]

Blocking VEGF could be beneficial for DC-based immunotherapy as VEGF has demonstrated the ability to suppress the activation and function of myeloid-derived DCs.[148–150] VEGF could induce the expression of PD-L1 on myeloid DCs, thereby suppressing the activation of T cells via this mechanism.[151] Blocking VEGF had resulted in improved induction of CTLs with a DC vaccine, as well as a reduction in Treg cell numbers.[152] In a clinical study, DCs pulsed with prostatic acid phosphatase (Sipuleucil-T) used in combination with bevacizumab led to a higher response rate in the patients.[153] In a recent study, we demonstrated that OC patients treated with DCs pulsed with HOCl-oxidized autologous WTL, bevacizumab, and CY experienced a longer progression-free survival and elicitation of T-cell responses against tumor neoantigens.[10] We further evaluated this combination with acetylsalicylic acid and low-dose IL-2 and found that OC patients treated with this treatment regimen showed significantly prolonged overall survival, as well as increased elicitation of polyfunctional $CD8^+$ T cells.[154] Analysis in the ID8 murine OC tumor model revealed that this combinatorial treatment regimen led to an increase infiltration of perforin-expressing $CD8^+$ T cells and a positive correlation with reduced tumor burden.[154]

Several VEGFR inhibitors have been developed and evaluated in clinics. Sorafenib, which is a multikinase inhibitor that targets VEGFR and platelet-derived growth factor receptor, has demonstrated antitumor effects against tumors.[155] Other antiangiogenesis inhibitors that target all VEGFRs, including sunitinib,[156] lenvatinib,[157] axitinib,[158] regorafenib,[159] cabozantinib,[160] pazopanib,[161] and vandetanib,[162] have been approved by FDA for treating different cancers. Ramucirumab, a fully humanized monoclonal antibody that targets the extracellular domain of VEGFR, has also been approved by FDA.[163]

Conclusion

Targeting tumors with WTL cancer vaccines is an attractive approach as it allows the simultaneous activation of $CD4^+$ and $CD8^+$ T-cell responses to prevent tumor escape variants. Vaccination with personalized autologous WTL is preferred as it could help to induce antitumor responses against both TAAs and private tumor neoantigens for more efficient tumor control. The immunogenicity and overall efficacy of WTL vaccine could be enhanced via different steps, that is, vaccine preparation, administration, and use in combinatorial treatment regimen with different FDA-approved immunostimulatory adjuvants.

References

1. Sasieni PD, Shelton J, Ormiston-Smith N, Thomson CS, Silcocks PB. What is the lifetime risk of developing cancer?: the effect of adjusting for multiple primaries. *Br J Cancer*. 2011;105(3):460–465. Available from: https://doi.org/10.1038/bjc.2011.250.

2. Smittenaar CR, Petersen KA, Stewart K, Moitt N. Cancer incidence and mortality projections in the UK until 2035. *Br J Cancer*. 2016;115(9):1147–1155. Available from: https://doi.org/10.1038/bjc.2016.304.
3. Tobias J, Hochhauser D. Systemic treatment for cancer. In: Tobias J, Hochhauser D, eds. *Cancer and Its Management*. Hoboken, NJ: John Wiley & Sons, Ltd; 2014:81–112. doi.org/10.1002/9781118468753.ch6.
4. Vasan N, Baselga J, Hyman DM. A view on drug resistance in cancer. *Nature*. 2019;575(7782):299–309. Available from: https://doi.org/10.1038/s41586-019-1730-1.
5. Kaplan DH, Shankaran V, Dighe AS, Stockert E, Aguet M, Old LJ, et al. Demonstration of an interferon gamma-dependent tumor surveillance system in immunocompetent mice. *Proc Natl Acad Sci USA*. 1998;95(13):7556–7561. Available from: https://doi.org/10.1073/pnas.95.13.7556.
6. Shankaran V, Ikeda H, Bruce AT, White JM, Swanson PE, Old LJ, et al. IFN-gamma and lymphocytes prevent primary tumour development and shape tumour immunogenicity. *Nature*. 2001;410(6832):1107–1111. Available from: https://doi.org/10.1038/35074122.
7. Dunn GP, Old LJ, Schreiber RD. The three Es of cancer immunoediting. *Ann Rev Immunol*. 2004;22:329–360. Available from: https://doi.org/10.1146/annurev.immunol.22.012703.104803.
8. Mini E, Nobili S. Pharmacogenetics: implementing personalized medicine. *Clin Cases Miner Bone Metab*. 2009;6(1):17–24.
9. Chiang CL, Kandalaft LE, Tanyi J, Hagemann AR, Motz GT, Svoronos N, et al. A dendritic cell vaccine pulsed with autologous hypochlorous acid-oxidized ovarian cancer lysate primes effective broad antitumor immunity: from bench to bedside. *Clin Cancer Res*. 2013;19(17):4801–4815. Available from: https://doi.org/10.1158/1078-0432.Ccr-13-1185.
10. Tanyi JL, Bobisse S, Ophir E, Tuyaerts S, Roberti A, Genolet R, et al. Personalized cancer vaccine effectively mobilizes antitumor T cell immunity in ovarian cancer. *Sci Transl Med*. 2018;10(436). Available from: https://doi.org/10.1126/scitranslmed.aao5931.
11. Toes RE, Ossendorp F, Offringa R, Melief CJ. CD4 T cells and their role in antitumor immune responses. *J Exp Med*. 1999;189(5):753–756. Available from: https://doi.org/10.1084/jem.189.5.753.
12. Neller MA, López JA, Schmidt CW. Antigens for cancer immunotherapy. *Semin Immunol*. 2008;20(5):286–295. Available from: https://doi.org/10.1016/j.smim.2008.09.006.
13. Sosman JA, Sondak VK. Melacine®: an allogeneic melanoma tumor cell lysate vaccine. *Expert Rev Vaccines*. 2003;2(3):353–368. Available from: https://doi.org/10.1586/14760584.2.3.353.
14. Greenman C, Stephens P, Smith R, Dalgliesh GL, Hunter C, Bignell G, et al. Patterns of somatic mutation in human cancer genomes. *Nature*. 2007;446:153. Available from: https://doi.org/10.1038/nature05610.
15. Wolfel T, Hauer M, Schneider J, Serrano M, Wolfel C, Klehmann-Hieb E, et al. A p16INK4a-insensitive CDK4 mutant targeted by cytolytic T lymphocytes in a human melanoma. *Science (New York, NY)*. 1995;269(5228):1281.
16. Zhou J, Dudley ME, Rosenberg SA, Robbins PF. Persistence of multiple tumor-specific T-cell clones is associated with complete tumor regression in a melanoma patient receiving adoptive cell transfer therapy. *J Immunother*. 2005;28(1):53–62.
17. Verdegaal EME, de Miranda NFCC, Visser M, Harryvan T, van Buuren MM, Andersen RS, et al. Neoantigen landscape dynamics during human melanoma–T cell interactions. *Nature*. 2016;536:91–95. Available from: https://doi.org/10.1038/nature18945.

18. Balachandran VP, Łuksza M, Zhao JN, Makarov V, Moral JA, Remark R, et al. Identification of unique neoantigen qualities in long-term survivors of pancreatic cancer. *Nature*. 2017;551:512. Available from: https://doi.org/10.1038/nature24462.
19. Linnemann C, van Buuren MM, Bies L, Verdegaal EM, Schotte R, Calis JJ, et al. High-throughput epitope discovery reveals frequent recognition of neo-antigens by $CD4^{+}$ T cells in human melanoma. *Nat Med*. 2015;21:81–85. Available from: https://doi.org/10.1038/nm.3773.
20. Lauber K, Blumenthal SG, Waibel M, Wesselborg S. Clearance of apoptotic cells: getting rid of the corpses. *Mol Cell*. 2004;14(3):277–287. Available from: https://doi.org/10.1016/S1097-2765(04)00237-0.
21. Truman LA, Ford CA, Pasikowska M, Pound JD, Wilkinson SJ, Dumitriu IE, et al. CX3CL1/fractalkine is released from apoptotic lymphocytes to stimulate macrophage chemotaxis. *Blood*. 2008;112(13):5026–5036. Available from: https://doi.org/10.1182/blood-2008-06-162404.
22. Lauber K, Bohn E, Kröber SM, Xiao Y-j, Blumenthal SG, Lindemann RK, et al. Apoptotic cells induce migration of phagocytes via caspase-3-mediated release of a lipid attraction signal. *Cell*. 2003;113(6):717–730. Available from: https://doi.org/10.1016/S0092-8674(03)00422-7.
23. Elliott MR, Chekeni FB, Trampont PC, Lazarowski ER, Kadl A, Walk SF, et al. Nucleotides released by apoptotic cells act as a find-me signal to promote phagocytic clearance. *Nature*. 2009;461(7261):282–286. Available from: https://doi.org/10.1038/nature08296.
24. Gude DR, Alvarez SE, Paugh SW, Mitra P, Yu J, Griffiths R, et al. Apoptosis induces expression of sphingosine kinase 1 to release sphingosine-1-phosphate as a "come-and-get-me" signal. *FASEB J*. 2008;22(8):2629–2638. Available from: https://doi.org/10.1096/fj.08-107169.
25. Marguet D, Luciani M-F, Moynault A, Williamson P, Chimini G. Engulfment of apoptotic cells involves the redistribution of membrane phosphatidylserine on phagocyte and prey. *Nat Cell Biol*. 1999;1(7):454–456. Available from: https://doi.org/10.1038/15690.
26. Fadok VA, Bratton DL, Rose DM, Pearson A, Ezekewitz RAB, Henson PM. A receptor for phosphatidylserine-specific clearance of apoptotic cells. *Nature*. 2000;405(6782):85–90. Available from: https://doi.org/10.1038/35011084.
27. Obeid M, Tesniere A, Ghiringhelli F, Fimia GM, Apetoh L, Perfettini JL, et al. Calreticulin exposure dictates the immunogenicity of cancer cell death. *Nat Med*. 2007;13(1):54–61. Available from: https://doi.org/10.1038/nm1523.
28. Kinchen JM, Ravichandran KS. Phagosome maturation: going through the acid test. *Nat Rev Mol Cell Biol*. 2008;9(10):781–795. Available from: https://doi.org/10.1038/nrm2515.
29. Kinchen JM, Ravichandran KS. Identification of two evolutionarily conserved genes regulating processing of engulfed apoptotic cells. *Nature*. 2010;464(7289):778–782. Available from: https://doi.org/10.1038/nature08853.
30. Tian J, Avalos AM, Mao SY, Chen B, Senthil K, Wu H, et al. Toll-like receptor 9-dependent activation by DNA-containing immune complexes is mediated by HMGB1 and RAGE. *Nat Immunol*. 2007;8(5):487–496. Available from: https://doi.org/10.1038/ni1457.
31. Rovere P, Peri G, Fazzini F, Bottazzi B, Doni A, Bondanza A, et al. The long pentraxin PTX3 binds to apoptotic cells and regulates their clearance by antigen-presenting dendritic cells. *Blood*. 2000;96(13):4300–4306.
32. Baruah P, Propato A, Dumitriu IE, Rovere-Querini P, Russo V, Fontana R, et al. The pattern recognition receptor PTX3 is recruited at the synapse between dying and dendritic

cells, and edits the cross-presentation of self, viral, and tumor antigens. *Blood*. 2006;107 (1):151–158. Available from: https://doi.org/10.1182/blood-2005-03-1112.

33. Gallo PM, Gallucci S. The dendritic cell response to classic, emerging, and homeostatic danger signals. Implications for autoimmunity. *Front Immunol*. 2013;4:138. Available from: https://doi.org/10.3389/fimmu.2013.00138.
34. Shi Y, Evans JE, Rock KL. Molecular identification of a danger signal that alerts the immune system to dying cells. *Nature*. 2003;425(6957):516–521. Available from: https://doi.org/10.1038/nature01991.
35. Apetoh L, Ghiringhelli F, Tesniere A, Obeid M, Ortiz C, Criollo A, et al. Toll-like receptor 4-dependent contribution of the immune system to anticancer chemotherapy and radiotherapy. *Nat Med*. 2007;13(9):1050–1059. Available from: https://doi.org/10.1038/nm1622.
36. Andersson U, Wang H, Palmblad K, Aveberger A-C, Bloom O, Erlandsson-Harris H, et al. High Mobility Group 1 Protein (Hmg-1) stimulates proinflammatory cytokine synthesis in human monocytes. *J Exp Med*. 2000;192(4):565–570. Available from: https://doi.org/10.1084/jem.192.4.565.
37. Asea A, Rehli M, Kabingu E, Boch JA, Bare O, Auron PE, et al. Novel signal transduction pathway utilized by extracellular HSP70: role of toll-like receptor (TLR) 2 and TLR4. *J Biol Chem*. 2002;277(17):15028–15034. Available from: https://doi.org/10.1074/jbc.M200497200.
38. Basu S, Binder RJ, Suto R, Anderson KM, Srivastava PK. Necrotic but not apoptotic cell death releases heat shock proteins, which deliver a partial maturation signal to dendritic cells and activate the NF-kappa B pathway. *Int Immunol*. 2000;12(11):1539–1546. Available from: https://doi.org/10.1093/intimm/12.11.1539.
39. Asea A, Kraeft SK, Kurt-Jones EA, Stevenson MA, Chen LB, Finberg RW, et al. HSP70 stimulates cytokine production through a CD14-dependant pathway, demonstrating its dual role as a chaperone and cytokine. *Nat Med*. 2000;6(4):435–442. Available from: https://doi.org/10.1038/74697.
40. Shi Y, Rock KL. Cell death releases endogenous adjuvants that selectively enhance immune surveillance of particulate antigens. *Eur J Immunol*. 2002;32(1):155–162. Available from: https://doi.org/10.1002/1521-4141(200201)32:1 < 155::Aid-immu155 > 3.0.Co;2-p.
41. Shi Y, Zheng W, Rock KL. Cell injury releases endogenous adjuvants that stimulate cytotoxic T cell responses. *Proc Natl Acad Sci USA*. 2000;97(26):14590–14595. Available from: https://doi.org/10.1073/pnas.260497597.
42. Shi Y, Galusha SA, Rock KL. Cutting edge: elimination of an endogenous adjuvant reduces the activation of CD8 T lymphocytes to transplanted cells and in an autoimmune diabetes model. *J Immunol*. 2006;176(7):3905–3908. Available from: https://doi.org/10.4049/jimmunol.176.7.3905.
43. Chiang CL, Ledermann JA, Rad AN, Katz DR, Chain BM. Hypochlorous acid enhances immunogenicity and uptake of allogeneic ovarian tumor cells by dendritic cells to cross-prime tumor-specific T cells. *Cancer Immunol Immunother*. 2006;55(11):1384–1395. Available from: https://doi.org/10.1007/s00262-006-0127-9.
44. Chiang CL, Ledermann JA, Aitkens E, Benjamin E, Katz DR, Chain BM. Oxidation of ovarian epithelial cancer cells by hypochlorous acid enhances immunogenicity and stimulates T cells that recognize autologous primary tumor. *Clin Cancer Res*. 2008;14 (15):4898–4907. Available from: https://doi.org/10.1158/1078-0432.Ccr-07-4899.
45. Prokopowicz ZM, Arce F, Biedroń R, Chiang CL, Ciszek M, Katz DR, et al. Hypochlorous acid: a natural adjuvant that facilitates antigen processing, cross-priming, and the induction of adaptive immunity. *J Immunol*. 2010;184(2):824–835. Available from: https://doi.org/10.4049/jimmunol.0902606.

46. Takaku S, Terabe M, Ambrosino E, Peng J, Lonning S, McPherson JM, et al. Blockade of TGF-beta enhances tumor vaccine efficacy mediated by CD8(+) T cells. *Int J Cancer*. 2010;126(7):1666–1674. Available from: https://doi.org/10.1002/ijc.24961.
47. Loercher AE, Nash MA, Kavanagh JJ, Platsoucas CD, Freedman RS. Identification of an IL-10-producing HLA-DR-negative monocyte subset in the malignant ascites of patients with ovarian carcinoma that inhibits cytokine protein expression and proliferation of autologous T cells. *J Immunol*. 1999;163(11):6251–6260.
48. Chiang CL, Hagemann AR, Leskowitz R, Mick R, Garrabrant T, Czerniecki BJ, et al. Day-4 myeloid dendritic cells pulsed with whole tumor lysate are highly immunogenic and elicit potent anti-tumor responses. *PLoS One*. 2011;6(12):e28732. Available from: https://doi.org/10.1371/journal.pone.0028732.
49. Wollenberg A, Wagner M, Günther S, Towarowski A, Tuma E, Moderer M, et al. Plasmacytoid dendritic cells: a new cutaneous dendritic cell subset with distinct role in inflammatory skin diseases. *J Invest Dermatol*. 2002;119(5):1096–1102. Available from: https://doi.org/10.1046/j.1523-1747.2002.19515.x.
50. Stary G, Bangert C, Tauber M, Strohal R, Kopp T, Stingl G. Tumoricidal activity of TLR7/8-activated inflammatory dendritic cells. *J Exp Med*. 2007;204(6):1441–1451. Available from: https://doi.org/10.1084/jem.20070021.
51. Urosevic M, Dummer R, Conrad C, Beyeler M, Laine E, Burg G, et al. Disease-independent skin recruitment and activation of plasmacytoid predendritic cells following imiquimod treatment. *J Natl Cancer Inst*. 2005;97(15):1143–1153. Available from: https://doi.org/10.1093/jnci/dji207.
52. Le Bon A, Etchart N, Rossmann C, Ashton M, Hou S, Gewert D, et al. Cross-priming of CD8 + T cells stimulated by virus-induced type I interferon. *Nat Immunol*. 2003;4(10):1009–1015. Available from: https://doi.org/10.1038/ni978.
53. Jego G, Palucka AK, Blanck JP, Chalouni C, Pascual V, Banchereau J. Plasmacytoid dendritic cells induce plasma cell differentiation through type I interferon and interleukin 6. *Immunity*. 2003;19(2):225–234. Available from: https://doi.org/10.1016/s1074-7613(03)00208-5.
54. Le Bon A, Schiavoni G, D'Agostino G, Gresser I, Belardelli F, Tough DF. Type i interferons potently enhance humoral immunity and can promote isotype switching by stimulating dendritic cells in vivo. *Immunity*. 2001;14(4):461–470. Available from: https://doi.org/10.1016/s1074-7613(01)00126-1.
55. Dudek AZ, Yunis C, Harrison LI, Kumar S, Hawkinson R, Cooley S, et al. First in human phase I trial of 852A, a novel systemic toll-like receptor 7 agonist, to activate innate immune responses in patients with advanced cancer. *Clin Cancer Res*. 2007;13(23):7119–7125. Available from: https://doi.org/10.1158/1078-0432.Ccr-07-1443.
56. Dummer R, Hauschild A, Becker JC, Grob JJ, Schadendorf D, Tebbs V, et al. An exploratory study of systemic administration of the toll-like receptor-7 agonist 852A in patients with refractory metastatic melanoma. *Clin Cancer Res*. 2008;14(3):856–864. Available from: https://doi.org/10.1158/1078-0432.Ccr-07-1938.
57. Chang BA, Cross JL, Najar HM, Dutz JP. Topical resiquimod promotes priming of CTL to parenteral antigens. *Vaccine*. 2009;27(42):5791–5799. Available from: https://doi.org/10.1016/j.vaccine.2009.07.062.
58. Aucouturier J, Dupuis L, Deville S, Ascarateil S, Ganne V. Montanide ISA 720 and 51: a new generation of water in oil emulsions as adjuvants for human vaccines. *Expert Rev Vaccines*. 2002;1(1):111–118. Available from: https://doi.org/10.1586/14760584.1.1.111.
59. Sabado RL, Pavlick A, Gnjatic S, Cruz CM, Vengco I, Hasan F, et al. Resiquimod as an immunologic adjuvant for NY-ESO-1 protein vaccination in patients with high-risk

melanoma. *Cancer Immunol Res*. 2015;3(3):278–287. Available from: https://doi.org/10.1158/2326-6066.Cir-14-0202.
60. Hemmi H, Takeuchi O, Kawai T, Kaisho T, Sato S, Sanjo H, et al. A toll-like receptor recognizes bacterial DNA. *Nature*. 2000;408(6813):740–745. Available from: https://doi.org/10.1038/35047123.
61. Bauer S, Kirschning CJ, Häcker H, Redecke V, Hausmann S, Akira S, et al. Human TLR9 confers responsiveness to bacterial DNA via species-specific CpG motif recognition. *Proc Natl Acad Sci USA*. 2001;98(16):9237–9242. Available from: https://doi.org/10.1073/pnas.161293498.
62. Kumagai Y, Takeuchi O, Akira S. TLR9 as a key receptor for the recognition of DNA. *Adv Drug Deliv Rev*. 2008;60(7):795–804. Available from: https://doi.org/10.1016/j.addr.2007.12.004.
63. Karbach J, Gnjatic S, Bender A, Neumann A, Weidmann E, Yuan J, et al. Tumor-reactive CD8+ T-cell responses after vaccination with NY-ESO-1 peptide, CpG 7909 and Montanide ISA-51: association with survival. *Int J Cancer*. 2010;126(4):909–918. Available from: https://doi.org/10.1002/ijc.24850.
64. Valmori D, Souleimanian NE, Tosello V, Bhardwaj N, Adams S, O'Neill D, et al. Vaccination with NY-ESO-1 protein and CpG in montanide induces integrated antibody/Th1 responses and CD8 T cells through cross-priming. *Proc Natl Acad Sci USA*. 2007;104(21):8947–8952. Available from: https://doi.org/10.1073/pnas.0703395104.
65. Fourcade J, Kudela P, Andrade Filho PA, Janjic B, Land SR, Sander C, et al. Immunization with analog peptide in combination with CpG and montanide expands tumor antigen-specific CD8+ T cells in melanoma patients. *J Immunother*. 2008;31(8):781–791. Available from: https://doi.org/10.1097/CJI0.0b013e318183af0b.
66. Speiser DE, Liénard D, Rufer N, Rubio-Godoy V, Rimoldi D, Lejeune F, et al. Rapid and strong human CD8+ T cell responses to vaccination with peptide, IFA, and CpG oligodeoxynucleotide 7909. *J Clin Invest*. 2005;115(3):739–746. Available from: https://doi.org/10.1172/jci23373.
67. Flacher V, Bouschbacher M, Verronèse E, Massacrier C, Sisirak V, Berthier-Vergnes O, et al. Human Langerhans cells express a specific TLR profile and differentially respond to viruses and Gram-positive bacteria. *J Immunol*. 2006;177(11):7959–7967. Available from: https://doi.org/10.4049/jimmunol.177.11.7959.
68. Klechevsky E, Morita R, Liu M, Cao Y, Coquery S, Thompson-Snipes L, et al. Functional specializations of human epidermal Langerhans cells and CD14+ dermal dendritic cells. *Immunity*. 2008;29(3):497–510. Available from: https://doi.org/10.1016/j.immuni.2008.07.013.
69. Oh S, Perera LP, Burke DS, Waldmann TA, Berzofsky JA. IL-15/IL-15Ralpha-mediated avidity maturation of memory CD8+ T cells. *Proc Natl Acad Sci USA*. 2004;101(42):15154–15159. Available from: https://doi.org/10.1073/pnas.0406649101.
70. Rasheed MA, Latner DR, Aubert RD, Gourley T, Spolski R, Davis CW, et al. Interleukin-21 is a critical cytokine for the generation of virus-specific long-lived plasma cells. *J Virol*. 2013;87(13):7737–7746. Available from: https://doi.org/10.1128/jvi0.00063-13.
71. Kuchen S, Robbins R, Sims GP, Sheng C, Phillips TM, Lipsky PE, et al. Essential role of IL-21 in B cell activation, expansion, and plasma cell generation during CD4+ T cell-B cell collaboration. *J Immunol*. 2007;179(9):5886–5896. Available from: https://doi.org/10.4049/jimmunol.179.9.5886.
72. Jongbloed SL, Kassianos AJ, McDonald KJ, Clark GJ, Ju X, Angel CE, et al. Human CD141+ (BDCA-3)+ dendritic cells (DCs) represent a unique myeloid DC subset that

cross-presents necrotic cell antigens. *J Exp Med.* 2010;207(6):1247–1260. Available from: https://doi.org/10.1084/jem.20092140.

73. Haniffa M, Shin A, Bigley V, McGovern N, Teo P, See P, et al. Human tissues contain CD141hi cross-presenting dendritic cells with functional homology to mouse CD103 + nonlymphoid dendritic cells. *Immunity.* 2012;37(1):60–73. Available from: https://doi.org/10.1016/j.immuni.2012.04.012.
74. Caskey M, Lefebvre F, Filali-Mouhim A, Cameron MJ, Goulet JP, Haddad EK, et al. Synthetic double-stranded RNA induces innate immune responses similar to a live viral vaccine in humans. *J Exp Med.* 2011;208(12):2357–2366. Available from: https://doi.org/10.1084/jem.20111171.
75. Salazar AM, Levy HB, Ondra S, Kende M, Scherokman B, Brown D, et al. Long-term treatment of malignant gliomas with intramuscularly administered polyinosinic-polycytidylic acid stabilized with polylysine and carboxymethylcellulose: an open pilot study. *Neurosurgery.* 1996;38(6):1096–1103. discussion 1103–4.
76. Dhodapkar MV, Sznol M, Zhao B, Wang D, Carvajal RD, Keohan ML, et al. Induction of antigen-specific immunity with a vaccine targeting NY-ESO-1 to the dendritic cell receptor DEC-205. *Sci Transl Med.* 2014;6(232):232ra51. Available from: https://doi.org/10.1126/scitranslmed.3008068.
77. Gowen BB, Wong MH, Jung KH, Sanders AB, Mitchell WM, Alexopoulou L, et al. TLR3 is essential for the induction of protective immunity against Punta Toro Virus infection by the double-stranded RNA (dsRNA), poly(I:C12U), but not Poly(I:C): differential recognition of synthetic dsRNA molecules. *J Immunol.* 2007;178(8):5200–5208. Available from: https://doi.org/10.4049/jimmunol.178.8.5200.
78. Navabi H, Jasani B, Reece A, Clayton A, Tabi Z, Donninger C, et al. A clinical grade poly I:C-analogue (Ampligen) promotes optimal DC maturation and Th1-type T cell responses of healthy donors and cancer patients in vitro. *Vaccine.* 2009;27(1):107–115. Available from: https://doi.org/10.1016/j.vaccine.2008.10.024.
79. Martin M, Michalek SM, Katz J. Role of innate immune factors in the adjuvant activity of monophosphoryl lipid A. *Infect Immun.* 2003;71(5):2498–2507. Available from: https://doi.org/10.1128/iai.71.5.2498-2507.2003.
80. Mata-Haro V, Cekic C, Martin M, Chilton PM, Casella CR, Mitchell TC. The vaccine adjuvant monophosphoryl lipid A as a TRIF-biased agonist of TLR4. *Science.* 2007;316(5831):1628–1632. Available from: https://doi.org/10.1126/science.1138963.
81. Okemoto K, Kawasaki K, Hanada K, Miura M, Nishijima M. A potent adjuvant monophosphoryl lipid A triggers various immune responses, but not secretion of IL-1β or activation of caspase-1. *J Immunol.* 2006;176(2):1203–1208. Available from: https://doi.org/10.4049/jimmunol.176.2.1203.
82. Fransen F, Boog CJ, van Putten JP, van der Ley P. Agonists of toll-like receptors 3, 4, 7, and 9 are candidates for use as adjuvants in an outer membrane vaccine against Neisseria meningitidis serogroup B. *Infect Immun.* 2007;75(12):5939–5946. Available from: https://doi.org/10.1128/iai.00846-07.
83. Didierlaurent AM, Morel S, Lockman L, Giannini SL, Bisteau M, Carlsen H, et al. AS04, an aluminum salt- and TLR4 agonist-based adjuvant system, induces a transient localized innate immune response leading to enhanced adaptive immunity. *J Immunol.* 2009;183(10):6186–6197. Available from: https://doi.org/10.4049/jimmunol.0901474.
84. Rhee EG, Kelley RP, Agarwal I, Lynch DM, La Porte A, Simmons NL, et al. TLR4 ligands augment antigen-specific CD8 + T lymphocyte responses elicited by a viral vaccine vector. *J Virol.* 2010;84(19):10413–10419. Available from: https://doi.org/10.1128/JVI.00928-10.

85. Lu H, Dietsch GN, Matthews MA, Yang Y, Ghanekar S, Inokuma M, et al. VTX-2337 is a novel TLR8 agonist that activates NK cells and augments ADCC. *Clin Cancer Res.* 2012;18(2):499–509. Available from: https://doi.org/10.1158/1078-0432.Ccr-11-1625.
86. Northfelt DW, Ramanathan RK, Cohen PA, Von Hoff DD, Weiss GJ, Dietsch GN, et al. A phase I dose-finding study of the novel toll-like receptor 8 agonist VTX-2337 in adult subjects with advanced solid tumors or lymphoma. *Clin Cancer Res.* 2014;20(14):3683–3691. Available from: https://doi.org/10.1158/1078-0432.Ccr-14-0392.
87. Chow LQM, Morishima C, Eaton KD, Baik CS, Goulart BH, Anderson LN, et al. Phase Ib trial of the toll-like receptor 8 agonist, motolimod (VTX-2337), combined with cetuximab in patients with recurrent or metastatic SCCHN. *Clin Cancer Res.* 2017;23(10):2442–2450. Available from: https://doi.org/10.1158/1078-0432.Ccr-16-1934.
88. Garg AD, Coulie PG, Van den Eynde BJ, Agostinis P. Integrating next-generation dendritic cell vaccines into the current cancer immunotherapy landscape. *Trends Immunol.* 2017;38(8):577–593. Available from: https://doi.org/10.1016/j.it.2017.05.006.
89. Grouard G, Rissoan MC, Filgueira L, Durand I, Banchereau J, Liu YJ. The enigmatic plasmacytoid T cells develop into dendritic cells with interleukin (IL)-3 and CD40-ligand. *J Exp Med.* 1997;185(6):1101–1111. Available from: https://doi.org/10.1084/jem.185.6.1101.
90. Tel J, Aarntzen EH, Baba T, Schreibelt G, Schulte BM, Benitez-Ribas D, et al. Natural human plasmacytoid dendritic cells induce antigen-specific T-cell responses in melanoma patients. *Cancer Res.* 2013;73(3):1063–1075. Available from: https://doi.org/10.1158/0008-5472.Can-12-2583.
91. Romano E, Rossi M, Ratzinger G, de Cos MA, Chung DJ, Panageas KS, et al. Peptide-loaded Langerhans cells, despite increased IL15 secretion and T-cell activation in vitro, elicit antitumor T-cell responses comparable to peptide-loaded monocyte-derived dendritic cells in vivo. *Clin Cancer Res.* 2011;17(7):1984–1997. Available from: https://doi.org/10.1158/1078-0432.Ccr-10-3421.
92. Anguille S, Smits EL, Lion E, van Tendeloo VF, Berneman ZN. Clinical use of dendritic cells for cancer therapy. *Lancet Oncol.* 2014;15(7):e257–e267. Available from: https://doi.org/10.1016/S1470-2045(13)70585-0.
93. Ueno H, Schmitt N, Klechevsky E, Pedroza-Gonzalez A, Matsui T, Zurawski G, et al. Harnessing human dendritic cell subsets for medicine. *Immunol Rev.* 2010;234(1):199–212. Available from: https://doi.org/10.1111/j.0105-2896.2009.00884.x.
94. Ardavín C, Amigorena S, Reis e Sousa C. Dendritic cells: immunobiology and cancer immunotherapy. *Immunity.* 2004;20(1):17–23. Available from: https://doi.org/10.1016/s1074-7613(03)00352-2.
95. de Vries IJ, Eggert AA, Scharenborg NM, Vissers JL, Lesterhuis WJ, Boerman OC, et al. Phenotypical and functional characterization of clinical grade dendritic cells. *J Immunother.* 2002;25(5):429–438. Available from: https://doi.org/10.1097/00002371-200209000-00007.
96. Jonuleit H, Kühn U, Müller G, Steinbrink K, Paragnik L, Schmitt E, et al. Pro-inflammatory cytokines and prostaglandins induce maturation of potent immunostimulatory dendritic cells under fetal calf serum-free conditions. *Eur J Immunol.* 1997;27(12):3135–3142. Available from: https://doi.org/10.1002/eji.1830271209.
97. Trabanelli S, Lecciso M, Salvestrini V, Cavo M, Očadlíková D, Lemoli RM, et al. PGE2-induced IDO1 inhibits the capacity of fully mature DCs to elicit an in vitro antileukemic immune response. *J Immunol Res.* 2015;2015:253191. Available from: https://doi.org/10.1155/2015/253191.

98. Munn DH, Mellor AL. Indoleamine 2,3 dioxygenase and metabolic control of immune responses. *Trends Immunol.* 2013;34(3):137–143. Available from: https://doi.org/10.1016/j.it.2012.10.001.
99. Akiyama Y, Oshita C, Kume A, Iizuka A, Miyata H, Komiyama M, et al. α-type-1 polarized dendritic cell-based vaccination in recurrent high-grade glioma: a phase I clinical trial. *BMC Cancer.* 2012;12:623. Available from: https://doi.org/10.1186/1471-2407-12-623.
100. Mailliard RB, Wankowicz-Kalinska A, Cai Q, Wesa A, Hilkens CM, Kapsenberg ML, et al. alpha-type-1 polarized dendritic cells: a novel immunization tool with optimized CTL-inducing activity. *Cancer Res.* 2004;64(17):5934–5937. Available from: https://doi.org/10.1158/0008-5472.Can-04-1261.
101. Hughes GA. Nanostructure-mediated drug delivery. *Nanomedicine.* 2005;1(1):22–30. Available from: https://doi.org/10.1016/j.nano.2004.11.009.
102. Sindhwani S, Chan WCW. Nanotechnology for modern medicine: next steps towards clinical translation. *J Intern Med.* 2021;. Available from: https://doi.org/10.1111/joim.13254.
103. Davis ME, Chen Z, Shin DM. Nanoparticle therapeutics: an emerging treatment modality for cancer. *Nat Rev Drug Discov.* 2008;7(9):771–782. Available from: https://doi.org/10.1038/nrd2614.
104. Yoon HY, Selvan ST, Yang Y, Kim MJ, Yi DK, Kwon IC, et al. Engineering nanoparticle strategies for effective cancer immunotherapy. *Biomaterials.* 2018;178:597–607. Available from: https://doi.org/10.1016/j.biomaterials.2018.03.036.
105. Anselmo AC, Mitragotri S. Nanoparticles in the clinic: an update. *Bioeng Transl Med.* 2019;4(3):e10143. Available from: https://doi.org/10.1002/btm2.10143.
106. Solbrig CM, Saucier-Sawyer JK, Cody V, Saltzman WM, Hanlon DJ. Polymer nanoparticles for immunotherapy from encapsulated tumor-associated antigens and whole tumor cells. *Mol Pharm.* 2007;4(1):47–57. Available from: https://doi.org/10.1021/mp060107e.
107. Won JE, Byeon Y, Wi TI, Lee JM, Kang TH, Lee JW, et al. Enhanced antitumor immunity using a tumor cell lysate-encapsulated CO_2-generating liposomal carrier system and photothermal irradiation. *ACS Appl Bio Mater.* 2019;2(6):2481–2489. Available from: https://doi.org/10.1021/acsabm.9b00183.
108. Shi GN, Zhang CN, Xu R, Niu JF, Song HJ, Zhang XY, et al. Enhanced antitumor immunity by targeting dendritic cells with tumor cell lysate-loaded chitosan nanoparticles vaccine. *Biomaterials.* 2017;113:191–202. Available from: https://doi.org/10.1016/j.biomaterials.2016.10.047.
109. Heße C, Kollenda S, Rotan O, Pastille E, Adamczyk A, Wenzek C, et al. A tumor-peptide–based nanoparticle vaccine elicits efficient tumor growth control in antitumor immunotherapy. *Mol Cancer Ther.* 2019;18(6):1069. Available from: https://doi.org/10.1158/1535-7163.MCT-18-0764.
110. Rosalia RA, Cruz LJ, van Duikeren S, Tromp AT, Silva AL, Jiskoot W, et al. CD40-targeted dendritic cell delivery of PLGA-nanoparticle vaccines induce potent anti-tumor responses. *Biomaterials.* 2015;40:88–97. Available from: https://doi.org/10.1016/j.biomaterials.2014.10.053.
111. Topalian SL, Drake CG, Pardoll DM. Immune checkpoint blockade: a common denominator approach to cancer therapy. *Cancer Cell.* 2015;27(4):450–461. Available from: https://doi.org/10.1016/j.ccell.2015.03.001.
112. Wilgenhof S, Corthals J, Heirman C, van Baren N, Lucas S, Kvistborg P, et al. Phase II study of autologous monocyte-derived mRNA electroporated dendritic cells (TriMixDC-

MEL) plus ipilimumab in patients with pretreated advanced melanoma. *J Clin Oncol.* 2016;34(12):1330–1338. Available from: https://doi.org/10.1200/jco0.2015.63.4121.

113. Ribas A, Comin-Anduix B, Chmielowski B, Jalil J, de la Rocha P, McCannel TA, et al. Dendritic cell vaccination combined with CTLA4 blockade in patients with metastatic melanoma. *Clin Cancer Res.* 2009;15(19):6267–6276. Available from: https://doi.org/10.1158/1078-0432.CCR-09-1254.
114. Antonios JP, Soto H, Everson RG, Orpilla J, Moughon D, Shin N, et al. PD-1 blockade enhances the vaccination-induced immune response in glioma. *JCI Insight.* 2016;1(10): e87059. Available from: https://doi.org/10.1172/jci.insight.87059.
115. Kadam P, Sharma S. PD-1 immune checkpoint blockade promotes therapeutic cancer vaccine to eradicate lung cancer. *Vaccines (Basel).* 2020;8(2):317. Available from: https://doi.org/10.3390/vaccines8020317.
116. Belkaid Y, Rouse BT. Natural regulatory T cells in infectious disease. *Nat Immunol.* 2005;6(4):353–360. Available from: https://doi.org/10.1038/ni1181.
117. Sakaguchi S, Yamaguchi T, Nomura T, Ono M. Regulatory T cells and immune tolerance. *Cell.* 2008;133(5):775–787. Available from: https://doi.org/10.1016/j.cell.2008.05.009.
118. Miyara M, Yoshioka Y, Kitoh A, Shima T, Wing K, Niwa A, et al. Functional delineation and differentiation dynamics of human CD4 + T cells expressing the FoxP3 transcription factor. *Immunity.* 2009;30(6):899–911. Available from: https://doi.org/10.1016/j.immuni.2009.03.019.
119. Ito T, Hanabuchi S, Wang YH, Park WR, Arima K, Bover L, et al. Two functional subsets of FOXP3 + regulatory T cells in human thymus and periphery. *Immunity.* 2008;28(6):870–880. Available from: https://doi.org/10.1016/j.immuni.2008.03.018.
120. Chen ML, Pittet MJ, Gorelik L, Flavell RA, Weissleder R, von Boehmer H, et al. Regulatory T cells suppress tumor-specific CD8 T cell cytotoxicity through TGF-beta signals in vivo. *Proc Natl Aatl Sci USA.* 2005;102(2):419–424. Available from: https://doi.org/10.1073/pnas.0408197102.
121. Fahlén L, Read S, Gorelik L, Hurst SD, Coffman RL, Flavell RA, et al. T cells that cannot respond to TGF-beta escape control by $CD4^+CD25^+$ regulatory T cells. *J Exp Med.* 2005;201(5):737–746. Available from: https://doi.org/10.1084/jem.20040685.
122. O'Garra A, Vieira PL, Vieira P, Goldfeld AE. IL-10-producing and naturally occurring CD4 + Tregs: limiting collateral damage. *J Clin Invest.* 2004;114(10):1372–1378. Available from: https://doi.org/10.1172/jci23215.
123. Collison LW, Workman CJ, Kuo TT, Boyd K, Wang Y, Vignali KM, et al. The inhibitory cytokine IL-35 contributes to regulatory T-cell function. *Nature.* 2007;450(7169):566–569. Available from: https://doi.org/10.1038/nature06306.
124. Wing K, Onishi Y, Prieto-Martin P, Yamaguchi T, Miyara M, Fehervari Z, et al. CTLA-4 control over Foxp3 + regulatory T cell function. *Science.* 2008;322(5899):271–275. Available from: https://doi.org/10.1126/science.1160062.
125. Onishi Y, Fehervari Z, Yamaguchi T, Sakaguchi S. Foxp3 + natural regulatory T cells preferentially form aggregates on dendritic cells in vitro and actively inhibit their maturation. *Proc Natl Aatl Sci USA.* 2008;105(29):10113–10118. Available from: https://doi.org/10.1073/pnas.0711106105.
126. Grohmann U, Orabona C, Fallarino F, Vacca C, Calcinaro F, Falorni A, et al. CTLA-4–Ig regulates tryptophan catabolism in vivo. *Nat Immunol.* 2002;3(11):1097–1101. Available from: https://doi.org/10.1038/ni846.
127. Dejean AS, Beisner DR, Ch'en IL, Kerdiles YM, Babour A, Arden KC, et al. Transcription factor Foxo3 controls the magnitude of T cell immune responses by

modulating the function of dendritic cells. *Nat Immunol.* 2009;10(5):504–513. Available from: https://doi.org/10.1038/ni.1729.

128. Ghiringhelli F, Larmonier N, Schmitt E, Parcellier A, Cathelin D, Garrido C, et al. CD4 + CD25 + regulatory T cells suppress tumor immunity but are sensitive to cyclophosphamide which allows immunotherapy of established tumors to be curative. *Eur J Immunol.* 2004;34(2):336–344. Available from: https://doi.org/10.1002/eji.200324181.
129. Wada S, Yoshimura K, Hipkiss EL, Harris TJ, Yen H-R, Goldberg MV, et al. Cyclophosphamide augments antitumor immunity: studies in an autochthonous prostate cancer model. *Cancer Res.* 2009;69(10):4309–4318. Available from: https://doi.org/10.1158/0008-5472.CAN-08-4102.
130. Scurr MJ, Pembroke T, Bloom A, Roberts DJ, Thomson A, Smart K, et al. Low-dose cyclophosphamide induces anti-tumor T-cell responses which associate with survival in metastatic colorectal cancer. *Clin Cancer Res.* 2017;23(22):6771–6780. Available from: https://doi.org/10.1158/1078-0432.CCR-17-0895.
131. van der Most RG, Currie AJ, Mahendran S, Prosser A, Darabi A, Robinson BWS, et al. Tumor eradication after cyclophosphamide depends on concurrent depletion of regulatory T cells: a role for cycling TNFR2-expressing effector-suppressor T cells in limiting effective chemotherapy. *Cancer Immunol Immunother.* 2008;58(8):1219. Available from: https://doi.org/10.1007/s00262-008-0628-9.
132. Herman AE, Freeman GJ, Mathis D, Benoist C. CD4 + CD25 + T regulatory cells dependent on ICOS promote regulation of effector cells in the prediabetic lesion. *J Exp Med.* 2004;199(11):1479–1489. Available from: https://doi.org/10.1084/jem.20040179.
133. Chen X, Subleski JJ, Kopf H, Howard OM, Männel DN, Oppenheim JJ. Cutting edge: expression of TNFR2 defines a maximally suppressive subset of mouse CD4 + CD25 + FoxP3 + T regulatory cells: applicability to tumor-infiltrating T regulatory cells. *J Immunol.* 2008;180(10):6467–6471. Available from: https://doi.org/10.4049/jimmunol.180.10.6467.
134. Strauss L, Bergmann C, Szczepanski MJ, Lang S, Kirkwood JM, Whiteside TL. Expression of ICOS on human melanoma-infiltrating CD4 + CD25highFoxp3 + T regulatory cells: implications and impact on tumor-mediated immune suppression. *J Immunol.* 2008;180(5):2967–2980. Available from: https://doi.org/10.4049/jimmunol.180.5.2967.
135. Cornelissen R, Hegmans JPJJ, Maat APWM, Kaijen-Lambers MEH, Bezemer K, Hendriks RW, et al. Extended tumor control after dendritic cell vaccination with low-dose cyclophosphamide as adjuvant treatment in patients with malignant pleural mesothelioma. *Am J Respir Crit Care Med.* 2016;193(9):1023–1031. Available from: https://doi.org/10.1164/rccm.201508-1573OC.
136. Bota DA, Chung J, Dandekar M, Carrillo JA, Kong X-T, Fu BD, et al. Phase II study of ERC1671 plus bevacizumab vs bevacizumab plus placebo in recurrent glioblastoma: interim results and correlations with CD4(+) T-lymphocyte counts. *CNS Oncol.* 2018;7(3):CNS22. Available from: https://doi.org/10.2217/cns-2018-0009.
137. Beyer M, Kochanek M, Darabi K, Popov A, Jensen M, Endl E, et al. Reduced frequencies and suppressive function of CD4 + CD25hi regulatory T cells in patients with chronic lymphocytic leukemia after therapy with fludarabine. *Blood.* 2005;106(6):2018–2025. Available from: https://doi.org/10.1182/blood-2005-02-0642.
138. Folkman J. Angiogenesis. In: Brenner S, Miller JH, eds. *Encyclopedia of Genetics*. New York: Academic Press; 2001:66–73.
139. Ferrara N, Gerber HP, LeCouter J. The biology of VEGF and its receptors. *Nat Med.* 2003;9(6):669–676. Available from: https://doi.org/10.1038/nm0603-669.

140. Jain RK. Antiangiogenesis strategies revisited: from starving tumors to alleviating hypoxia. *Cancer Cell.* 2014;26(5):605–622. Available from: https://doi.org/10.1016/j.ccell.2014.10.006.
141. Rivera LB, Bergers G. CANCER. Tumor angiogenesis, from foe to friend. *Science.* 2015;349(6249):694–695. Available from: https://doi.org/10.1126/science.aad0862.
142. Ferrara N, Adamis AP. Ten years of anti-vascular endothelial growth factor therapy. *Nat Rev Drug Discov.* 2016;15(6):385–403. Available from: https://doi.org/10.1038/nrd.2015.17.
143. Garcia J, Hurwitz HI, Sandler AB, Miles D, Coleman RL, Deurloo R, et al. Bevacizumab (Avastin®) in cancer treatment: a review of 15 years of clinical experience and future outlook. *Cancer Treat Rev.* 2020;86:102017. Available from: https://doi.org/10.1016/j.ctrv.2020.102017.
144. Hurwitz H, Fehrenbacher L, Novotny W, Cartwright T, Hainsworth J, Heim W, et al. Bevacizumab plus irinotecan, fluorouracil, and leucovorin for metastatic colorectal cancer. *N Engl J Med.* 2004;350(23):2335–2342. Available from: https://doi.org/10.1056/NEJMoa032691.
145. Kabbinavar FF, Hambleton J, Mass RD, Hurwitz HI, Bergsland E, Sarkar S. Combined analysis of efficacy: the addition of bevacizumab to fluorouracil/leucovorin improves survival for patients with metastatic colorectal cancer. *J Clin Oncol.* 2005;23(16):3706–3712. Available from: https://doi.org/10.1200/jco.2005.00.232.
146. Sandler A, Gray R, Perry MC, Brahmer J, Schiller JH, Dowlati A, et al. Paclitaxel-carboplatin alone or with bevacizumab for non-small-cell lung cancer. *N Engl J Med.* 2006;355(24):2542–2550. Available from: https://doi.org/10.1056/NEJMoa061884.
147. Tewari KS, Sill MW, Long 3rd HJ, Penson RT, Huang H, Ramondetta LM, et al. Improved survival with bevacizumab in advanced cervical cancer. *N Engl J Med.* 2014;370(8):734–743. Available from: https://doi.org/10.1056/NEJMoa1309748.
148. Oyama T, Ran S, Ishida T, Nadaf S, Kerr L, Carbone DP, et al. Vascular endothelial growth factor affects dendritic cell maturation through the inhibition of nuclear factor-kappa B activation in hemopoietic progenitor cells. *J Immunol.* 1998;160(3):1224–1232.
149. Dikov MM, Ohm JE, Ray N, Tchekneva EE, Burlison J, Moghanaki D, et al. Differential roles of vascular endothelial growth factor receptors 1 and 2 in dendritic cell differentiation. *J Immunol.* 2005;174(1):215–222. Available from: https://doi.org/10.4049/jimmunol.174.1.215.
150. Takahashi A, Kono K, Ichihara F, Sugai H, Fujii H, Matsumoto Y. Vascular endothelial growth factor inhibits maturation of dendritic cells induced by lipopolysaccharide, but not by proinflammatory cytokines. *Cancer Immunol Immunother.* 2004;53(6):543–550. Available from: https://doi.org/10.1007/s00262-003-0466-8.
151. Curiel TJ, Wei S, Dong H, Alvarez X, Cheng P, Mottram P, et al. Blockade of B7-H1 improves myeloid dendritic cell-mediated antitumor immunity. *Nat Med.* 2003;9(5):562–567. Available from: https://doi.org/10.1038/nm863.
152. Li MO, Wan YY, Sanjabi S, Robertson AK, Flavell RA. Transforming growth factor-beta regulation of immune responses. *Ann Rev Immunol.* 2006;24:99–146. Available from: https://doi.org/10.1146/annurev.immunol.24.021605.090737.
153. Rini BI, Weinberg V, Fong L, Conry S, Hershberg RM, Small EJ. Combination immunotherapy with prostatic acid phosphatase pulsed antigen-presenting cells (provenge) plus bevacizumab in patients with serologic progression of prostate cancer after definitive local therapy. *Cancer.* 2006;107(1):67–74. Available from: https://doi.org/10.1002/cncr.21956.

154. Tanyi JL, Chiang L-LC, Chiffelle J, Thierry A-H, Baumgartener P, et al. A novel personalized dendritic cell vaccine strategy elicits a potent polyfunctional T cell response and demonstrates clinical benefits in recurrent ovarian cancer. *NPJ Vaccines*. 2021;. Available from: https://doi.org/10.1038/s41541-021-00297-5. in press.
155. Llovet JM, Ricci S, Mazzaferro V, Hilgard P, Gane E, Blanc J-F, et al. Sorafenib in advanced hepatocellular carcinoma. *N Engl J Med*. 2008;359(4):378–390. Available from: https://doi.org/10.1056/NEJMoa0708857.
156. Motzer RJ, Hutson TE, Tomczak P, Michaelson MD, Bukowski RM, Rixe O, et al. Sunitinib vs interferon alfa in metastatic renal-cell carcinoma. *N Engl J Med*. 2007;356(2):115–124. Available from: https://doi.org/10.1056/NEJMoa065044.
157. Schlumberger M, Tahara M, Wirth LJ, Robinson B, Brose MS, Elisei R, et al. Lenvatinib vs placebo in radioiodine-refractory thyroid cancer. *N Engl J Med*. 2015;372(7):621–630. Available from: https://doi.org/10.1056/NEJMoa1406470.
158. Motzer RJ, Escudier B, Tomczak P, Hutson TE, Michaelson MD, Negrier S, et al. Axitinib vs sorafenib as second-line treatment for advanced renal cell carcinoma: overall survival analysis and updated results from a randomised phase 3 trial. *Lancet Oncol*. 2013;14(6):552–562. Available from: https://doi.org/10.1016/s1470-2045(13)70093-7.
159. Grothey A, Van Cutsem E, Sobrero A, Siena S, Falcone A, Ychou M, et al. Regorafenib monotherapy for previously treated metastatic colorectal cancer (CORRECT): an international, multicentre, randomised, placebo-controlled, phase 3 trial. *Lancet*. 2013;381(9863):303–312. Available from: https://doi.org/10.1016/s0140-6736(12)61900-x.
160. Elisei R, Schlumberger MJ, Müller SP, Schöffski P, Brose MS, Shah MH, et al. Cabozantinib in progressive medullary thyroid cancer. *J Clin Oncol*. 2013;31(29):3639–3646. Available from: https://doi.org/10.1200/JCO.2012.48.4659.
161. Sternberg CN, Davis ID, Mardiak J, Szczylik C, Lee E, Wagstaff J, et al. Pazopanib in locally advanced or metastatic renal cell carcinoma: results of a randomized phase III trial. *J Clin Oncol*. 2010;28(6):1061–1068. Available from: https://doi.org/10.1200/jco.2009.23.9764.
162. Wells Jr. SA, Robinson BG, Gagel RF, Dralle H, Fagin JA, Santoro M, et al. Vandetanib in patients with locally advanced or metastatic medullary thyroid cancer: a randomized, double-blind phase III trial. *J Clin Oncol*. 2012;30(2):134–141. Available from: https://doi.org/10.1200/jco.2011.35.5040.
163. Fuchs CS, Tomasek J, Yong CJ, Dumitru F, Passalacqua R, Goswami C, et al. Ramucirumab monotherapy for previously treated advanced gastric or gastro-oesophageal junction adenocarcinoma (REGARD): an international, randomised, multicentre, placebo-controlled, phase 3 trial. *Lancet*. 2014;383(9911):31–39. Available from: https://doi.org/10.1016/s0140-6736(13)61719-5.

Chapter 3

Tumor antigens for preventative cancer vaccines

Camille Jacqueline and Olivera J. Finn
Department of Immunology, University of Pittsburgh School of Medicine, Pittsburgh, PA, United States

Introduction

Prevention of tumor onset and progression is a concern for all multicellular organisms.[1] On one hand, cell-intrinsic controls, such as apoptosis, effective DNA repair, epigenetic modifications, cell cycle checkpoints, telomere shortening, and specific tissue architecture,[2] prevent unbridled cell proliferation and invasion of other tissues; on the other hand, extrinsic mechanisms, such as the immune system, ensure the elimination of abnormal cells. Determining to what extent the immune system plays a role in the natural prevention of cancer has been a highly active field of research since the 1960s, when cancer immunosurveillance was first defined as the ability of the immune system to identify and destroy nascent tumors. Since then, the process of immunosurveillance has become better understood and shown to be fully involved from the appearance of earliest premalignant lesions. Tumor–immune system interaction can yield one of at least three potential outcomes: elimination of the tumor, or more likely the premalignant lesion, considered to be the most frequent outcome; equilibrium, where premalignant lesion survives over a long period of time through constant interaction with the immune system but without overt evidence of clinical disease; and escape, where the tumor escapes immune control and becomes clinical disease.[3] Therefore all cancer prevention mechanisms, from DNA repair to immunosurveillance, limit cancer development to some extent, but none are 100% effective. Many of these mechanisms, including the immune system, can be modulated, which opens up the opportunity for the development of strategies that aim to strengthen the natural cancer prevention.

Cancer immunoprevention

Immunoprevention is one of many interventional modalities designed to reduce the risk of developing cancer over a lifetime or to reduce progression

Cancer Vaccines as Immunotherapy of Cancer. DOI: https://doi.org/10.1016/B978-0-12-823901-8.00006-6
© 2022 Elsevier Inc. All rights reserved.

from precancer to invasive and ultimately lethal disease. The main tool in the immunoprevention tool kit is vaccines that have been the highly successful approach for the prevention of infectious diseases. With the increasing knowledge of how immune responses are activated and regulated, other immune system-directed interventions, such as specific cytokines and checkpoint inhibitors, may in the future be added to the cancer immunoprevention tool kit. The work in the field of cancer prevention has been carried out along three separate goals—primary prevention, secondary prevention, and tertiary prevention.

Primary prevention aims to intervene before cancer is initiated. There are public health interventions to mediate cancer risk in the general population by controlling exposure to carcinogens and reduce chronic inflammation by promoting healthy habits, such as healthy diet, exercise, and stress reduction.[4] Some of these could easily be considered immunoprevention as it is known that these interventions boost immune responses. Primary prevention is also directed toward individuals known to have a strong genetic predisposition to develop certain cancers. There are, for example, chemotherapy drugs developed for the prevention of estrogen receptor-positive breast cancer or prostate cancer.[5,6] A more focused public health immunoprevention approach is vaccines against cancer causing viruses, such as Epstein–Barr virus (EBV), hepatitis B virus (HBV), and human papillomavirus (HPV) that prevent cancers by preventing the initial viral infection. Whereas all these strategies are already well-established in humans, the use of preventative vaccines against cancers of nonviral origin is only beginning to receive a well-deserved attention.

Secondary cancer prevention refers to some signs of a disease having been detected and using various interventions to stop disease progression. Prevention here can involve a simple intervention, such as careful surveillance for signs of progression or surgical removal of the precancerous lesions. This particular prevention setting, often also referred to as "cancer interception," is an ideal setting for testing preventative vaccines. It has already been shown that the HPV vaccine can be effective in clearing premalignant cervical lesions.[7] The question now is whether vaccines developed against nonviral cancers could have the same effect, and a limited number of recent vaccine efforts in this setting appears promising.

Tertiary prevention applies to individuals who have already been treated for their primary malignancy but are now susceptible to developing secondary malignancies. The aim is to achieve long-term remission or prolong the period to disease recurrence. While vaccines have never been tested in this postdiagnosis setting in pathogen-caused diseases, there have been over three decades of vaccine efforts in tertiary cancer prevention. Unfortunately, trials undertaken so far have suggested that as a lone intervention, therapeutic vaccines have limited clinical efficacy and that the future for tertiary prevention will likely be to combine vaccines with other preventative drugs and measures, including various immunomodulatory agents, such as checkpoint inhibitors.

Targeted tumor antigens

Based on the knowledge acquired in the field of infectious diseases where live-attenuated pathogens were successfully used in vaccines, the first cancer vaccines were composed of whole tumor cells irradiated or inactivated.[8] However, as this work transitioned to using spontaneous tumors as a source, whole tumor cell vaccines proved to be nonimmunogenic or weakly immunogenic. In addition, whole tumor cell vaccines presented risk for autoimmunity as dendritic cells (DCs) would also process and present many normal antigens. In parallel with this approach, researchers started to develop vaccines that targeted defined antigens.

Recognition of cancer by the immune effectors relies on the expression of proteins called tumor antigens, which can be organized in two large categories. The first category is tumor-specific antigens (TSA) that includes mutated neoantigens unique to each tumor,[9] products of oncogene mutations, such as K-ras and N-ras, or gene translocations and fusions, such as BCR-ABL, shared by many tumors,[10] and oncofetal antigens, shared by many tumors but not expressed in normal adult tissues.[11] In this category, we can also find the antigens derived from viral proteins. Viral proteins are produced inside the tumor cells and therefore give rise to antigenic peptides that can be detected by T cells. Yet another source of TSA is products of genes expressed in cancer cells and germ cells but not in other cell types. They include the melanoma antigens MAGE[12] and GAGE[13] and the broadly expressed antigen NY-ESO-1.[14]

The second category is tumor-associated antigens (TAA), which includes differentiation antigens and overexpressed and posttranslationally abnormally modified antigens. Differentiation antigens are encoded by genes with a tissue-specific expression and found therefore on specific tumor types and the corresponding healthy tissues. They include CEA,[15] PSA,[16] tyrosinase,[17] mammoglobin A,[18] gp100,[19] and melanA/MART-1.[20] Overexpressed antigens include a large variety of molecules, such as MUC1,[21] cyclin B1,[22] HER-2/neu, hTERT,[23] survivin,[24] and mesothelin,[25] which are overexpressed in many epithelial tumors due to tumor-specific increase in their gene transcription, gene amplification, and/or increased protein stability.[26] Today, hundreds of these types of molecules are in the databases of confirmed tumor antigens.[27]

A series of seminal observations on melanoma tumors suggested that targeting mutated TSA was crucial for clinically effective antimelanoma T-cell responses.[28,29] While this still needs to be confirmed, it may also be unique to melanomas as recent studies showed that mutated TSA were rare in the antigenic repertoire of nonmelanoma tumor types[30] and that instead, the nonmutated TSA or TAA were the major targets of antitumor T cells.[31,32]

Because the work on the role of mutated TSA is still relatively new and inconclusive, and because this chapter discusses preventative vaccines that are more likely to require shared, broadly applicable tumor antigens, we will focus on vaccines targeting nonmutated TSA or TAA and their advantages and benefits at different stages of the disease (Fig. 3.1).

Primary prevention

Prevention of cancers caused by viruses

A growing number of pathogens are recognized to be oncogenic where the infection is a prerequisite for maintaining or initiating the growth of cancer cells.[33] The World Health Organization acknowledges that at least 20% of cancers have an infectious origin.[34] HPV, HBV, hepatitis C virus (HCV), EBV, Kaposi's sarcoma-associated herpes virus, human immunodeficiency virus, and human T-cell lymphoma virus are well-established contributors to human cancers. Several HBV immunization programs showed that preventative cancer vaccines given in infancy provide protection against hepatocellular carcinoma in adulthood.[35,36] A similar observation was made after the introduction of a vaccine against HPV type 16 or 18 that successfully reduced the prevalence of the virus in vaccinated women but also protected against cervical neoplasia and cancer.[37]

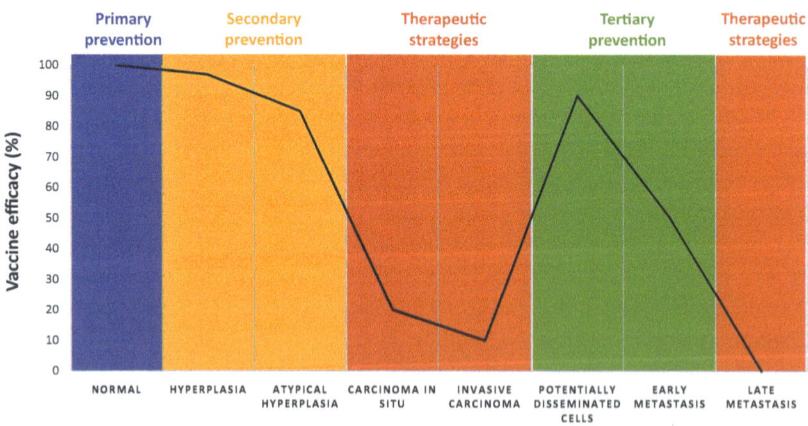

FIGURE 3.1 Hypothesized efficacy of preventative cancer vaccines based on tumor-specific antigens and tumor-associated antigens given at successive phases of cancer development. The predicted best strategies are showed by different colors. The bottom axis shows different clinical stages of the disease.

Prevention of nonviral cancers

The possibility of applying prophylactic vaccines based on TSA or TAA to prevent the occurrence of cancers with unknown etiology was proposed almost two decades ago.[38,39] The rationale for the implementation of immunoprevention in nonviral cancers is strong because, in disease-free individuals, the immune system is not yet impaired by tumor- and treatment-induced suppression nor has it had a chance to become tolerant to tumor antigens encountered in the tumor microenvironment in the absence of costimulatory/danger signals.

There has been a strong pushback on using vaccines based on TSA and TAA in healthy individuals with increased cancer risk. This is in great part due to the assumption that nonmutated TSA and TAA are self-antigens that in the first place might not elicit strong enough protective immunity, or if they did, this would result in autoimmunity. Fortunately, this perception is beginning to change due to the efforts of several laboratories that are accumulating data in murine models supporting both efficacy and safety of tumor antigen-based preventative vaccines for nonviral cancers.[40,41]

Nonmutated tumor-specific antigens (TSA) suitable for primary prevention

TSA are likely to be good candidate antigens for preventative cancer vaccines as their expression in certain tumor types is highly predictable and because they are expressed only in tumors, which reduces the risk of autoimmunity. Oncofetal antigens, such as alpha-fetoprotein (AFP), have been investigated in preclinical models of primary prevention. DCs were engineered to express AFP and used as a vaccine to induce a strong T-cell response. This completely protected mice from a lethal dose of an AFP^+ tumor challenge.[42] The use of fetal materials beyond peptides has also been suggested to prevent cancer. For example, vaccination with undifferentiated, pluripotent embryonic stem cells has been shown to delay tumor growth in mouse models of transplantable colon and lung cancers.[43,44] It was hypothesized that this protection relies on the expression of similar molecules by cancer-initiating cells and embryonic stem cells, such as OCT4, NANOG, LY6G, and BMI1, which could open opportunities for vaccination using peptides derived from these molecules.[45]

Cancer testis antigens (CTA) could also be appropriate for primary prevention as they are expressed in diverse types of tumors but not in normal adult tissues except for testes and ovaries. Among all CTA found so far, NY-ESO-1 is the most immunogenic and therefore has been widely studied for its potential in therapeutic vaccination.[46,47] It was only recently considered in the development of a preventative vaccine for multiple myeloma. In a murine model, the authors found that vaccination with NY-ESO-1 peptides increased survival and reduced tumor growth.[48] Another CTA of interest belongs to the MAGE family of molecules expressed in many tumor types. Intradermal administration of a vaccine composed of a DNA plasmid encoding MAGE-1 induced anti-MAGE-1-specific antibody in BALB/c mice and was associated with tumor challenge rejection.[49]

Tumor-associated antigens (TAA) suitable for primary prevention

Transgenic mouse models have been extensively used over the years to test cancer prevention potential and safety of vaccines against the well-known TAA HER-2/neu, CEA, MUC1, mTERT, and mesothelin among others.[39] Decades of research showed that the outcomes of vaccination can vary between models and vaccine formulations. They can cause as little as a slight delay in tumor onset to completely prevent tumor initiation.[50] Vaccine antitumor activity was shown to be principally driven by T cells producing cytokines, such as interferon (IFN)-γ, and antibodies that mediated antibody-dependent cellular cytotoxicity and complement-dependent cytotoxicity.[51–53] A lot of older work has been extensively reviewed;[38,54–56] therefore in the following sections, we will focus on the most recent examples of TAA-based preventative vaccines tested in mouse models.

A vaccine based on an adenovirus expressing the extracellular and transmembrane domains of Her-2 was used to immunize BALB-*neu*T mice and prevent the development of spontaneous mammary tumors.[52] This protection was linked to the elicitation of anti-Her-2 antibodies rather than antigen-specific T cells. In the same mouse model, virus-like particles carrying human Her-2/neu were used as a vaccine.[57] After a single immunization, the vaccine induced a significant delay in tumor outgrowth. More recently, a nanoliposomal vaccine containing Her-2/neu peptides promoted high IFN-γ production by Her-2/neu-specific CD8$^+$ T cells and increased survival of BALB/c mice challenged with TUBO tumor cells expressing rat Her-2/neu.[58,59] Two interesting findings support the use and the expected benefit of similar strategies for conducting clinical studies in patients. First, a study showed that the vaccination of BALB-neuT mice before the onset of the spontaneous tumor with a Her-2/neu peptide vaccine induced a delay in tumor growth and an increase in survival. The vaccinated mice exhibited long-lasting CTL responses, not only against the immunizing peptide but also against other peptides derived from rat Her-2/neu product, suggesting epitope spreading.[60] Second, some evidence was obtained that the protection conferred by cancer vaccines could be transmitted vertically from mother to offspring. Indeed, Barutello et al. found that the vaccination against Her-2/neu of BALB-*neu*T transgenic mice protects not only the mothers but also their offspring from spontaneous mammary tumor development.[61] They linked this protection to the transfer through milk of maternal antineu antibodies together with immune complexes formed between the *neu* extracellular domain and antineu IgG. These complexes could be taken up by DCs in the intestinal mucosa via their Fc receptors, which would also activate them and promote their migration to a draining lymph node and presentation of Her-2/neu to the T cells of the offspring. Maternal immunization has already been shown to be efficient in reducing pathogen-induced diseases in newborns.[62] If confirmed, transmission of protective immunity from mother to offspring after cancer vaccination could delay cancer development in genetically predisposed offspring and offer a new approach for neonatal cancer prevention.

MUC1 is a highly glycosylated, transmembrane protein expressed apically on luminal epithelium and abnormally glycosylated in cancer cells that has been extensively investigated for immunoprevention applications. A fusion protein consisting of glutathione-*S*-transferase and 10 mouse MUC1 tandem repeats was used to vaccinate C3H/HeOuj mice, which resulted in a decreased incidence of spontaneous breast cancer.[63] Immunized mice had high levels of anti-MUC1 antibodies and MUC1-specific T-cell responses. In human MUC1 transgenic mice, a vaccine consisting of BCG transformed with a plasmid containing the human interleukin (IL)-2 and truncated MUC1 cDNA was able to stimulate a T-cell-mediated immune response, to decrease inflammation and inhibit the cancer growth of engrafted MUC1$^+$ tumor, which demonstrated its efficacy as a prevention strategy for breast cancer.[64] A synthetic glycopeptide vaccine containing a 22-amino-acid-long sequence from the variable number tandem repeats region of human MUC1 coupled to tetanus toxoid was also tested to prevent breast cancer in human MUC1-transgenic mice. Vaccination resulted in an MUC1-specific humoral immune response, a slowdown of tumor progression, and an increase in survival of breast tumor-bearing mice.[65] More recently, colorectal cancer stem cells (CCSC) were engineered to express different levels of MUC1 and used as vaccines in wild-type mice. The authors found that the CCSC-based vaccine significantly reduced the growth of the tumor used to challenge the immunized mice and increased the cytotoxicity of natural killer cells and antibody production and that MUC1 overexpression dramatically enhanced the CCSC vaccine immunogenicity.[66]

In addition to MUC1 and Her-2/neu, other TAA are attractive targets for preventative vaccination strategies. The prostate stem cell antigen (PSCA) is a membrane glycoprotein expressed constitutively at low levels in prostate tissue but overexpressed in cancer. A study with the human PSCA transgenic mouse (TRAMP), a model for prostate adenocarcinoma, showed that immune memory induced by the PSCA vaccine has a prophylactic effect, as animals receiving the vaccine were protected against TRAMP-C2 tumor challenge.[67] This protection was linked to an increase in tumor-infiltrating CD4 and CD8 cells and enhanced intratumoral production of IL-2, IFN-γ, granulocyte-macrophage colony-stimulating factor, and tumor necrosis factor. Importantly, this effective tumor protection was not accompanied by autoimmunity.

Using both transplantable and spontaneous melanoma models, it was found that lentiviral vector expressing human gp100, a melanoma TAA, was able to induce potent CD8 T-cell responses and to prevent tumor establishment.[68] The prophylactic potential of a vaccine based on α-lactalbumin, a protein normally expressed on lactating breast ducts but overexpressed in breast cancer, was also assessed in a transgenic model recapitulating human Her-2/neu overexpressing breast cancer. In this model, where 50% of mice develop cancer by day 205, vaccination against α-lactalbumin totally prevented spontaneous development of breast cancer as well as growth of implanted breast cancer cells.[69] Finally, three WT1-derived peptides were used to vaccinate wild-type mice that were later challenged with a lethal dose of WT1-expressing FBL3 cells. The peptide with the highest binding affinity for H-2D^b molecules induced high cytotoxic T-cell responses specific to WT1. Vaccinated mice also rejected WT1-exppressing tumors and had an increased survival with no signs of autoimmunity.[70]

Disease-associated antigens (DAA) suitable for primary cancer prevention

Primary cancer immunoprevention poses the strictest requirements on candidate vaccine antigens limiting the number of already existing promising targets. One desired characteristic is that their expression needs to be predictable, expressed early in cancer development and on many different cancers or all cancers of the same tissue type. An other important characteristic is an abnormal expression on affected cells compared to healthy counterparts. Therefore research efforts have recently been directed toward either confirmation of these characteristics for existing TAA or identification of new molecular targets that undergo changes in expression and posttranslation modifications. This has led to a completely new category of potential vaccine antigens that transiently change expression to levels similar to what is observed in cancer, in other conditions, such as inflammation and acute infections. Such molecules were defined as disease-associated antigens (DAA), and evidence accumulated so far suggested that this change in expression, or vaccination with these molecules, elicits antigen-specific immunity and long-term immune memory. Indeed, the presence of antibodies against DAA molecules, known to be expressed in cancer cells as TAA, was found to decrease cancer risk in healthy individuals without any evidence of autoimmunity. For instance, Cramer et al. showed a decrease in ovarian cancer risk correlated with the number of inflammatory conditions, such as mastitis, pelvic surgery, and intrauterine device use.[71] The same study showed that the likelihood of having anti-MUC1 antibodies increased from 24% for women with 0 or 1 inflammatory condition to 51% for women with five or more conditions. These results were then confirmed by a prospective nested case–control study showing that anti-MUC1 antibodies could be used as predictive markers of subsequent ovarian cancer.[72] The role of inflammation in inducing changes in the expression of many self-antigens in normal cells was recently confirmed. Acute inflammation mimicked in vitro was sufficient to induce the transient expression of DAA MUC1, Her-2/neu, CEA, EEF2, SerpinB1, SOD2, and NAMPT in normal mammary epithelial cells.[73] These molecules were found to be expressed as early as in grade 2 ductal carcinoma in situ (DCIS) lesions (our unpublished data) and are now investigated for the development of preventative breast cancer vaccines.

The presence of an immune memory against DAA has also been observed after infections. It was found that women who experienced mumps infection early in life had higher serum levels of anti-MUC1 antibodies, which was associated with a drastically reduced risk of ovarian cancer later in life.[74,75] These observations in humans have been modeled in mice. In C57BL6 mice, a repeated infection with the influenza virus yielded numerous postinfection IgG antibodies and T cells directed against antigens present on a mouse lung tumor cell line 3LL and protection from 3LL challenge 3 months post last influenza infection. These antigens were found to be nonmutated self-molecules constitutively overexpressed on 3LL cells and transiently overexpressed on infected mouse lungs and therefore can be considered as DAA.[76] Vaccination with peptides derived from these DAA could be the best and safest strategy to reduce the risk of developing cancer.[74] Indeed, in the study on influenza mentioned above, it was shown that vaccination based on a cocktail of DAA protected mice from tumor challenge by stimulating immune memory.[76]

A large body of literature in the field of cancer epidemiology shows that strong febrile infections, especially those that characterize common childhood diseases, can profoundly reduce life-time risk of various cancers and opens even more opportunity for the identification of DAA for cancer prevention. For example, case–control studies in lymphoma, stomach, colorectal, breast, and ovarian cancers found that childhood diseases, such as chicken pox and pertussis, as well as repeated cold and influenza infections throughout life, significantly decreased life-time risk for these cancers.[77,78] In addition, a large case–control study in patients with melanoma from six European countries showed reduction of melanoma risk with increasing numbers of febrile viral infections experienced early in life.[79,80] Once the DAA expressed in each of the acute infections are characterized, a global preventative strategy could be envisioned where the lack of DAA-specific immune memory due to limited early exposure to infectious inflammatory events could be compensated with a vaccine composed of a cocktail of these DAA.[55] As these DAA are expressed in other conditions, such as viral infections or inflammatory diseases, it could also open a new opportunity for the development of vaccines that confer "universal" protection.

Secondary prevention: preventing progression to invasive cancer

Primary prevention is particularly challenging to test in humans as the absence of toxicity is a prerequisite in healthy individuals but more importantly, because the type of cancer that will arise later in life is not always predictable. However, targeting earlier stage of disease to prevent the later insurmountable tumor-intrinsic mechanisms of immune regulation is a unique opportunity to enhance the utility of vaccination. Therefore researchers have focused on vaccinating patients with chronic diseases and premalignant lesions as a means to avoid progression to life-threatening diseases. Although preventative vaccines have principally been studied in mouse models of human chronic inflammation or premalignant lesions, several clinical trials have also been completed. TSA and TAA were found to be expressed in various premalignant lesions.[81] For instance, 60% of DCIS highly expressed CTA, which support their potential as targets for vaccines to prevent the development of cancer in women harboring those premalignant lesions.[82] MAGE-A1, A3–4 and NY-ESO-1 were also found on premalignant lesions[83] and antibodies against NY-ESO-1 were detected in patients with premalignant lesions, including benign prostate hyperplasia.[84] The tumor form of MUC1 is expressed in many preneoplastic lesions, such as pancreatic intraepithelial neoplasia, precursors of pancreatic cancer, and adenomatous polyps, precursors of colon cancer.[85,86]

Preclinical studies in secondary prevention

A few TAA have been investigated for their protective effects in this setting. MUC1 peptide plus TLR4 agonist adjuvant vaccine was administrated to $MUC1^{+}/IL10^{-/-}$ transgenic mice that develop MUC1-positive inflammatory bowel disease (IBD). Vaccination against MUC1 before any evidence of IBD, induced MUC1-specific CTLs and anti-MUC1 antibodies that eliminated inflamed colonocytes and prevented IBD progression to colitis-associated colon cancer (CACC) without causing autoimmune reactions against $MUC1^{+}$ healthy cells.[87] In the dextran sulfate sodium mouse model of CACC, the same vaccine prevented the development of dysplasia in the colon and supported increased survival.[88]

The same Her-2/neu peptide vaccine used for primary prevention described above was used to vaccinate 15-week-old Balb-NeuT mice when unvaccinated mice harbor their first measurable tumors. When administrated together with CpG and after three doses, the vaccine delayed by 5 weeks the time required for these mice to develop tumors and elicited tumor-specific T-cell responses.[60] The authors also showed that the efficacy of the vaccine could be increased by the concomitant injection of anti-CD25 antibody suggesting that regulatory T cells could be involved in the tumor escape at such early stages.

Finally, in a TRAMP mouse model of prostate intraepithelial neoplasia, the murine PSCA (mPSCA) cDNA vaccine, followed by a boost with a Venezuelan equine encephalitis virus carrying mPSCA, was able to efficiently induce antitumor immune responses[89] with vaccinated animals showing 90% survival compared to 20% for control mice.[67] The tumors of vaccinated TRAMP mice were smaller and well-differentiated focal adenocarcinomas with higher immune infiltration compared to unvaccinated mice. However, the same vaccine given to mice after they developed prostate cancer had no effect on survival, strengthening the idea that vaccination in premalignant settings represents a unique opportunity for cancer prevention before the onset of multiple immune evasion mechanisms.

Human clinical trials of secondary immunoprevention

Only a few preventative vaccines based on tumor antigen were tested in a premalignant setting in individuals at high risk for cancer. For instance, a pilot clinical trial assessed anti-MUC1 vaccine immunogenicity in patients at high risk of developing MUC1$^+$ colon cancer after the removal of advanced polyps.[90] This vaccine elicited strong immunity in 43% of the patients without any side effects, and a long-term immune memory as measured by antibody responses to a booster injection at 1 year.[90] Later, it was shown that antibodies made in response to the vaccine were highly tumor specific with a range of affinities no different from the one reported for antiviral responses.[91] When the nonresponders to the vaccine were studied, it was found that they had increased numbers of circulating myeloid-derived suppressor cells (MDSCs). Similarly to what was observed in mice after Her-2/neu vaccination, this suggested that, in some of these patients, the immunosuppressive environment was already in place at this early stage of cancer development, limiting the efficacy of the vaccine.[90] More importantly, in vaccine responders, there were no autoimmune problems reported after more than 14 years from the first vaccination. A multicenter placebo-controlled efficacy trial of the same vaccine and in the same premalignant setting was recently completed. Results were similar to the results of the pilot study with vaccinated individuals divided into responders and nonresponders and the nonresponders having higher levels of circulating MDSC than responders. The vaccine also appeared to protect from new polyps as determined by follow-up colonoscopy at year 3 postvaccine (manuscript in preparation).

Efforts to prevent recurrence of DCIS by immunizing against Her-2/neu have also been made. A vaccine composed of DCs pulsed with 6 Her-2/neu peptides was administrated to patients with DCIS postbiopsy but before surgical resection. At surgery, some patients had no more evidence of disease.[92] Another trial using the same vaccine showed effective activation of Her-2/neu-specific CD4$^+$ and CD8$^+$ T cells, their accumulation in the DCIS lesion, and a decrease in Her-2/neu expression in surgical tumor specimens.[93] A follow-up randomized trial found that the pathological response rate was higher in DCIS patients than in patients with more advanced disease, reinforcing the idea that targeting early premalignant lesions could be an efficient approach to prevent cancer.[94] Finally, an ongoing phase II clinical trial is investigating the safety and the immunogenicity of a vaccine based on a Her-2/neu-derived peptide in DCIS patients before surgery.

MUC1 and Her-2/neu are not the only TAA expressed in premalignant lesions that could be targeted to avoid progression to invasive disease. In fact, mesothelin, cyclin B1, SOX-2, and EGFR have all been found to be expressed in premalignant diseases and still need to be investigated as targets for the development of vaccines in patients at high risk of developing cancer.[81]

Tertiary immunoprevention: avoiding recurrence

Each year, more than 700,000 people undergo cancer surgery in the United State. However, more than 40% of those patients develop recurrence and have a poor outcome. Vaccination was for a long time considered to be a promising approach in patients with high risk of local recurrence after surgical removal of the primary tumor. In the absence of the primary tumor, the vaccine was expected to boost ongoing antitumor responses or generate new ones that would go on to eradicate residual or disseminated tumor cells by targeting them specifically, and establish long-lasting immunity. The enthusiasm for this approach had not yet been dampened by knowledge that came much later about the immunosuppressive tumor microenvironment. Indeed, tumor cells debilitate the antitumor responses by establishing T-cell tolerance against tumor antigens, secreting immunosuppressive factors, and attracting immunosuppressive cells, such as Tregs, tumor-associated macrophages, and regulatory DCs, among other mechanisms.[95–97] It has been shown that after surgery, the lingering immunosuppressive cells can repopulate small pockets of residual disease quickly limiting the efficacy of vaccine and increasing the risk of recurrence.[98] That might explain why most clinical studies described below have shown none or only limited vaccine efficacy, although many different vaccine formulations and many different TSA and TAA were tried.

Given its ability to induce both humoral and cellular immune responses, NY-ESO-1 has been considered a suitable antigen for a cancer vaccine. In a phase I study, a vaccine based on an NY-ESO-1 peptide combined with two adjuvants was used on patients with resected high-risk stage III/IV malignancies. The trial showed that the vaccine could induce NY-ESO-1-specific T-cell responses, which were associated with a favorable clinical outcome and an increase in overall survival.[99] A phase I/II clinical trial was conducted in high-risk resected melanoma and the NY-ESO-1 vaccine was well tolerated and capable of inducing antibody and T-cell responses but the impact on survival or disease recurrence was not evaluated.[100] Another phase II clinical trial administrated a vaccine based on the recombinant MAGE-A3 protein with or without adjuvant to patients with nonsmall-cell lung cancer after tumor resection. The vaccine was safe and immunogenic and induced high peptide-specific antibody titer and cytotoxic T cells.[101] A randomized clinical phase III trial was conducted in patients with lung cancer after complete resection to assess the efficiency of the same MAGE-A3 vaccine. After 13 intramuscular injections, there was no increase in disease-free survival and further development of this vaccine was stopped.[102]

MUC1 peptide loaded onto autochthonous DCs was administered intradermally or subcutaneously in patients with pancreatic cancer after surgical resection of the tumor.[103] This early phase I/II clinical trial was a proof of principle that vaccination was feasible and well tolerated. The authors also reported a significant improvement in disease-free survival compared to historical controls; however, the immunological mechanisms responsible for protection were not clear. It was quickly followed in 2005 by a phase I/II MUC1 trial testing the MUC1 peptide vaccine in patients with surgically resected pancreatic cancer and no other chemotherapy.[104] The vaccine was administered at 3-week intervals and was well tolerated with a significant increase in the number of $CD8^+$ cells in the peripheral blood. However, the trial was relatively small and the clinical outcome had to be compared with historical controls. Two of 15 patients were alive and disease free at 32- and 61-month follow-ups, which were not different from controls. Later, another MUC1 vaccine was developed composed of autologous DCs modified with a recombinant vaccinia virus to express CEA, MUC1, and three costimulatory molecules B7.1, ICAM-1, and LFA-3. In a randomized phase II clinical trial on patients with colorectal cancer after complete resection of the primary tumor, the authors showed that it could be beneficial in tertiary prevention as vaccinated patients had a superior survival compared to the unvaccinated group.[105] Recombinant CEA without MUC1 was also used to design a vaccine. A randomized, dose ranging study on 24 patients with a resected colorectal cancer found that vaccination was safe and induce long-lasting CEA-specific T-cell and IgG responses.[106] The titers of anti-CEA IgG were associated with increased survival especially for patients with high level of CEA at the start of vaccination.

Vaccines based on three Her-2/neu-derived peptides (E75, GP2, and AE37) were developed and used with different adjuvants in clinical trials to prevent the recurrence of breast cancer. In a phase I/II clinical trial, vaccinated patients positive for HLA-A2/A3 that are restricting elements for E75 peptide showed a trend toward improved disease-free survival with a 48% reduction in the relative risk of recurrence.[107,108] A randomized prospective phase II trial of the AE37 vaccine reported an enhanced disease-free survival in patients with Her-2 expression with a 32% reduction of recurrence.[107,108] Another phase II clinical trial was conducted to evaluate the efficacy of a vaccine based on GP2 administrated to patients with disease-free, node-positive, and high-risk node-negative breast cancer to prevent recurrence.[109] The study suggested that the vaccine was safe and could be effective in decreasing recurrence in patients with tumors expressing high levels of

Her-2/neu. Recently, three clinical trials were conducted to examine the potential of Her-2/neu vaccine to reduce recurrence. In a phase I clinical trial, women with resected Her2^{+} cancer were vaccinated six times with four Her2-derived peptides.[110] Ninety percent of the patients generated Her-2-specific T cells and antibodies that lasted for at least 2 years after the last vaccination. A randomized phase III clinical trial used a vaccine based on E75 in combination with trastuzumab on clinically disease-free patients after they undergo standard therapy. The vaccine was safe[111] but did not significantly increase disease-free survival at 17 months' median follow-up and the trial was stopped.[112]

Concluding remarks

Our systematic review of the literature on cancer vaccines based on defined tumor antigens showed that 88% of them were tested in a therapeutic setting. Even though these vaccines are used in a large spectrum of cancers (Fig. 3.2A), they are often used in late stage or metastatic cancers in patients with highly immunosuppressive tumor microenvironments or impaired immune functions related to previous treatments, such as chemotherapy and radiotherapy.[113] Therefore therapeutic vaccines showed limited results in clinical trials and only 26 of 189 trials we analyzed reported an increase in survival for patients with tumor antigen-specific humoral or cellular immunity (Fig. 3.2B). We hypothesize that vaccines may instead reach their full potential in individuals without malignant lesions where low levels of immune activation will eliminate precancer cells without autoimmune side effects, especially in primary or secondary preventative settings. In contrast, the future for vaccine-based approaches in tertiary prevention and therapeutic setting may belong to combinations of vaccines with other immunomodulatory interventions, for example, checkpoint inhibitors, depletion of regulatory T cells or MDSCs, inhibition of suppressive cytokines and chemokines, or addition of immunostimulatory cytokines and chemokines, among others. Clinical trials are ongoing to assess the efficacy of therapeutic vaccines in combination with checkpoint inhibitors[114] while toxicities associated with checkpoint inhibitors in cancer-free patients continue to be assessed.[115]

The field of preventative cancer vaccines still faces major challenges in identifying the formulation, timing, and antigens that will maximize the immune responses to future cancer cells. In the tertiary prevention and the therapeutic setting, our systematic review showed that tumor antigen vaccines were predominantly based on peptides or pulsed DC (Fig. 3.2C). As immune responses may be stronger in healthy individuals, further studies will need to compare methods in primary and secondary prevention settings to identify those most relevant.

Theoretically, every individual above 25 years of age harbors some premalignant lesions, and this is why the determination of the best timing for

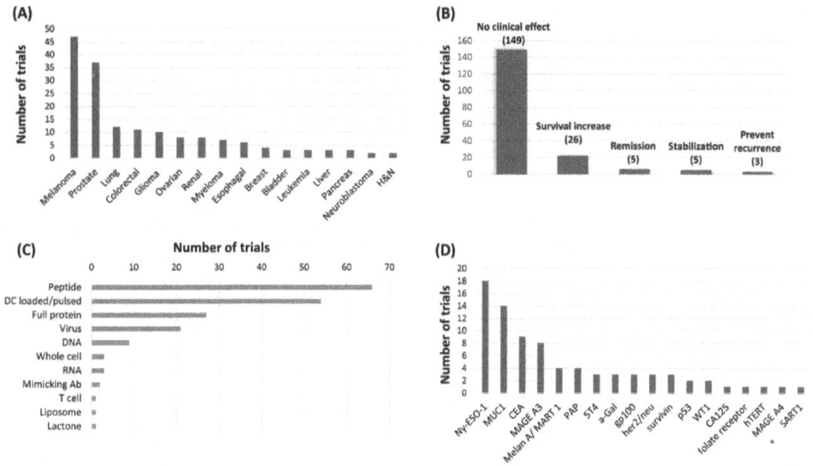

FIGURE 3.2 Qualitative synthesis of clinical trials of therapeutic cancer vaccines based on tumor-specific antigens and tumor-associated antigens. We identified 4178 records in the PubMed database using the MeSH terms "cancer vaccine" AND "tumor antigen" from year 1995 to year 2019. We excluded review articles ($n = 2128$) and mouse experimental studies ($n = 1534$). We then manually screened 516 records and excluded trials with an earlier date than 2009 and publications for which the full text was not available. A total of 236 full-text articles were then assessed for eligibility and we excluded case reports ($n = 13$) and trials that did not test vaccines based on defined tumor antigens ($n = 34$). A total of 189 studies were included in our final analysis. (A) Number of clinical trials in patients affected by different cancers. *H&N*, head and neck. (B) Trials reporting different clinical outcomes. (C) Number of clinical trials using various vaccination methods. *Ab*, antibody. (D) Number of clinical trials using various tumor antigens. *NY-ESO-1*, New York esophageal squamous cell carcinoma 1; *MUC1*, mucin 1; *CEA*, carcinoembryonic antigen; *MAGE-A3*, melanoma-associated antigen 3; *MelanA/MART-1*, melanoma antigen recognized by T cell 1; *PAP*, prostatic acid phosphatase; *α-Gal*, galactose-alpha-1,3-galactose; *gp100*, glycoprotein 100; *WT1*, Wilms' tumor suppressor gene 1; *CA125*, cancer antigen 125; *hTERT*, human telomerase reverse transcriptase; *SART1*, spliceosome associated factor 1.

vaccination is an extremely important factor. For primary prevention, vaccination in childhood might be a promising strategy as it might contribute to educating the immune system to eliminate premalignant cells when they appear, similarly to what was observed in the context of childhood viral or bacterial diseases.[116] For secondary prevention, vaccination might be most efficient in early premalignancy before immunoediting decreases tumor antigen expression and increases the population of immunosuppressive cells, such as regulatory T cells and MDSCs.[3] However, the use of these vaccines will depend on the continued development of better detection methods for these early stages of disease.

The suitability of a tumor antigen for preventative vaccine depends on the proportion of tumors that express it, its abundance on tumor cells and the predictability of its expression. Our meta-analysis showed that clinical trials so far have mostly focused on four tumor antigens, NY-ESO, MUC1, CEA, and MAGE-A3 to develop cancer vaccines in tertiary prevention and therapeutic settings (Fig. 3.2D). These antigens might also be promising in primary and secondary prevention strategies. The choice of antigens for preventative vaccines has to satisfy specific requirements: (1) the antigens have to have an essential role in tumor growth or progression to avoid escape of antigen loss variants but play a marginal role in healthy adult tissues; (2) be a good immune target that elicits long-term immunity; and (3) be accessible for direct recognition (cell surface protein) to continue to be recognized even if major histocompatibility complex molecules are downregulated during transformation.[39] Antigens that are either biochemically or structurally different in tumor cells than in normal cells could be especially suitable for prevention vaccine antigens. Good examples are posttranscriptionally modified protein, such as the hypoglycosylated form of MUC1.

Cancer vaccines can be cost-effective, off-the shelf reagents that can be made broadly available around the world. In this chapter, we highlighted some innovative ways in which vaccines can prevent cancer altogether or slow down its progression and therefore could decrease world cancer incidence and mortality. With the fast-accumulating knowledge in cancer immunology, cancer biology, and new technological advances in medical diagnostics, the time has come to focus more attention on cancer immunoprevention through the use of tumor antigen-based vaccines. In the distant future, we can hope that safe and effective cancer vaccines will be administrated to the general population, but until then, preventative vaccines should be developed for specific groups of people who are at an increased risk for developing specific cancers.

References

1. Aktipis CA, Nesse RM. Evolutionary foundations for cancer biology. *Evol Appl.* 2013;6:144–159. Available from: https://doi.org/10.1111/eva.12034.
2. DeGregori J. Evolved tumor suppression: why are we so good at not getting cancer? *Cancer Res.* 2011;71:3739–3744. Available from: https://doi.org/10.1158/0008-5472.CAN-11-0342.
3. Dunn GP, Old LJ, Schreiber RD. The three Es of cancer immunoediting. *Annu Rev Immunol.* 2004;22:329–360. Available from: https://doi.org/10.1146/annurev.immunol.22.012703.104803.
4. Anand P, Kunnumakara AB, Sundaram C, Harikumar KB, Tharakan ST, Lai OS, et al. Cancer is a preventable disease that requires major lifestyle changes. *Pharm Res.* 2008. Available from: https://doi.org/10.1007/s11095-008-9661-9.
5. Narod SA, Brunet JS, Ghadirian P, Robson M, Heimdal K, Neuhausen SL, et al. Tamoxifen and risk of contralateral breast cancer in BRCA1 and BRCA2 mutation carriers: a case-control study. *Lancet.* 2000;356:1876–1881. Available from: https://doi.org/10.1016/S0140-6736(00)03258-X.
6. Stephenson AJ, Abouassaly R, Klein EA. Chemoprevention of prostate cancer. *Urol Clin North Am.* 2010. Available from: https://doi.org/10.1016/j.ucl.2009.11.003.

7. Karimi-Zarchi M, Allahqoli L, Nehmati A, Kashi AM, Taghipour-Zahir S, Alkatout I. Can the prophylactic quadrivalent HPV vaccine be used as a therapeutic agent in women with CIN? A randomized trial. *BMC Public Health*. 2020;20. Available from: https://doi.org/10.1186/s12889-020-8371-z.
8. Ward S, Casey D, Labarthe MC, Whelan M, Dalgleish A, Pandha H, et al. Immunotherapeutic potential of whole tumour cells. *Cancer Immunol Immunother*. 2002;51:351–357. Available from: https://doi.org/10.1007/s00262-002-0286-2.
9. Chen F, Zou Z, Du J, Su S, Shao J, Meng F, et al. Neoantigen identification strategies enable personalized immunotherapy in refractory solid tumors. *J Clin Invest*. 2019;129:2056–2070. Available from: https://doi.org/10.1172/JCI99538.
10. Lee EYHP, Muller WJ. Oncogenes and tumor suppressor genes. *Cold Spring Harb Perspect Biol*. 2010. Available from: https://doi.org/10.1101/cshperspect.a003236.
11. Wepsic HT. Overview of oncofetal antigens in cancer. *Ann Clin Lab Sci*. 1983;13:261–266.
12. Chomez P, De Backer O, Bertrand M, De Plaen E, Boon T, Lucas S. An overview of the MAGE gene family with the identification of all human members of the family. *Cancer Res*. 2001;61:5544–5551.
13. De Backer O, Arden KC, Boretti M, Vantomme V, De Smet C, Czekay S, et al. Characterization of the GAGE genes that are expressed in various human cancers and in normal testis. *Cancer Res*. 1999;59:3157–3165.
14. Chen YT, Scanlan MJ, Sahin U, Türeci Ö, Gure AO, Tsang S, et al. A testicular antigen aberrantly expressed in human cancers detected by autologous antibody screening. *Proc Natl Acad Sci USA*. 1997;94:1914–1918. Available from: https://doi.org/10.1073/pnas.94.5.1914.
15. Nollau P, Scheller H, Kona-Horstmann M, Rohde S, Hagenmüller F, Wagener C, et al. Expression of CD66a (human C-CAM) and other members of the carcinoembryonic antigen gene family of adhesion molecules in human colorectal adenomas. *Cancer Res*. 1997;57:2354–2357.
16. Salman JW, Schoots IG, Carlsson SV, Jenster G, Roobol MJ. *Prostate specifi c antigen as a tumor marker in prostate cancer: biochemical and clinical aspects. Advances in Experimental Medicine and Biology*. New York LLC: Springer; 2015:93–114. Available from: http://doi.org/10.1007/978-94-017-7215-0-7.
17. Chi DD, Merchant RE, Rand R, Conrad AJ, Garrison D, Turner R, et al. Molecular detection of tumor-associated antigens shared by human cutaneous melanomas and gliomas. *Am J Pathol*. 1997;150:2143–2152.
18. Watson MA, Darrow C, Zimonjic DB, Popescu NC, Fleming TP. Structure and transcriptional regulation of the human mammaglobin gene, a breast cancer associated member of the uteroglobin gene family localized to Chromosome 11q13. *Oncogene*. 1998;16:817–824. Available from: https://doi.org/10.1038/sj.onc.1201597.
19. Kawakami Y, Eliyahu S, Jennings C, Sakaguchi K, Kang X, Southwood S, et al. Recognition of multiple epitopes in the human melanoma antigen gp100 by tumor-infiltrating T lymphocytes associated with in vivo tumor regression. *J Immunol*. 1995;154:3961–3968.
20. Seiter S, Monsurro V, Nielsen M-B, Wang E, Provenzano M, Wunderlich JR, et al. Frequency of MART-1/MelanA and gp100/PMel17-specific T cells in tumor metastases and cultured tumor-infiltrating lymphocytes. *J Immunother*. 1997;25:252–263.
21. Vlad AM, Kettel JC, Alajez NM, Carlos CA, Finn OJ. MUC1 immunobiology: from discovery to clinical applicationsadvances in immunology. *Adv Immunol*. 2004;249–293. Available from: https://doi.org/10.1016/S0065-2776(04)82006-6.

22. Kao H, Marto JA, Hoffmann TK, Shabanowitz J, Finkelstein SD, Whiteside TL, et al. Identification of cyclin B1 as a shared human epithelial tumor-associated antigen recognized by T cells. *J Exp Med.* 2001;194:1313.
23. Hiyama E, Hiyama K. Telomerase as tumor marker. *Cancer Lett.* 2003;194:221–233. Available from: https://doi.org/10.1016/s0304-3835(02)00709-7.
24. Schmidt SM, Schag K, Müller MR, Weck MM, Appel S, Kanz L, et al. Survivin is a shared tumor-associated antigen expressed in a broad variety of malignancies and recognized by specific cytotoxic T cells. *Blood.* 2003;102:571–576. Available from: https://doi.org/10.1182/blood-2002-08-2554.
25. Argani P, Iacobuzio-Donahue C, Ryu B, Rosty C, Goggins M, Wilentz RE, et al. Mesothelin is overexpressed in the vast majority of ductal adenocarcinomas of the pancreas. *Clin Cancer Res.* 2001;7:3862–3868.
26. Slamon DJ, Godolphin W, Jones LA, Holt JA, Wong SG, Keith DE, et al. Studies of the HER-2/neu proto-oncogene in human breast and ovarian cancer. *Science.* 1989;244:707–712.
27. Novellino L, Castelli C, Parmiani G. A listing of human tumor antigens recognized by T cells: March 2004 update. *Cancer Immunol Immunother.* 2005. Available from: https://doi.org/10.1007/s00262-004-0560-6.
28. Zhou J, Dudley ME, Rosenberg SA, Robbins PF. Persistence of multiple tumor-specific T-cell clones is associated with complete tumor regression in a melanoma patient receiving adoptive cell transfer therapy. *J Immunother.* 2005;28:53–62. Available from: https://doi.org/10.1097/00002371-200501000-00007.
29. Lennerz V, Fatho M, Gentilini C, Frye RA, Lifke A, Ferel D, et al. The response of autologous T cells to a human melanoma is dominated by mutated neoantigens. *Proc Natl Acad Sci USA.* 2005;102:16013–16018. Available from: https://doi.org/10.1073/pnas.0500090102.
30. Löffler MW, Mohr C, Bichmann L, Freudenmann LK, Walzer M, Schroeder CM, et al. Multi-omics discovery of exome-derived neoantigens in hepatocellular carcinoma. *Genome Med.* 2019;11:28. Available from: https://doi.org/10.1186/s13073-019-0636-8.
31. Bilich T, Nelde A, Bichmann L, Roerden M, Salih HR, Kowalewski DJ, et al. The HLA ligandome landscape of chronic myeloid leukemia delineates novel T-cell epitopes for immunotherapy. *Blood.* 2019;133:550–565. Available from: https://doi.org/10.1182/blood-2018-07-866830.
32. Laumont CM, Vincent K, Hesnard L, Audemard É, Bonneil É, Laverdure JP, et al. Noncoding regions are the main source of targetable tumor-specific antigens. *Sci Transl Med.* 2018;10. Available from: https://doi.org/10.1126/scitranslmed.aau5516.
33. Zur Hausen H. The search for infectious causes of human cancers: where and why. *Virology.* 2009;392:1–10. Available from: https://doi.org/10.1016/j.virol.2009.06.001.
34. de Martel C, Ferlay J, Franceschi S, Vignat J, Bray F, Forman D, et al. Global burden of cancers attributable to infections in 2008: a review and synthetic analysis. *Lancet Oncol.* 2012;13:607–615. Available from: https://doi.org/10.1016/S1470-2045(12)70137-7.
35. Chang MH, You SL, Chen CJ, Liu CJ, Lai MW, Wu TC, et al. Long-term effects of hepatitis B immunization of infants in preventing liver cancer. *Gastroenterology.* 2016;151:472–480.e1. Available from: https://doi.org/10.1053/j.gastro.2016.05.048.
36. Posuwan N, Wanlapakorn N, Sa-nguanmoo P, Wasitthankasem R, Vichaiwattana P, Klinfueng S, et al. The success of a universal hepatitis B immunization program as part of Thailand's EPI after 22 years' implementation. *PLoS One.* 2016;11:e0150499. Available from: https://doi.org/10.1371/journal.pone.0150499.

37. Harper DM, DeMars LR. HPV vaccines—a review of the first decade. *Gynecol Oncol.* 2017. Available from: https://doi.org/10.1016/j.ygyno.2017.04.004.
38. Finn OJ. Cancer vaccines: between the idea and the reality. *Nat Rev Immunol.* 2003. Available from: https://doi.org/10.1038/nri1150.
39. Finn OJ, Forni G. Prophylactic cancer vaccines. *Curr Opin Immunol.* 2002. Available from: https://doi.org/10.1016/S0952-7915(02)00317-5.
40. Finn OJ. The dawn of vaccines for cancer prevention. *Nat Rev Immunol.* 2018;18:183–194. Available from: https://doi.org/10.1038/nri.2017.140.
41. Finn OJ. Human tumor antigens yesterday, today, and tomorrow. *Cancer Immunol Res.* 2017;5:347–354. Available from: https://doi.org/10.1158/2326-6066.CIR-17-0112.
42. Vollmer Jr. CM, Eilber FC, Butterfield LH, Ribas A, Dissette VB, Koh A, et al. α-Fetoprotein-specific genetic immunotherapy for hepatocellular carcinoma. *Cancer Res.* 1999. Available from: https://cancerres.aacrjournals.org/content/59/13/3064.
43. Dong W, Du J, Shen H, Gao D, Li Z, Wang G, et al. Administration of embryonic stem cells generates effective antitumor immunity in mice with minor and heavy tumor load. *Cancer Immunol Immunother.* 2010;59:1697–1705. Available from: https://doi.org/10.1007/s00262-010-0899-9.
44. Li Y, Zeng H, Xu R-H, Liu B, Li Z. Vaccination with human pluripotent stem cells generates a broad spectrum of immunological and clinical response against colon cancer. *Stem Cells.* 2009;27. Available from: https://doi.org/10.1002/stem.234.
45. Tuohy VK, Johnson JM, Mazumder S. Primary immunoprevention of adult onset cancers by vaccinating against retired tissue-specific self-proteins. *Semin Immunol.* 2020. Available from: https://doi.org/10.1016/j.smim.2020.101392.
46. Fujita S, Wada H, Jungbluth AA, Sato S, Nakata T, Noguchi Y, et al. NY-ESO-1 expression and immunogenicity in esophageal cancer. *Clin Cancer Res.* 2004;10:6551–6558. Available from: https://doi.org/10.1158/1078-0432.CCR-04-0819.
47. Sugita Y, Wada H, Fujita S, Nakata T, Sato S, Noguchi Y, et al. NY-ESO-1 expression and immunogenicity in malignant and benign breast tumors. *Cancer Res.* 2004;64:2199–2204. Available from: https://doi.org/10.1158/0008-5472.CAN-03-3070.
48. Wang H, Huang W, Gao H, Liu TT. NY-ESO-1 protein vaccine combining alum, CpG ODN, and HH2 complex adjuvant induces protective and therapeutic anti-tumor responses in murine multiple myeloma. *Onco Targets Ther.* 2020;13:8069–8077. Available from: https://doi.org/10.2147/ott.s255713.
49. Park JH, Kim CJ, Lee JH, Shin SH, Chung GH, Jang YS. Effective immunotherapy of cancer by DNA vaccination. *Mol Cells.* 1999;9:384–391.
50. Lollini P-L, Nicoletti G, Landuzzi L, Cavallo F, Forni G, De Giovanni C, et al. Vaccines and other immunological approaches for cancer immunoprevention. *Curr Drug Targets.* 2011;12:1957–1973. Available from: https://doi.org/10.2174/138945011798184146.
51. Dhodapkar MV, Sexton R, Das R, Dhodapkar KM, Zhang L, Sundaram R, et al. Prospective analysis of antigen-specific immunity, stem-cell antigens, and immune checkpoints in monoclonal gammopathy. *Blood.* 2015;126:2475–2478. Available from: https://doi.org/10.1182/blood-2015-03-632919.
52. Park JM, Terabe M, Sakai Y, Munasinghe J, Forni G, Morris JC, et al. Early role of CD4 + Th1 cells and antibodies in HER-2 adenovirus vaccine protection against autochthonous mammary carcinomas. *J Immunol.* 2005;174:4228–4236. Available from: https://doi.org/10.4049/jimmunol.174.7.4228.
53. Spadaro M, Ambrosino E, Iezzi M, Di Carlo E, Sacchetti P, Curcio C, et al. Cure of mammary carcinomas in Her-2 transgenic mice through sequential stimulation of innate

(neoadjuvant interleukin-12) and adaptive (DNA vaccine electroporation) immunity. *Clin Cancer Res*. 2005;11:1941–1952. Available from: https://doi.org/10.1158/1078-0432.CCR-04-1873.

54. Farkas AM, Finn OJ. Vaccines based on abnormal self-antigens as tumor-associated antigens: immune regulation. *Semin Immunol*. 2010. Available from: https://doi.org/10.1016/j.smim.2010.03.003.
55. Finn OJ. Vaccines for cancer prevention: a practical and feasible approach to the cancer epidemic. *Cancer Immunol Res*. 2014;2:708–713. Available from: https://doi.org/10.1158/2326-6066.CIR-14-0110.
56. Lollini PL, Cavallo F, Nanni P, Quaglino E. The promise of preventive cancer vaccines. *Vaccines*. 2015. Available from: https://doi.org/10.3390/vaccines3020467.
57. Andreasson K, Tegerstedt K, Eriksson M, Curcio C, Cavallo F, Forni G, et al. Murine pneumotropic virus chimeric Her2/ neu virus-like particles as prophylactic and therapeutic vaccines against Her2/ neu expressing tumors. *Int J Cancer*. 2009;124:150–156. Available from: https://doi.org/10.1002/ijc.23920.
58. Farzad N, Barati N, Momtazi-Borojeni AA, Yazdani M, Arab A, Razazan A, et al. P435 HER2/neu-derived peptide conjugated to liposomes containing DOPE as an effective prophylactic vaccine formulation for breast cancer. *Artif Cells Nanomed Biotechnol*. 2019;47:664–672. Available from: https://doi.org/10.1080/21691401.2019.1576702.
59. Shariat S, Badiee A, Jalali SA, Mansourian M, Yazdani M, Mortazavi SA, et al. P5 HER2/neu-derived peptide conjugated to liposomes containing MPL adjuvant as an effective prophylactic vaccine formulation for breast cancer. *Cancer Lett*. 2014;355:54–60. Available from: https://doi.org/10.1016/j.canlet.2014.09.016.
60. Nava-Parada P, Forni G, Knutson KL, Pease LR, Celis E. Peptide vaccine given with a toll-like receptor agonist is effective for the treatment and prevention of spontaneous breast tumors. *Cancer Res*. 2007;67:1326–1334. Available from: https://doi.org/10.1158/0008-5472.CAN-06-3290.
61. Barutello G, Curcio C, Spadaro M, Arigoni M, Trovato R, Bolli E, et al. Antitumor immunization of mothers delays tumor development in cancer-prone offspring. *Oncoimmunology*. 2015;4:e1005500. Available from: https://doi.org/10.1080/2162402X.2015.1005500.
62. Albrecht M, Arck PC. Vertically transferred immunity in neonates: mothers, mechanisms and mediators. *Front Immunol*. 2020;11:555. Available from: https://doi.org/10.3389/fimmu.2020.00555.
63. Xing PX, Poulos G, McKenzie IFC. Breast cancer in mice: effect of murine muc-1 immunization on tumor incidence in C3H/HeOuj mice. *J Immunother*. 2001;24:10–18. Available from: https://doi.org/10.1097/00002371-200101000-00002.
64. Chung, et al. Development and preclinical evaluation of a Bacillus Calmette-Guérin-MUC1-based novel breast cancer vaccine. *Cancer Research*. 2003.
65. Stergiou N, Gaidzik N, Heimes AS, Dietzen S, Besenius P, Jäkel J, et al. Reduced breast tumor growth after immunization with a tumor-restricted MUC1 glycopeptide conjugated to tetanus toxoid. *Cancer Immunol Res*. 2019;7:113–122. Available from: https://doi.org/10.1158/2326-6066.CIR-18-0256.
66. Guo M, Luo B, Pan M, Li M, Xu H, Zhao F, et al. Colorectal cancer stem cell vaccine with high expression of MUC1 serves as a novel prophylactic vaccine for colorectal cancer. *Int Immunopharmacol*. 2020;88:106850. Available from: https://doi.org/10.1016/j.intimp.2020.106850.
67. Garcia-Hernandez MDLL, Gray A, Hubby B, Klinger OJ, Kast WM. Prostate stem cell antigen vaccination induces a long-term protective immune response against prostate

cancer in the absence of autoimmunity. *Cancer Res*. 2008;68:861–869. Available from: https://doi.org/10.1158/0008-5472.CAN-07-0445.

68. Xiao H, Peng Y, Hong Y, Liu Y, Guo ZS, Bartlett DL, et al. Lentivector prime and vaccinia virus vector boost generate high-quality CD8 memory T cells and prevent autochthonous mouse melanoma. *J Immunol*. 2011;187:1788–1796. Available from: https://doi.org/10.4049/jimmunol.1101138.
69. Jaini R, Kesaraju P, Johnson JM, Altuntas CZ, Jane-Wit D, Tuohy VK. An autoimmune-mediated strategy for prophylactic breast cancer vaccination. *Nat Med*. 2010;16:799–803. Available from: https://doi.org/10.1038/nm.2161.
70. Oka Y, Udaka K, Tsuboi A, Elisseeva OA, Ogawa H, Aozasa K, et al. Cancer immunotherapy targeting Wilms' tumor gene WT1 product. *J Immunol*. 2000;164:1873–1880. Available from: https://doi.org/10.4049/jimmunol.164.4.1873.
71. Cramer DW, Titus-Ernstoff L, McKolanis JR, Welch WR, Vitonis AF, Berkowitz RS, et al. Conditions associated with antibodies against the tumor-associated antigen MUC1 and their relationship to risk for ovarian cancer. *Cancer Epidemiol Biomarkers Prev*. 2005;14:1125–1131. Available from: https://doi.org/10.1158/1055-9965.EPI-05-0035.
72. Pinheiro SP, Hankinson SE, Tworoger SS, Rosner BA, McKolanis JR, Finn OJ, et al. Anti-MUC1 antibodies and ovarian cancer risk: prospective data from the nurses' health studies. *Cancer Epidemiol Biomarkers Prev*. 2010;19:1595–1601. Available from: https://doi.org/10.1158/1055-9965.EPI-10-0068.
73. Jacqueline C, Lee A, Frey N, Minden JS, Finn OJ. Inflammation-induced abnormal expression of self-molecules on epithelial cells: targets for tumor immunoprevention. *Cancer Immunol Res*. 2020. Available from: https://doi.org/10.1158/2326-6066.CIR-19-0870.
74. Cramer DW, Finn OJ. Epidemiological perspective on immune-surveillance in cancer. *Curr Opin Immunol*. 2012;23:265–271. Available from: https://doi.org/10.1016/j.coi.2011.01.002.Epidemiologic.
75. Cramer DW, Vitonis AF, Pinheiro SP, McKolanis JR, Fichorova RN, Brown KE, et al. Mumps and ovarian cancer: moder interpretation of an historic association. *Cancer Causes Control*. 2011;21:1193–1201. Available from: https://doi.org/10.1007/s10552-010-9546-1.Mumps.
76. Iheagwara UK, Beatty PL, Van PT, Ross TM, Minden JS, Finn OJ. Influenza virus infection elicits protective antibodies and T cells specific for host cell antigens also expressed as tumor associated antigens: a new view of cancer immunosurveillance. *Cancer Immunol Res*. 2015;2:263–273. Available from: https://doi.org/10.1158/2326-6066.CIR-13-0125.Influenza.
77. Abel U, Becker N, Angerer R, Frentzel-beyme R, Kaufmann M, Wysoeki S, et al. Common infections in the history of cancer patients and controls. *J Cancer Res Clin Oncol*. 1991;117:339–344.
78. Parodi S, Crosignani P, Miligi L, Nanni O, Ramazzotti V, Rodella S, et al. Childhood infectious diseases and risk of leukaemia in an adult population. *Int J cancer*. 2013;133:1892–1899. Available from: https://doi.org/10.1002/ijc.28205.
79. Kölmel KF, Pfahlberg A, Mastrangelo G, Niin M, Botev IN, Seebacher C, et al. Infections and melanoma risk: results of a multicentre EORTC case-control study. European Organization for Research and Treatment of Cancer. *Melanoma Res*. 1999;9:511–519.
80. Krone B, Kölmel KF, Grange JM, Mastrangelo G, Henz BM, Botev IN, et al. Impact of vaccinations and infectious diseases on the risk of melanoma—evaluation of an EORTC case-control study. *Eur J Cancer*. 2003;39:2372–2378. Available from: https://doi.org/10.1016/s0959-8049(03)00625-7.

81. Finn OJ, Beatty PL. Cancer immunoprevention. *Curr Opin Immunol.* 2016;39:52–58. Available from: https://doi.org/10.1016/j.coi.2016.01.002.
82. Adams S, Greeder L, Reich E, Shao Y, Fosina D, Hanson N, et al. Expression of cancer testis antigens in human BRCA-associated breast cancers: potential targets for immunoprevention? *Cancer Immunol Immunother.* 2011;60:999–1007. Available from: https://doi.org/10.1007/s00262-011-1005-7.
83. Hudolin T, Juretic A, Spagnoli GC, Pasini J, Bandic D, Heberer M, et al. Immunohistochemical expression of tumor antigens MAGE-A1, MAGE-A3/4, and NY-ESO-1 in cancerous and benign prostatic tissue. *Prostate.* 2006;66:13–18. Available from: https://doi.org/10.1002/pros.20312.
84. Jung V, Fischer E, Imig J, Kleber S, Nuber N, Reinshagen F, et al. Yeast-based identification of prostate tumor antigens provides an effective vaccine platform. *Anticancer Res.* 2010;30:895–902.
85. van der Vegt B, de Roos MAJ, Peterse JL, Patriarca C, Hilkens J, de Bock GH, et al. The expression pattern of MUC1 (EMA) is related to tumour characteristics and clinical outcome of invasive ductal breast carcinoma. *Histopathology.* 2007;51:322–335. Available from: https://doi.org/10.1111/j.1365-2559.2007.02757.x.
86. Yonezawa S, Higashi M, Yamada N, Goto M. Precursor lesions of pancreatic cancer. *Gut Liver.* 2008;2:137–154. Available from: https://doi.org/10.5009/gnl.2008.2.3.137.
87. Beatty PL, Narayanan S, Gariépy J, Ranganathan S, Finn OJ. Vaccine against MUC1 antigen expressed in inflammatory bowel disease and cancer lessens colonic inflammation and prevents progression to colitis-associated colon cancer. *Cancer Prev Res.* 2010;3:438–446. Available from: https://doi.org/10.1158/1940-6207.CAPR-09-0194.
88. Beatty P, Ranganathan S, Finn OJ. Prevention of colitis-associated colon cancer using a vaccine to target abnormal expression of the MUC1 tumor antigen. *Oncoimmunology.* 2012;1:263–270. Available from: https://doi.org/10.4161/onci.18950.
89. Wang X, Wang JP, Maughan MF, Lachman LB. Alphavirus replicon particles containing the gene for HER2/neu inhibit breast cancer growth and tumorigenesis. *Breast Cancer Res.* 2005;7. Available from: https://doi.org/10.1186/bcr962.
90. Kimura T, McKolanis JR, Dzubinski LA, Islam K, Potter MD, Salazar AM, et al. MUC1 vaccine for individuals with advanced adenoma of the colon: a cancer immunoprevention feasibility study. *Cancer Prev Res (Phila).* 2012;100:130–134. Available from: https://doi.org/10.1016/j.pestbp.2011.02.012.Investigations.
91. Lohmuller, et al. Antibodies elicited by the first non-viral prophylactic cancer vaccine show tumor-specificity and immunotherapeutic potential. *Scientific Reports.* 2016.
92. Fracol M, Xu S, Mick R, Fitzpatrick E, Nisenbaum H, Roses R, et al. Response to HER-2 Pulsed DC1 vaccines is predicted by both HER-2 and estrogen receptor expression in DCIS. *Ann Surg Oncol.* 2013;20:3233–3239. Available from: https://doi.org/10.1245/s10434-013-3119-y.
93. Czerniecki BJ, Koski GK, Koldovsky U, Xu S, Cohen PA, Mick R, et al. Targeting HER-2/neu in early breast cancer development using dendritic cells with staged interleukin-12 burst secretion. *Cancer Res.* 2007;67:1842–1852. Available from: https://doi.org/10.1158/0008-5472.CAN-06-4038.
94. Lowenfeld L, Mick R, Datta J, Xu S, Fitzpatrick E, Fisher CS, et al. Dendritic cell vaccination enhances immune responses and induces regression of HER2pos DCIS independent of route: results of randomized selection design trial. *Clin Cancer Res.* 2017;23:2961–2971. Available from: https://doi.org/10.1158/1078-0432.CCR-16-1924.

95. Whiteside TL. The tumor microenvironment and its role in promoting tumor growth. *Oncogene*. 2008. Available from: https://doi.org/10.1038/onc.2008.271.
96. Labani-Motlagh A, Ashja-Mahdavi M, Loskog A. The tumor microenvironment: a milieu hindering and obstructing antitumor immune responses. *Front Immunol*. 2020. Available from: https://doi.org/10.3389/fimmu.2020.00940.
97. Rabinovich GA, Gabrilovich D, Sotomayor EM. Immunosuppressive strategies that are mediated by tumor cells. *Annu Rev Immunol*. 2007;25:267–296. Available from: https://doi.org/10.1146/annurev.immunol.25.022106.141609.
98. Predina J, Eruslanov E, Judy B, Kapoor V, Cheng G, Wang LC, et al. Changes in the local tumor microenvironment in recurrent cancers may explain the failure of vaccines after surgery. *Proc Natl Acad Sci USA*. 2013;110. Available from: https://doi.org/10.1073/pnas.1211850110.
99. Karbach J, Gnjatic S, Bender A, Neumann A, Weidmann E, Yuan J, et al. Tumor-reactive CD8 + T-cell responses after vaccination with NY-ESO-1 peptide, CpG 7909 and Montanide® ISA-51: association with survival. *Int J Cancer*. 2009;126. Available from: https://doi.org/10.1002/ijc.24850.
100. Pavlick A, Blazquez AB, Meseck M, Lattanzi M, Ott PA, Marron TU, et al. Combined vaccination with NY-ESO-1 protein, poly-ICLC, and montanide improves humoral and cellular immune responses in patients with high-risk melanoma. *Cancer Immunol Res*. 2020;8:70–80. Available from: https://doi.org/10.1158/2326-6066.CIR-19-0545.
101. Atanackovic D, Altorki NK, Cao Y, Ritter E, Ferrara CA, Ritter G, et al. Booster vaccination of cancer patients with MAGE-A3 protein reveals long-term immunological memory or tolerance depending on priming. *Proc Natl Acad Sci USA*. 2008;105:1650–1655. Available from: https://doi.org/10.1073/pnas.0707140104.
102. Vansteenkiste JF, Cho BC, Vanakesa T, De Pas T, Zielinski M, Kim MS, et al. Efficacy of the MAGE-A3 cancer immunotherapeutic as adjuvant therapy in patients with resected MAGE-A3-positive non-small-cell lung cancer (MAGRIT): a randomised, double-blind, placebo-controlled, phase 3 trial. *Lancet Oncol*. 2016;17:822–835. Available from: https://doi.org/10.1016/S1470-2045(16)00099-1.
103. Lepisto AJ, Whiteside T, McKolanis J, Moser AJ, Zeh H, Lee K, et al. A phase 1B study of a MUC1 pulsed autologous dendritic cell (DC) vaccine as adjuvant therapy in patients (Pts) with resected pancreatic or biliary tumors. *J Clin Oncol*. 2004;22. Available from: https://doi.org/10.1200/jco.2004.22.90140.2578. 2578–2578.
104. Ramanathan RK, Lee KM, McKolanis J, Hitbold E, Schraut W, Moser AJ, et al. Phase I study of a MUC1 vaccine composed of different doses of MUC1 peptide with SB-AS2 adjuvant in resected and locally advanced pancreatic cancer. *Cancer Immunol Immunother*. 2005;54:254–264. Available from: https://doi.org/10.1007/s00262-004-0581-1.
105. Morse MA, Niedzwiecki D, Marshall JL, Garrett C, Chang DZ, Aklilu M, et al. A randomized phase II study of immunization with dendritic cells modified with poxvectors encoding CEA and MUC1 compared with the same poxvectors plus GM-CSF for resected metastatic colorectal cancer. *Ann Surg*. 2013;258:879–886. Available from: https://doi.org/10.1097/SLA.0b013e318292919e.
106. Ullenhag GJ, Frödin JE, Jeddi-Tehrani M, Strigård K, Eriksson E, Samanci A, et al. Durable carcinoembryonic antigen (CEA)-specific humoral and cellular immune responses in colorectal carcinoma patients vaccinated with recombinant CEA and granulocyte/macrophage colony-stimulating factor. *Clin Cancer Res*. 2004;10:3273–3281. Available from: https://doi.org/10.1158/1078-0432.CCR-03-0706.

107. Mittendorf E, Clifton G. *Final Report of the Phase I/II Clinical Trial of the E75 (Nelipepimut-S) Vaccine with Booster Inoculations to Prevent Disease Recurrence in High-Risk Breast*. Oxford University Press; 2014.
108. Mittendorf EA, Schneble EJ, Perez SA, Symanowski JT, Patil R, Vreeland TJ, et al. Primary analysis of the prospective, randomized, single-blinded phase II trial of AE37 vaccine versus GM-CSF alone administered in the adjuvant setting to high-risk breast cancer patients. *J Clin Oncol*. 2014;32:638. Available from: https://doi.org/10.1200/jco.2014.32.15-suppl.638.
109. Mittendorf EA, Ardavanis A, Litton JK, Shumway NM, Hale DF, Murray JL, et al. Primary analysis of a prospective, randomized, single-blinded phase II trial evaluating the HER2 peptide GP2 vaccine in breast cancer patients to prevent recurrence. *Oncotarget*. 2016;7:66192–66201. Available from: https://doi.org/10.18632/oncotarget.11751.
110. Knutson KL, Block MS, Norton N, Erskine CL, Hobday TJ, Dietz AB, et al. Rapid generation of sustainable HER2-specific T-cell immunity in patients with HER2 breast cancer using a degenerate HLA class II epitope vaccine. *Clin Cancer Res*. 2020;26:1045–1053. Available from: https://doi.org/10.1158/1078-0432.CCR-19-2123.
111. Clifton GT, Peace KM, Holmes JP, Vreeland TJ, Hale DF, Herbert GS, et al. Initial safety analysis of a randomized phase II trial of nelipepimut-S + GM-CSF and trastuzumab compared to trastuzumab alone to prevent recurrence in breast cancer patients with HER2 low-expressing tumors. *Clin Immunol*. 2019;201:48–54. Available from: https://doi.org/10.1016/j.clim.2019.02.011.
112. Mittendorf E, Lu B, Melisko M, Hiller JP, Bondarenko I, Brunt AM, et al. Efficacy and safety analysis of nelipepimut-S vaccine to prevent breast cancer recurrence: a randomized, multicenter, phase III clinical trial. *Clin Cancer Res*. 2019;25:4248–4254. Available from: https://doi.org/10.1158/1078-0432.CCR-18-2867.
113. Kang DH, Weaver MT, Park NJ, Smith B, McArdle T, Carpenter J. Significant impairment in immune recovery after cancer treatment. *Nurs Res*. 2009;58:105–114. Available from: https://doi.org/10.1097/NNR.0b013e31818fcecd.
114. Maeng H, Terabe M, Berzofsky JA. Cancer vaccines: translation from mice to human clinical trials. *Curr Opin Immunol*. 2018. Available from: https://doi.org/10.1016/j.coi.2018.03.001.
115. Westdorp H, Kolders S, Hoogerbrugge N, de Vries IJM, Jongmans MCJ, Schreibelt G. Immunotherapy holds the key to cancer treatment and prevention in constitutional mismatch repair deficiency (CMMRD) syndrome. *Cancer Lett*. 2017. Available from: https://doi.org/10.1016/j.canlet.2017.06.018.
116. Simon AK, Hollander GA, McMichael A. Evolution of the immune system in humans from infancy to old age. *Proc R Soc B Biol Sci*. 2015;282:20143085. Available from: https://doi.org/10.1098/rspb.2014.3085.

Chapter 4

Regulatory aspects of cancer immunotherapy in Europe

Filomena Nappi and Maria Cristina Galli
National Centre for the Control and Evaluation of Medicines, Italian National Institute of Health, Rome, Italy

Introduction

Immunotherapy is becoming a fast growing area of new medicinal products in oncology for treating patients along with or alternatively to chemo- or radio-therapy, aiming at fighting the disease through strengthening host anti-tumor immunity. These new agents have considerably changed the landscape of treatment options for patients with cancer. Due to advancements in cancer immunology, different strategies are currently under development, for example, those based on different types of medicinal products, such as monoclonal antibodies, peptides, dendritic cells (DCs), natural killer cells, recombinant vectors, or chimeric antigen receptor (CAR) T cells.[1]

Cancer vaccines are based on tumor antigens able to elicit a strong tumor-specific cellular immune response, aiming at tumor cells killing and at cancer regression. Cancer vaccines may contain the tumor antigens in the form of purified proteins/peptides or nucleic acid sequences or cells or recombinant vectors.

In Europe monoclonal antibodies and purified proteins/peptides are regulated as biological medicinal products within the frame of EU Regulation 726/2004.[2] They are out of the scope of this chapter. Medicinal products, such as DCs and CAR T cells, are classified as advanced therapy medicinal products (ATMP) and regulated within the frame of EU Regulation 1394/2007.[3] Medicinal products containing nucleic acid sequences or recombinant vectors can be classified in different way, as it will be described in this chapter, according to the intended use.

The focus of this chapter is on so-called *cancer vaccines* containing cells or recombinant vectors, thus falling under the definitions of ATMP as clarified below.

An overview of the European regulatory landscape applicable to such cancer vaccines is provided, describing the differences between traditional

Cancer Vaccines as Immunotherapy of Cancer. DOI: https://doi.org/10.1016/B978-0-12-823901-8.00009-1
© 2022 Elsevier Inc. All rights reserved.

vaccines and so-called cancer vaccines from a regulatory perspective. Regulatory issues for classification, marketing authorization, production, applicable scientific requirements, and legal requirements for traceability, pharmacovigilance, and batch release of so-called cancer vaccines as opposed to traditional vaccines are discussed. Those differences in legal and scientific requirements between so-called cancer vaccines and prophylactic vaccines that exist in the European regulatory landscape need to be taken into careful consideration when planning and carrying out the entire product development.

Definition of vaccines within the EU legislation

Traditional vaccines, that is, vaccines for infectious diseases, such as those caused by Ebola or Covid-19 viruses, are *prophylactic* vaccines because they intend to prevent those pathologies.

Live attenuated vaccines have been so far among the most effective vaccines against infectious disease. Live viral vectors expressing heterologous antigens (also known as live recombinant vectored vaccines) have also been developed, based on for example, pox viruses, adenoviruses, alphaviruses, measles virus, yellow fever virus, and vesicular stomatitis virus. The expressed heterologous antigens derive from those infectious agents against which the vaccine is directed, such as Coronavirus and Ebola virus.[4]

European Directive 2001/83 (Article I, Section 4, Annex I, Part I—3.2.1.1) does not contain a specific definition of vaccines but instead it gives a definition of immunological medicinal product: "any medicinal product consisting of vaccines, toxins, serums or allergen products".[5]

According to European Pharmacopoeia (Ph. Eur.), vaccines for human use are "preparations containing antigens capable of inducing a specific and active immunity in man against an infecting agent or the toxin or antigen elaborated by it".[6]

It is evident that the above mentioned legal definitions pertain to *prophylactic* vaccines.

The wording *cancer vaccines* is widely used in the scientific community. Cancer vaccines are therapeutic medicinal products because they are used to treat cancer after diagnosis aiming at eliciting a specific immune response against tumor antigens or amplifying an existing antitumor immune response.

The wording *cancer vaccines* is not accepted in Europe when used in a regulatory context, to keep *therapeutic* vaccines separated from *prophylactic* vaccines. Cancer vaccines therefore when containing nucleic acid sequences or cells or recombinant vectors as described above, are referred to as ATMP within the European regulatory frame. ATMP, as described in the EU Regulation 1394/2007, are gene therapy medicinal products (GTMP), cell

TABLE 4.1 Examples of EMA CAT classification of a medicinal product based on biological properties and intended use.

Medicinal product	Composition	Intended use	CAT classification
Recombinant viral vector	Lentiviral vector encoding HIV epitopes	Therapeutic HIV vaccination	Not an ATMP (infectious disease)
Recombinant viral vector	AdV containing a mutated HCV structural region	Prevention and treatment of HCV and HCV-induced hepatocellular carcinoma	Not an ATMP (infectious disease)
Recombinant viral vector	HSV-1 containing human GM-CSF	Melanoma therapy	GTMP
Dendritic cells	Autologous dendritic cells, pulsed with allogeneic tumor cell lysate	Malignant mesothelioma immunotherapy	CTMP
NK cells	Allogeneic NK cells	Multiple myeloma immunotherapy	CTMP
Lymphocytes	Autologous tumor-infiltrating lymphocytes	Melanoma immunotherapy	CTMP
Lymphocytes	Allogeneic T lymphocytes specific for EBV or CMV	Posttransplant infections in cancer patients	CTMP
Lymphocytes	CAR T cells	Cancer immunotherapy	GTMP
Tumor cells	Autologous tumor-derived cells	Cancer immunotherapy	CTMP
Liposomes/ virosomes/ nanoparticles	mRNA derived from autologous tumor	Cancer immunotherapy	GTMP
Vesicles/ nanoparticles	Tumor-derived protein mixtures	Cancer immunotherapy	Not an ATMP (biological)
Plasmids	DNA plasmid encoding hIL12	Melanoma therapy	GTMP
RNA	In vitro transcribed RNA encoding tumor antigen	Cancer immunotherapy	GTMP

AdV, adenovirus; *ATMP*, advanced therapy medicinal products; *CAR*, chimeric antigen receptor; *CAT*, Committee for Advanced Therapies; *CMV*, cytomegalovirus; *CTMP*, cell therapy medicinal products; *EBV*, Epstein Barr virus; *GM-CSF*, granulocyte-macrophage colony-stimulating factor; *GTMP*, gene therapy medicinal products; *HCV*, hepatitis C virus; *HIV*, human immunodeficiency virus; *HSV*, herpes simplex virus; *NK*, natural killer.
Source: EMA website/Advanced Therapies/Classification.

therapy medicinal products (CTMP), and tissue engineering medicinal products. This terminology will be used throughout the chapter.[3]

Classification

The EU decision to keep therapeutic vaccines separated from prophylactic vaccines is very well shown in the definition of GTMP given in EU Directive 2009/120: "2.1. Gene therapy medicinal product......Gene therapy medicinal products shall not include vaccines against infectious diseases".[7]

Therefore in EU, as an example, a recombinant vector that contains genetic sequences coding for Covid-19 virus antigens is considered a vaccine and not a GTMP, while the same vector containing a tumor antigen aimed at cancer therapy is considered an ATMP (see Table 4.1).

Provenge, indicated for castration-resistant prostate cancer and approved for EU market in 2013, is an example of so-called *cancer vaccine*. Provenge contains autologous peripheral blood mononuclear cells activated with prostatic acid phosphatase granulocyte colony-stimulating factor.[8] Since it fulfills the European definition of cell therapy product,[7] Provenge is classified in Europe as an ATMP.

Classification of a medicinal product as an ATMP or not is a task that EU Regulation 1394/2007 gives to the Committee for Advanced Therapies (CAT) within the European Medicine Agency (EMA).[3]

CAT classification is based on the current legal European definitions given in EU Regulation 1394/2007[3] and in Directive 2009/120.[7]

Criteria for classification are the product composition, its therapeutic indication, and its target population. Live recombinant viral vectors delivering genes that encode tumor-associated antigen sequences fulfill the definition of ATMP when used for cancer immunotherapy, but similar products would not be classified as ATMP when intended to be used as prophylactic products against infectious disease.[9] CAT has classified so far many products.[10] Table 4.1 provides some examples of CAT classification.

Consequences of medicinal product classification

Classification of a medicinal product as a vaccine or as an ATMP has important consequences on its development pathway with a view to European market, since many legal and/or scientific requirements are similar but many others are different.

The areas impacted by the product classification are the following:

1. traceability,
2. marketing authorization procedure,
3. production,
4. quality/preclinical/clinical studies,

5. pharmacovigilance, and
6. batch release.

Traceability

EU Regulation 1394/2007 in its Article 15 requires that the ATMP market authorization holder as well as the hospital or clinics where the ATMP is used to establish and maintain a traceability system that ensures tracking for 30 years after expiry date of the product and its production process, as well as the patient and the ATMP received.[3] In other words, traceability must enable to track ATMP to patient and vice versa.

This requirement is not legally applicable to vaccines for infectious diseases, independently from their composition, and to purified proteins/peptides.

Marketing authorization

EU Regulation 726/2004 contains the European legal frame for marketing the authorization of medicinal products.

The European centralized procedure, that is, authorization at central level by EMA, is mandatory for the following types of medicinal products for human use[2]:

1. medicinal products developed by means of recombinant DNA technology,
2. monoclonal antibodies,
3. ATMP,[3]
4. medicinal products for human use containing a new active substance not authorized in the community, indicated for the therapy of AIDS, cancer, neurodegenerative disorder, diabetes, autoimmune diseases, and viral diseases, and
5. medicinal products that are designated as orphan medicinal products.

Thus medicinal products aiming at cancer immunotherapy and containing purified proteins/peptides or nucleic acid sequences or cells or recombinant vectors are always subjected to centralized marketing authorization procedure because they fall under the above categories.

For prophylactic vaccines, unless produced through recombinant DNA technology or designed as orphan drugs, EMA-centralized procedure is not mandatory. Market authorization of prophylactic vaccines can be carried out at national level, for example, with mutual recognition procedure, or at central level if voluntarily chosen by applicant. In the case of centralized procedure for a vaccine (be it classical, or subunit, or live vectored), the application is handled directly by Committee for Medicinal Products for

Human Use (CHMP) without involving CAT, as opposed to ATMP application that is examined by CAT and then by CHMP.

Production

Production of any medicinal product, either investigational or market approved, must be authorized by the competent authority.[5] Thus production of a prophylactic vaccine and a so-called cancer vaccine must be authorized during the entire lifecycle.

Authorization is granted by means of a manufacturing and importation authorization (MIA) certificate.

In the MIA certificate, specific authorization is foreseen for each class of medicines, as follows:

1.3.1.1 biological products,
1.3.1.2 immunological products,
1.3.1.3 cell therapy, and
1.3.1.4 gene therapy.

It is evident that GTMP/CTMP production is not allowed in a facility certified only for prophylactic vaccines (i.e., immunological products) and conversely prophylactic vaccine production is not allowed if a facility is certified only for GTMP/CTMP. Each type of production should be covered in the MIA certificate.

Production should comply with the applicable Good Manufacturing Practice (GMP) guidelines.

Many production issues are shared between vaccines and ATMP, since there may be similarities in production process, and/or similar containment and segregation may be required due to the nature of the product.

Nevertheless, GMP guidelines address the production of ATMP in a specific part (Part IV), while production of all other types of medicinal product is addressed in Parts I and II and their annexes, of which Annex 2 "Biologicals" is applicable to vaccines.[11]

Therefore even though there might be areas of overlapping, it should be noted that GMP requirements given in Part IV are specific for ATMP and are not applicable to vaccines, while GMP requirements described in Annex 2 are applicable only to vaccines and not to ATMP. Examples of differences are the requirements for traceability and lot release procedure. Procedures to ensure compliance with traceability legal requirements as well as keeping the associated records for the required time period are mandatory for ATMP production (Ref. 11—Part IV, 6.6). On the contrary, this is not required for vaccines production. Two separated subsequent steps for lot release are acceptable for ATMP (Ref. 11—Part IV, 11.3) while this is not possible for vaccines.

Ph. Eur. provides requirements for GTMP quality in the general chapter "5.14 Gene transfer medicinal products for human use."[12] This chapter, currently under revision for updating, contains a series of texts on recombinant vectors, genetically modified cells, plasmid vectors, bacterial cells used for the manufacture of plasmid vectors, adenovirus vectors, poxviruses, retroviridae-derived vectors, and adenoassociated virus vectors, thus providing a framework of requirements applicable to the production and control of these GTMP.

Ph. Eur. also contains a chapter describing the quality requirements for raw materials used in ATMP manufacture.[13]

Moreover, Ph. Eur. provides several texts on the quality of traditional vaccines for human use that should be consulted for the production of such products including general texts[13–16] and monographs specific for individual vaccine types (as example, Hepatitis B vaccine).[17]

Quality requirements

The European requirements concerning data to be provided to EU Authorities with a view to bring any medicinal products into European market are described in Annex I of Directive 2001/83/EC.[5]

In Annex I Part III, a Vaccine Antigen Master File (VAMF), a specific stand-alone document, is mentioned for vaccines. VAMF contains quality data for every vaccine antigen in the medicinal product to which it is applied. If a vaccine to be market authorized contains an antigen for which a VAMF has been previously provided to authorities in the context of another vaccine, then for the new one the VAMF can be referred to.

The European requirements specific for ATMP are provided in Part IV, Annex I, as amended in Directive 2009/120.[7] At difference with traditional vaccines, an ATMP Master File is not yet considered in the requirements and therefore cannot be referred to in the marketing authorization application documentation.

In the case of a GTMP containing a recombinant viral vector, detailed information including nucleic acid sequence verification, genotypic stability, attenuation of virulence, and tropism for specific tissues and cell types is generally expected.

In the case of a traditional vaccine, such information is generally not required, unless it is a recombinant vectored vaccine, for which requirements similar to those for GTMP are described in a specific EMA guideline.[4]

Several European guidelines contain detailed specific requirements for the characterization and control of ATMP.[18–21]

Quality of traditional vaccines is generally described in Ph. Eur. texts, as stated above. In addition, many EMA guidelines are available for vaccines development.[22–24]

Preclinical and clinical studies

For classical vaccines, the required preclinical data include safety (single and repeated dose toxicity, mutagenesis) and efficacy studies to document protection against a challenge from the pathogenic organism (immune responses according to dose, target population, and vaccination schedule; levels and type of antibody production; duration of response; local tolerance). Studies on adjuvant are also expected to address the risk of immunotoxicty of new adjuvants (e.g., hypersensitivity) or toxicity of a known additive in a new combination with antigen.

In general, studies on biodistribution, germ-line transmission, insertional mutagenesis, tumorigenicity, and embryo/fetal and perinatal toxicity are not required for classical vaccines, while they are very important for ATMP, particularly for GTMP, for which on the contrary classical mutagenesis, studies are not applicable.[25,26]

In the case of live vectored vaccines, requirements are more similar to GTMP.[4]

As far as the clinical studies, the most important difference between the two types of medicinal product pertains to population addressed: patients for ATMP and healthy subjects for vaccines. Safety and efficacy clinical studies should then be shaped on those populations according to targeted disease.

Pharmacovigilance

Pharmacovigilance in EU is regulated by EU Regulation 1235/2010[27] and pertains safety.

At Article 1.2, the Regulation states that obligations to conduct post-authorization follow-up efficacy studies can be recommended for any medicinal product if deemed appropriate. Thus while safety follow-up is always required for any medicine including classical vaccines, follow-up of efficacy is not a legal requirement for classical vaccines and may be required on a case-by-case basis.[27]

A notable exception to this regulation is represented by ATMP since they are regulated by EU Regulation 1394/2007.[3]

Article 14 of EU Regulation 1394/2007 on ATMP states that, in addition to safety pharmacovigilance, efficacy pharmacovigilance should be carried out. An EU Risk Management Plan should be prepared by the ATMP market authorization holder, including safety and efficacy follow-up plans. The plan is authorized in the context of market authorization and its outcome should be periodically reported to EMA.[3]

Batch release

According to Article 114 of Directive 2001/83/EC,[5] as amended by Directive 2004/27/EC,[28] an EU Member State laboratory may test a batch of an immunological medicinal product or a medicinal product derived from human blood or plasma before it is placed on the European market. The tests conducted by an Official Medicine Control Laboratory, in accordance with Official Control Authority Batch Release (OCABR) guidelines show that the batch is in compliance with the approved specifications laid down in the relevant monographs of the Ph. Eur. and in the relevant marketing authorization. OCABR performed by an EU Member State is mutually recognized by all others requiring OCABR for that product and by any State having signed a formal agreement of OCABR recognition with the EU, such as Switzerland.

OCABR is not required for ATMP.

Conclusion

It is evident that a correct classification of the medicinal product is the first important step in development.

The differences in legal and scientific requirements between ATMP and prophylactic vaccines that exist in the European regulatory landscape need to be taken into careful consideration when planning and carrying out the product development.

Regulatory considerations should be applied along the entire process (development, authorization, market, update).

An effective regulatory strategy able to manage risks and to prevent problems is more convenient in terms of money and time savings but more importantly ultimately contributes to successfully reaching the goal of bringing to patients the most effective therapies.

References

1. Buonaguro L, Tagliamonte M. Selecting target antigens for cancer vaccine development. *Vaccines*. 2020;8:615.
2. Regulation (EC) No. 726/2004 of the European Parliament and of the Council laying down Community procedures for the authorisation and supervision of medicinal products for human and veterinary use and establishing a European Medicines Agency.
3. Regulation (EC) No. 1394/2007 of the European Parliament and of the Council on advanced therapy medicinal products and amending Directive 2001/83/EC and Regulation (EC) No. 726/2004.
4. Guideline on quality, non-clinical and clinical aspects of live recombinant viral vectored vaccines EMA/CHMP/VWP/141697/2009.
5. Directive 2001/83/EC of the European Parliament and of the Council on the Community code relating to medicinal products for human use.
6. Vaccines for human use. European Pharmacopoeia: 07/2021:0153.
7. Commission Directive 2009/120/EC amending Directive 2001/83/EC of the European Parliament and of the Council on the Community code relating to medicinal products for human use as regards advanced therapy medicinal products.

8. https://www.ema.europa.eu/en/medicines/human/EPAR/provenge.
9. Reflection paper on classification of advanced therapy medicinal products EMA/CAT/600280/2010 rev.1.
10. https://www.ema.europa.eu/en/human-regulatory/marketing-authorisation/advanced-therapies/advanced-therapy-classification.
11. EudraLex—Good Manufacturing Practice (GMP) guidelines.
12. Gene Transfer Medicinal Products for human use. European Pharmacopoeia: 04/2019: 51400.
13. Raw materials of biological origin for the production of cell-based and gene therapy medicinal products. European Pharmacopoeia: 01/2017:50212.
14. Chicken flocks free from specified pathogens for the production and quality control of vaccines. European Pharmacopoeia: 07/2010:50202.
15. Cell substrates for the production of vaccines for human use. European Pharmacopoeia: 01/2018:50203.
16. Tests for extraneous agents in viral vaccines for human use. European Pharmacopoeia: 07/2020:20616.
17. Hepatitis B vaccine (rDNA). European Pharmacopoeia: 01/2019:1056.
18. Guideline on quality, non-clinical and clinical requirements for investigational advanced therapy medicinal products in clinical trials—Draft. EMA/CAT/852602/2018.
19. Guideline on the quality, non-clinical and clinical aspects of gene therapy medicinal products—EMA/CAT/80183/2014.
20. Guideline on quality, non-clinical and clinical aspects of medicinal products containing genetically modified cells—EMA/CAT/GTWP/671639/2008.
21. Guideline on development and manufacture of lentiviral vectors - CHMP/BWP/2458/03.
22. Note for Guidance on Preclinical Pharmacological and toxicological testing of vaccines—CPMP/SWP/465/95.
23. Guideline on Clinical Evaluation of New Vaccines—EMEA/CHMP/VWP/164653/2005.
24. Guideline on adjuvants in vaccines for human use—EMEA/CHMP/VEG/134716/2004.
25. Guideline on non-clinical testing for inadvertent germline transmission of gene transfer vectors—EMEA/273974/2005.
26. Guideline on the non-clinical studies required before first clinical use of gene therapy medicinal products—EMEA/CHMP/GTWP/125459/2006.
27. Regulation (EU) No. 1235/2010 of the European Parliament and of the Council amending, as regards pharmacovigilance of medicinal products for human use, Regulation (EC) No. 726/2004 laying down Community procedures for the authorization and supervision of medicinal products for human and veterinary use and establishing a European Medicines Agency, and Regulation (EC) No. 1394/2007 on advanced therapy medicinal products.
28. Directive 2004/27/EC of the European Parliament and of the Council of 31 March 2004 amending Directive 2001/83/EC on the Community code relating to medicinal products for human use.

Chapter 5

Identification of tumor antigens for T-cell-mediated cancer immunotherapy

Hans-Georg Rammensee[1,2,3]

[1]*Institute for Cell Biology, Department of Immunology, University of Tübingen, Tübingen, Germany,* [2]*Cluster of Excellence iFIT (EXC2180) "Image-Guided and Functionally Instructed Tumor Therapies," University of Tübingen, Tübingen, Germany,* [3]*German Cancer Consortium (DKTK) and German Cancer Research Center (DKFZ), Partner Site Tübingen, Tübingen, Germany*

Introduction

The immune system has the power to attack and eliminate cancers, even in advanced stages. This is mostly mediated by T cells. The first tumor antigens recognized by T cells in patients, and in experimental mouse systems, were identified via T cells the patients, respectively, the mice, developed spontaneously.[1,2] The first successful clinical application of T-cell-mediated cancer immunotherapy was the so-called donor-derived lymphocyte transfusion, DLT, in the specific setting of leukemia patients after hematopoietic stem cell transplantation.[3] In cases of relapse, donor lymphocytes are transfused into the patient, resulting in graft-versus-host reactions frequently accompanied by a beneficial graft-versus-leukemia effect. However, the real breakthrough in T-cell-mediated cancer immunotherapy came with the introduction of checkpoint inhibition (CHI), as first practised using antibodies against CTLA-4 and shortly thereafter, PD1.[4–7] This approach is beneficial in several solid tumor entities, particularly in those with a high mutation load, such as metastatic melanoma. This has resulted in long-term remission for many patients with otherwise refractory disease, but a substantial proportion of patients fail to respond, or later relapse. For many other cancer entities, such as the most frequent tumors, such as colorectal, breast, or prostate cancer, with a few exceptions CHI, has not proven beneficial.

Nonetheless, because the power of the immune system to efficiently attack tumors and even to completely eliminate them in some patients has

Cancer Vaccines as Immunotherapy of Cancer. DOI: https://doi.org/10.1016/B978-0-12-823901-8.00010-8
© 2022 Elsevier Inc. All rights reserved.

been shown, approaches to extend the benefit of cancer immunotherapy to all cancer patients are being intensively investigated. One important consideration is to determine the reasons why CHI does not work in a given patient. Because recognition of tumor peptide antigens presented on human leukocyte antigen (HLA) molecules by T cells is the crux of their mode of action, identification of these is expected to lead to further progress. As noted earlier, CHI works best for cancers with a high mutation load, suggesting that CHI-induced T cells recognize mutation-derived neoepitopes. We have contributed to the identification of such mutated neoepitopes in an experimental model and in individual patients by mass spectrometry.[8] It has been tried to predict such neoepitopes[9,10] using algorithms based on sequence similarities (peptide motifs) we previously found by analyzing normal peptides from different HLA molecules.[11] This approach, although very useful for identifying virus epitopes, frequently fails for mutated neoepitopes in that such predicted peptides can be verified by mass spectrometry only in a small fraction of cases.[12,13] One obvious reason for such failure is the limited sensitivity of this physical method. However, other observations and considerations suggest a different reason, namely, that the number of mutation-derived neoantigens actually presented as peptides on HLA molecules is rather small, even in cancers with a high mutation load. This is impressively illustrated by simply counting the number of different HLA class I molecules on the cancer cells. In most cases, an individual is heterozygous at the HLA loci and thus expresses six different HLA class I molecules, two alleles for each of the three loci HLA A, B, and C. However, these can also be identical by chance, so that a person can be homozygous at one or more loci. Thus a person homozygous at all three loci expresses only three different HLA class I molecules. If a melanoma patient lacks just one of the alleles, thus having only five instead of six, the response to CHI is already reduced. If this is combined with lower mutation load, the difference, measured as overall survival times, is very pronounced.[14] This demonstrates that the number of neoepitopes actually presented on the cancer cells of a patient with clinically relevance is rather low. One reason for this is likely the negative selection of tumor cells with highly immunogenic HLA peptides by the patient's T cells (by the process known as immunosurveillance, including immuno-editing, as analyzed in depth by Lloyd Old and Robert Schreiber and colleagues), so that what remains in a clinically relevant tumor are those cancer cells with no or few neoantigens presented by HLA molecules. On the other hand, cancer cells generally present hundreds of cancer-specific peptides in germline sequences, which are not found on normal cells, and which can also potentially be recognized by T cells.[15,16] In most cases, the reason for this differential expression is unknown, but one clear case of nonmutated neoantigen expression by cancer cells is due to a TAP deficiency.[17]

Antigen processing and presentation

This brings us to a brief consideration of the manner in which HLA molecules are loaded with peptides inside the cell. Antigen processing for the two HLA classes, I and II, is fundamentally different, although in both cases, the two protein chains composing the HLA molecules are first brought into the ER lumen. Thereafter, the pathways differ. Nascent HLA class I molecules, consisting of the polymorphic alpha chain and the almost monomorphic beta-2-microglobulin, are decorated with peptides by an extremely complicated loading machinery, designated the peptide loading complex (PLC).[18] The complexity of this is relevant for our context, because any alteration in its components in cancer cells can give rise to new peptides from the cellular proteins in germline sequence. The transporter associated with antigen processing (TAP) mentioned above is one example of this, but alterations in the other components of the PLC, such as Calreticulin, ERp57, Tapasin, and ERAP, might also contribute to the large number of apparently cancer-specific HLA ligands that we find in clinical tumor samples. The peptides to be loaded, or rather initially longer versions of these, are protein fragments cut by proteasomes in the cytosol, either with mature proteins as substrates, or with unfinished proteins (DRIPS) or nonfunctional small translation products originating from different sources.[19,20] HLA ligands derived from the latter are commonly referred to as cryptic peptides. Thus the entire protein compartment of a cell, and many more of the genome-encoded sequences, including noncoding RNAs, do contribute to a cell's HLA class I ligandome, and a cancer cell's ligandome is usually much different from a normal cell's. However, it remains challenging to identify those useful for cancer immunotherapy among the plethora of possibilities. HLA class II molecules also begin their assembly in the ER; the two chains, the almost monomorphic alpha chain and the polymorphic beta chain, are first chaperoned by the invariant chain to build a nonamer with the peptide-binding groove blocked by a part of the invariant chain, the CLIP peptide. The nonamer is then translocated via the Golgi system to the MIIC/late endosome, the major histocompatibility complex (MHC) class II loading compartment, a process regulated by DM and DO molecules. Peptide loading of the HLA class II molecules can only begin in the MIIC. The invariant chain is degraded, the CLIP peptide released upon association of DM with the now-freed alpha/beta chain dimer, and replaced by a new peptide. This can come from several sources: from exogenous proteins entering the cell by phagocytosis, from the cell's own cell membrane proteins, or from cytosolic proteins entering the MIIC by autophagy. Presentation of cancer antigens by HLA class II molecules is relevant in two scenarios: first, many cancer cells start to express HLA class II molecules, even if the tissue of origin is HLA class II -negative, as we have found, for example, for ovarian carcinomas. Second, antigen-presenting cells in the tumor microenvironment or in draining lymph nodes can present tumor antigens released by tumor cells, or freed from dead tumor cells.

The human leukocyte antigen system

The HLA region is the most polymorphic present in the genome. The loci relevant for our consideration are the class I HLA-A, B, and C loci, and the class II HLA-Dr, DQ, and DP loci. They code for peptide receptors with defined specificities. HLA class I molecules present peptides of mostly nine amino acids, and each allelic product has its own peptide specificity, defined by particular amino acids, for example, at positions 2, or 3, or 5, and the C-terminus at nine. The evolutionary purpose of this is believed to be to cover all sequences that might occur in infectious agents, at least on a population level. As a consequence, we now have cataloged more than 20,000 different HLA class I alleles at the three loci.[21] Many of these differ in the amino acid sequence of the expressed proteins, and among these most of polymorphic differences are situated in the peptide-binding groove. Thus thousands of different HLA receptor specificities, known as HLA motifs, are being used in the human population. An illustration of this functional polymorphism can be found in.[22] For HLA class I, peptides consist, in most cases, of exactly nine amino acids. Two of these, often position 2 and the C-terminus are anchoring amino acids, interacting with complementary sites within the peptide-binding groove. In addition, amino acids at other positions can interact with the groove, increasing the variability of HLA motifs, so that HLA molecules with identical anchor residues can considerably differ in their peptide repertoires, as illustrated, for example, for HLA-B44 allomorphs, all sharing a basic residue at (R or K) at position 2 and an aliphatic one at 9, but with almost entirely different ligand repertoires.[23] The N- and C-termini of the peptides bound to HLA class I molecules are buried in the groove,[24] illustrating the strict length limitation.

For HLA class II molecules, there are also two or three anchor residues within a stretch of nine amino acids, building up the motif, whereas the N- and C-termini are free, so that class II molecules bind peptides with a larger length variability, whereby the interaction with the binding groove is maintained by the nine amino acid stretch in the middle.[24] Generally, the stringency of peptide-binding requirements for HLA class II molecules is lower than for HLA class I, so that it is easier to find promiscuous peptides, meaning ones that can bind to more than one allomorph.

With regard to the HLA ligands on cancer cells, one has to take into account that different peptides may be presented even if they are derived from the very same cancer antigen. This also contributes to the explanation why mutated neoantigens are so rarely presented by HLA. For HLA class II, the most polymorphic genes are those coding for the beta-chains, DRB, DQB, and DPB, with together more than 7000 different alleles, whereas the alpha-chains are almost monomorphic, like the beta-2-microglobulin in the case of HLA class I. Thus the variability of HLA peptide receptor specificities in the human population is enormous, with the consequence that each individual presents a different selection of HLA ligands to her or his T cells from the same antigen, be it from an infectious agent, or from a self antigen, or from a tumor antigen.

Isolation of human leukocyte antigen ligands from tumor tissues or other sources

Because the peptides are noncovalently bound to HLA molecules, they can be separated without destroying them. This is usually accomplished by acid elution of the peptides from HLA molecules immunoprecipitated from cell lysates, a process requiring HLA-specific antibodies.[25] Pan-specific antibodies precipitating all HLA A, B, and C molecules (W6/32) or all Dr molecules (L243) are most commonly employed for this purpose. Tumor tissues surgically resected for therapeutic reasons are the most suitable source because they will give insight into the universe of ligands that were presented by the cancer cells in vivo. If one uses tumor cell lines derived from a patient's tumor, and grown in vitro, it is likely that the HLA ligand repertoire changes with time because the cells adapt to the culture conditions and thus change their protein composition. Acid elution of HLA-presented peptides can also be done by simply incubating living cells in medium with a low pH; the peptides then can be harvested from the supernatant.[26] One advantage of this method is that one does not need an HLA-specific antibody, and another is that the cells may be maintained as growing cultures so that the HLA ligands can be harvested repeatedly. Disadvantages are that the harvested supernatant contains a lot of other materials from the cells and that the method is only suitable for cells that can be grown in vitro.

An especially interesting source of HLA ligands is the soluble HLA molecules regularly found in the blood of both healthy and diseased individuals.[27] Why these molecules are present there is not fully understood, but they might simply be a remnant of dead cells, or they might be lost from living cells or from exosomes. In any event, they can be purified by immunoprecipitation from serum or plasma. In cancer patients, such blood-derived soluble HLA molecules have also been shown to contain peptides from cancer cells and thus could be used for identifying cancer-specific peptides for personalized immunotherapies. This would be advantageous, for example, in the case of inoperable patients, or if the tumor cannot be located. In addition, the repertoire of soluble HLA-derived ligands might be of diagnostic value.

Identification of human leukocyte antigen ligands by mass spectrometry

Once the peptides have been brought into solution, they present as a mixture of usually thousands of different molecules, all of similar size; HLA class I ligands are mostly peptides of 8–12 amino acids, with a peak at 9, whereas class II ligands range between 8 and 20 amino acids or more. The complex peptide mixtures have to be separated before injection into the ionization chamber of the mass spectrometer; this is usually done by high-performance liquid chromatography with very narrow (nanometer range) long (e.g., 50 cm) columns. In the mass spectrometer, the ionized peptides are first brought into motion by ionic strength, sorted according to their mass-to-charge ratio in the mass filter, and then anlayzed in the mass detector for their actual mass. To obtain sequence information, the peptides are fragmented in the collision chamber and the resulting fragments are then subjected to mass detection again. From the mass-to-charge ratio of the fragments, the original sequence can be calculated because each amino acid and each peptide fragment consisting of two or more amino acids have a discrete mass and a discrete charge. However, the probability of incorrect assignment (false discovery rate) is large. Detailed methods and the different mass spectrometry detection principles (time of flight, ion intensity) are described elsewhere.[28–30] A principal challenge here is the sheer complexity of the data. Prior to applying the identified peptides for patient treatment, validation by comparison with synthetic, isotope-labeled peptides on the same equipment is essential.[31]

The impact of information derived from DNA and RNA sequencing

Because the order of the peptide fragments identified by the mass spectrometer within the original peptide cannot be immediately identified, possible sequences are usually compared to known sequences in protein sequence databases, such as Swissprot. However, if the particular HLA ligand is not derived from a known protein sequence in the data bank, as is the case for mutated neoepitopes, for example, it is required to sequence DNA or RNA from the tissue or cells of origin. The predicted protein sequences are then used for identifying the ligand in question.

Accumulating evidence now indicates the existence of another universe of sources for HLA ligands, derived not from expressed proteins but usually from not, or not completely, translated sequences. Such HLA ligands are commonly referred to as "cryptic peptides."[32] They can be derived from noncoding RNAs, from usually nontranscribed DNA sequences up- or downstream of coding sequences, from introns, or from short open reading frames.[20,33] Such cryptic peptides can be found on normal tissues, but a few of them appear to be cancer-specific. Thus, in future, the space of HLA ligands specific for cancer cells is likely to increase further.

If information on the potential protein sequences is not available, mass spectrometry-based de novo sequencing of ligands can be accomplished by comparing the peptide fragments and their charges to in silico libraries of all possible peptides that can be formed by the 20 amino acids. This increases the search space tremendously, but also the false discovery rate.

The impact of human leukocyte antigen ligand prediction tools

The identification of large numbers of ligands eluted from MHC molecules and their analysis for sequence similarities allowed the development of algorithms to predict HLA binding characteristics of a particular peptide. We initiated this approach with a basic prediction tool named after the first MHC ligand that we identified, SYFPEITHI, which is derived from JAK1. In the meantime, a number of more sophisticated prediction tools have been developed, which allow peptide predictions for new MHC molecules just based on the MHC sequence, and which also allow prediction of the ligand affinity. Of these tools, NetMHCpan is frequently used for MHC class I molecules and NetMHCIIpan for MHC class II.[34] Both cover many HLA alleles, and also MHC molecules from a number of different species. In addition to predicting the likelihood of presentation, these and other tools allow the prediction of binding affinity of the peptides to MHC molecules.

It is currently popular to use HLA ligand prediction tools for mutated neoepitopes, but in most cases, this approach is not very useful because the predicted peptides are not, or no longer, presented on a patient's tumor cells. For example, in the case of melanoma, Bassani-Sternberg et al. have analyzed the tumor samples of five patients, all with high mutation load of around 1000 nonsynonymous mutations on the DNA.[13] Using highly sensitive mass spectrometry, they could identify 11 mutated neoepitopes actually presented by HLA class I molecules in the cancer cells, which means that 0,2% of all mutations have actually found by mass spectrometry. The number of actually existing neoepitopes in the five patients was 8, 2, 1, 0, 0, meaning that in two of the patients not a single neoepitope could be detected. In an attempt to find neoepitopes in cancers of intermediate mutation load, we analyzed the tumor samples of 16 patients with hepatocellular carcinoma, presenting with 66 nonsynonymous mutations at average., and could detect not a single HLA-presented neoantigen by mass spectrometry.[35] Similarly, we analyzed tumor samples of 16 glioma patients, found 450 nonsysnonymous mutations in the DNA, but again not a single neoantigen presented on HLA class I molecules.[36]

HLA ligand prediction tools are essential for determining which peptide is derived from which HLA molecule, because as mentioned earlier, the standard approach for HLA purification is immunoprecipitation with a pan-HLA-A,B-C antibody, such as the widely used W6/32, or antibodies precipitating all HLA-Dr (L243).[28] Thus if a tumor tissue expresses six different HLA class I molecules, as in most cases, we analyze the list of peptides found by mass spectrometry for fitting to the six HLA class I motifs of that patient. This not only allows an estimate of which of the allomorphs present that peptide, but also serves as a quality control step: peptides not fitting to any of the relevant class I motifs are discarded from further consideration. The disadvantage of this step is that peptides with a noncanonical motif are lost, but the reduction of the data volume for processing outweighs this drawback.

How to find cancer-specific ligands?

For peptides representing mutated neoepitopes as predicted based on DNA or RNA sequence information from the patient's tumor and from their normal cells, answering this question is easy: If the mutated peptide is present, one can assume that it is cancer-specific. This situation, however, is the exception. In most cases, the cancer-specific ligands are in germline sequence. Cancer specificity then has to be assessed by the absence of the peptide in question from all healthy cells of that patient. This is an art, but approximately possible. Ideally, one would determine the HLA ligandome of all tissues in this patient, which is, of course, not possible. The analysis of adjacent benign tissue next to the tumor, resected together with the tumor, is frequently practiced, but we find that such peptides are already different as compared to tissue of the same organ from a healthy individual. A solution to this quandary is to systematically analyze the HLA ligandome of cancer-free individuals. We have attempted this by determining the HLA ligandomes of autopsy-derived tissues as well as from tissues resected from patients for reasons other than cancer. Thus we have a large database of the normal HLA ligandome,[37] publicly available at https://hla-ligand-atlas.org. If we now analyze the HLA ligandome of cancer tissue of a given patient, we can systematically disregard all peptides listed in this HLA-ligand atlas, reducing the data space from typically between 5000 and 10,000 down to a few hundred candidates that might be cancer-specific. These then can be prioritized according to a number of criteria,[38] for example, are there peptides already known from other cancer tissues in different patients? And already shown to be immunogenic? Derived from cancer related genes? Usually, we end up with a selection of 10 peptides that we deem suitable for the patient in question: 10, because this is the usual size of the peptide cocktails we use in patients, and this size is covered by our present drug production approval.

Use of information on tumor antigens for therapy

The most straightforward strategy is to vaccinate cancer patients with a selection of the cancer-specific antigens found on their cancer cells. A cocktail of antigens, and not only one antigen, should be used to avoid rapid immune selection of antigen loss variants and then resistant to attack. Ideally, the cancer-specific antigens actually presented by HLA molecules should be identified for each patient, and the most relevant ones should be formulated de novo for every patient as a personalized vaccine, which is a demanding approach, at least for peptides. Since some of the cancer-specific HLA ligands are recurrent in several cancer patients, with the same HLA molecules, of course, a less demanding approach relies on the warehouse concept: Those peptides found repeatedly on the tumor tissues of different patients and are already characterized for their immunogenicity, are produced in advance in GMP-grade, and used for incorporation in a peptide cocktail for any new patient, using only those peptides actually presented on that very patient's cancer. We have shown that this can be done (GAPVAC) and that such a warehouse concept can be combined with de novo produced peptides covering new ligands found in that patient.

The formulation of the antigens in the vaccine could be anything that can induce strong immune responses. Hence, the adjuvant used and the mode of application are critical. Synthetic peptides given in Montanide together with an efficient adjuvant, such as CpG, or our newly developed adjuvant XS15, can induce strong T-cell responses after one single injection, lasting for several years, as we have found for XS15.[39] Encoding the peptides on mRNA is an elegant possibility, because once a production facility has been established, mRNAs are easy to produce in GMP-grade, independent of the sequence (unlike peptides).[40,41] We think that combining the relevant peptides with a strong adjuvant might be useful.[42] Other means of application might be vector-based, using viral or bacterial vectors.

In particular, for patients with advanced cancer, and/or an already weakened immune system incapable of a strong response to a vaccine, adoptive T-cell transfer could be considered. However, this is challenging for many points of view, both when using unmanipulated autologous tumor-specific T cells (hard to access) and when using allogeneic T cells (cytotoxic due to cross reactivity on normal cells). Alternatively, the transfer of T-cell receptors (TCRs) specific for the tumor antigens transfected into normal cells could be useful; these could be of allogeneic origin from an HLA-matched donor. The production of such T cells requires many steps: the identification of the relevant TCR repertoire, the transfection of T cells with the TCR of choice recognizing a cancer-specific HLA ligand on the patient's tumor, the expansion of the TCR-transfected cells in vitro and their infusion into the patient. This can only done with HLA-matched TCRs and most optimally would require simultaneous treatment with T cells carrying TCRS for more than one antigen, to avoid rapid tumor escape by antigen downregulation. Another approach is the use of bispecific antibody-like constructs, with one arm against CD3 and the other consisting of a TCR against the cancer-specific HLA ligand. One could envisage the construction of a warehouse consisting of several hundred such bispecific constructs, to be applied to patients according to their HLA typing and their tumor HLA ligandome. The usefulness of such a warehouse would depend on the presence of cancer-specific ligands shared between different individual tumors, not unique to each one. These are rare but do exist.

Conclusion

T-cell-mediated antigen-specific cancer immunotherapy should have a bright future. All technical and logistic problems are in principle solvable. In case specific T cells are induced but do not lead to tumor regression or do not prevent cancer recurrence, combination therapies are likely to improve matter, for example, CHI, or, if the tumor is "cold," that is, not invaded by immune cells, by local application of immune modulators.

Acknowledgments

Many thanks go to Graham Pawelec for carefully editing this text.The support of many sources for the funding of our work on the identification of HLA-presented peptides in health and disease is acknowledged: DFG (SFBs 120, 510, and 685, Cluster of Excellence 2180 iFIT, Leibniz-Preis), Landesforschungspreis Baden-Württemberg, ERC, EU 7th framework program (GAPVAC and HEPAVAC), Ernst-Jung-Stiftung, Bayer-Stiftung, BMBF, Deutsche Krebshilfe, and many more. I also thank the numerous students, PhD students, postdocs, and senior scientists working with me on the issue of finding cancer antigens over the last decades.

References

1. Scanlan MJ, Gure AO, Jungbluth AA, Old LJ, Chen YT. Cancer/testis antigens: an expanding family of targets for cancer immunotherapy. *Immunol Rev*. 2002;188:22–32.
2. van der Bruggen P, Traversari C, Chomez P, Lurquin C, De Plaen E, Van den Eynde B, et al. A gene encoding an antigen recognized by cytolytic T lymphocytes on a human melanoma. *Science*. 1991;254:1643–1647.
3. Kolb HJ, Holler E. Adoptive immunotherapy with donor lymphocyte transfusions. *Curr Opin Oncol*. 1997;9:139–145.
4. Leach DR, Krummel MF, Allison JP. Enhancement of antitumor immunity by CTLA-4 blockade. *Science*. 1996;271:1734–1736.
5. Page DB, Postow MA, Callahan MK, Allison JP, Wolchok JD. Immune modulation in cancer with antibodies. *Annu Rev Med*. 2014;65:185–202.
6. Wolchok JD, Hodi FS, Weber JS, Allison JP, Urba WJ, Robert C, et al. Development of ipilimumab: a novel immunotherapeutic approach for the treatment of advanced melanoma. *Ann N Y Acad Sci*. 2013;1291:1–13.
7. Eggermont AMM, Blank CU, Mandala M, Long GV, Atkinson V, Dalle S, et al. Adjuvant pembrolizumab vs placebo in resected stage III melanoma. *N Engl J Med*. 2018;378:1789–1801.
8. Gubin MM, Zhang X, Schuster H, Caron E, Ward JP, Noguchi T, et al. Checkpoint blockade cancer immunotherapy targets tumour-specific mutant antigens. *Nature*. 2014;515:577–581.

9. Sahin U, Derhovanessian E, Miller M, Kloke BP, Simon P, Lower M, et al. Personalized RNA mutanome vaccines mobilize poly-specific therapeutic immunity against cancer. *Nature*. 2017;547:222–226.
10. Keskin DB, Anandappa AJ, Sun J, Tirosh I, Mathewson ND, Li S, et al. Neoantigen vaccine generates intratumoral T cell responses in phase Ib glioblastoma trial. *Nature*. 2019;565:234–239.
11. Rammensee H, Bachmann J, Emmerich NP, Bachor OA, Stevanovic S. SYFPEITHI: database for MHC ligands and peptide motifs. *Immunogenetics*. 1999;50:213–219.
12. Finn OJ, Rammensee HG. Is it possible to develop cancer vaccines to neoantigens, what are the major challenges, and how can these be overcome? Neoantigens: nothing new in spite of the name. *Cold Spring Harb Perspect Biol*. 2018;10.
13. Bassani-Sternberg M, Braunlein E, Klar R, Engleitner T, Sinitcyn P, Audehm S, et al. Direct identification of clinically relevant neoepitopes presented on native human melanoma tissue by mass spectrometry. *Nat Commun*. 2016;7:13404.
14. Chowell D, Morris LGT, Grigg CM, Weber JK, Samstein RM, Makarov V, et al. Patient HLA class I genotype influences cancer response to checkpoint blockade immunotherapy. *Science*. 2018;359:582–587.
15. Kowalewski DJ, Schuster H, Backert L, Berlin C, Kahn S, Kanz L, et al. HLA ligandome analysis identifies the underlying specificities of spontaneous antileukemia immune responses in chronic lymphocytic leukemia (CLL). *Proc Natl Acad Sci U S A*. 2015;112:E166–E175.
16. Schuster H, Peper JK, Bosmuller HC, Rohle K, Backert L, Bilich T, et al. The immunopeptidomic landscape of ovarian carcinomas. *Proc Natl Acad Sci U S A*. 2017;114:E9942–E9951.
17. Marijt KA, Blijleven L, Verdegaal EME, Kester MG, Kowalewski DJ, Rammensee HG, et al. Identification of non-mutated neoantigens presented by TAP-deficient tumors. *J Exp Med*. 2018;215:2325–2337.
18. Blees A, Januliene D, Hofmann T, Koller N, Schmidt C, Trowitzsch S, et al. Structure of the human MHC-I peptide-loading complex. *Nature*. 2017;551:525–528.
19. Chong C, Muller M, Pak H, Harnett D, Huber F, Grun D, et al. Integrated proteogenomic deep sequencing and analytics accurately identify non-canonical peptides in tumor immunopeptidomes. *Nat Commun*. 2020;11:1293.
20. Erhard F, Halenius A, Zimmermann C, L'Hernault A, Kowalewski DJ, Weekes MP, et al. Improved Ribo-seq enables identification of cryptic translation events. *Nat Methods*. 2018;15:363–366.
21. Marsh SGE. Nomenclature for factors of the HLA system, update January, February and March 2021. *Int J Immunogenet*. 2021;.
22. Ghosh M, Di Marco M, Stevanovic S. Identification of MHC ligands and establishing MHC class I peptide motifs. *Methods Mol Biol*. 2019;1988:137–147.
23. Hillen N, Mester G, Lemmel C, Weinzierl AO, Muller M, Wernet D, et al. Essential differences in ligand presentation and T cell epitope recognition among HLA molecules of the HLA-B44 supertype. *Eur J Immunol*. 2008;38:2993–3003.
24. Rammensee HG, Falk K, Rotzschke O. MHC molecules as peptide receptors. *Curr Opin Immunol*. 1993;5:35–44.
25. Falk K, Rotzschke O, Stevanovic S, Jung G, Rammensee HG. Allele-specific motifs revealed by sequencing of self-peptides eluted from MHC molecules. *Nature*. 1991;351:290–296.
26. Sturm T, Sautter B, Worner TP, Stevanovic S, Rammensee HG, Planz O, et al. Mild acid elution and MHC immunoaffinity chromatography reveal similar albeit not identical profiles of the HLA class I immunopeptidome. *J Proteome Res*. 2021;20:289–304.

27. Bassani-Sternberg M, Barnea E, Beer I, Avivi I, Katz T, Admon A. Soluble plasma HLA peptidome as a potential source for cancer biomarkers. *Proc Natl Acad Sci U S A.* 2010;107:18769–18776.
28. Nelde A, Kowalewski DJ, Stevanovic S. Purification and identification of naturally presented MHC class I and II ligands. *Methods Mol Biol.* 2019;1988:123–136.
29. Bichmann L, Nelde A, Ghosh M, Heumos L, Mohr C, Peltzer A, et al. MHCquant: automated and reproducible data analysis for immunopeptidomics. *J Proteome Res.* 2019;18:3876–3884.
30. Pak H, Michaux J, Huber F, Chong C, Stevenson BJ, Muller M, et al. Sensitive immunopeptidomics by leveraging available large-scale multi-HLA spectral libraries, data-independent acquisition, and MS/MS prediction. *Mol Cell Proteomics.* 2021;20:100080.
31. Ghosh M, Gauger M, Marcu A, Nelde A, Denk M, Schuster H, et al. Guidance document: validation of a high-performance liquid chromatography-tandem mass spectrometry immunopeptidomics assay for the identification of HLA class I ligands suitable for pharmaceutical therapies. *Mol Cell Proteomics.* 2020;19:432–443.
32. Erhard F, Dolken L, Schilling B, Schlosser A. Identification of the cryptic HLA-I immunopeptidome. *Cancer Immunol Res.* 2020;8:1018–1026.
33. Laumont CM, Vincent K, Hesnard L, Audemard E, Bonneil E, Laverdure JP, et al. Noncoding regions are the main source of targetable tumor-specific antigens. *Sci Transl Med.* 2018;10.
34. Reynisson B, Alvarez B, Paul S, Peters B, Nielsen M. NetMHCpan-4.1 and NetMHCIIpan-4.0: improved predictions of MHC antigen presentation by concurrent motif deconvolution and integration of MS MHC eluted ligand data. *Nucleic Acids Res.* 2020;48:W449–W454.
35. Loffler MW, Mohr C, Bichmann L, Freudenmann LK, Walzer M, Schroeder CM, et al. Multi-omics discovery of exome-derived neoantigens in hepatocellular carcinoma. *Genome Med.* 2019;11:28.
36. Hilf N, Kuttruff-Coqui S, Frenzel K, Bukur V, Stevanovic S, Gouttefangeas C, et al. Actively personalized vaccination trial for newly diagnosed glioblastoma. *Nature.* 2019;565:240–245.
37. Marcu A, Bichmann L, Kuchenbecker L, Kowalewski DJ, Freudenmann LK, Backert L, et al. H.L.A. Ligand Atlas: a benign reference of HLA-presented peptides to improve T-cell-based cancer immunotherapy. *J Immunother Cancer.* 2021;9.
38. Haen SP, Loffler MW, Rammensee HG, Brossart P. Towards new horizons: characterization, classification and implications of the tumour antigenic repertoire. *Nat Rev Clin Oncol.* 2020;17:595–610.
39. Rammensee HG, Wiesmuller KH, Chandran PA, Zelba H, Rusch E, Gouttefangeas C, et al. A new synthetic toll-like receptor 1/2 ligand is an efficient adjuvant for peptide vaccination in a human volunteer. *J Immunother Cancer.* 2019;7:307.
40. Sahin U, Oehm P, Derhovanessian E, Jabulowsky RA, Vormehr M, Gold M, et al. An RNA vaccine drives immunity in checkpoint-inhibitor-treated melanoma. *Nature.* 2020;585:107–112.
41. Weide B, Garbe C, Rammensee HG, Pascolo S. Plasmid DNA- and messenger RNA-based anti-cancer vaccination. *Immunol Lett.* 2008;115:33–42.
42. Nelde A, Rammensee HG, Walz JS. The peptide vaccine of the future. *Mol Cell Proteomics.* 2021;20:100022.

Chapter 6

Therapeutic cancer vaccines targeting viral antigens

Cornelis J.M. Melief, Anna-Sophia Wiekmeijer and Esmé T.I. van der Gracht
ISA Pharmaceuticals, Oegstgeest, The Netherlands

Introduction

Approximately, 16%–20% of all cancers are caused by infectious microorganisms, in particular the viruses listed in Table 6.1 and bacteria such as *Helicobacter pylori*.

As a matter of course, virus-induced cancers are in theory preventable by the application of preventive vaccines inducing neutralizing antibodies against the respective viruses. This has been accomplished for hepatitis B virus (HBV) and the most frequent types of high-risk human papilloma virus (hrHPV). One would think that widespread application of these preventive vaccines would have obliterated the cancers caused by HBV and hrHPV a long time ago. However, because of the long latency between initial infection with HBV or HPV and disease occurrence, because of vaccine hesitancy or scepsis, as well as through failure of health authorities in many countries, hundreds of millions of people are still persistently infected with HBV or hrHPV, running the risk of liver cirrhosis or hepatocellular cancer (HBV) or of cervical cancer, other genital cancers, anal cancer or head and neck cancer (hrHPV). Preventive vaccines are not yet available against Epstein–Barr virus (EBV), human T-cell lymphotropic virus type 1 (HTLV-1), hepatitis C virus (HCV), human herpes virus 8, and Merkel polyoma virus. Therefore therapeutic vaccines are not only required for cancers induced by HBV or hrHPV for the reasons listed, but also for eradication of all other persistent infections, premalignant diseases, or cancers caused by the other viruses listed in Table 6.1, except for Kaposi sarcoma, formerly frequently associated with AIDS caused by human immunodeficiency virus (HIV). In the case of HIV-induced AIDS-associated Kaposi sarcomas, tumor eradication is usually implemented by successful delivery of antiretroviral therapy. The resulting restoration of immunocompetence then suffices to eradicate the virus-induced sarcomas.

Cancer Vaccines as Immunotherapy of Cancer. DOI: https://doi.org/10.1016/B978-0-12-823901-8.00013-3
© 2022 Elsevier Inc. All rights reserved.

TABLE 6.1 Human cancer viruses.

***Epstein–Barr virus (EBV)**, a herpes virus, can cause *nasopharyngeal carcinoma, B-cell lymphomas,* particularly in immunosuppressed/ immunodeficient persons, *Hodgkin's disease,* and *stomach cancer*
***Human T-cell lymphotropic virus type 1 (HTLV-1)**, a retrovirus, can cause *adult T-cell leukemia/lymphoma*
***Hepatitis B virus (HBV)**[a], a Hepadna virus, can cause liver cirrhosis and *hepatocellular carcinoma (HCC)*
***Hepatitis C virus (HCV)**, a flavivirus, can cause liver cirrhosis and *hepatocellular carcinoma (HCC)*
***Human papilloma virus (HPV)**[a], a papilloma virus, can cause *anogenital cancers* and *head and neck cancers*
***Kaposi sarcoma virus**, a herpes virus, human herpes virus 8 (HHV8) can cause *sarcomas* in immunodeficient patients
***Merkel carcinoma virus**, a polyoma virus, can cause *skin carcinomas*

[a]*Against these viruses preventive vaccines are available.*

Immunotherapy, including therapeutic vaccination against cancers caused by Epstein–Barr virus

The large majority of people all over the world are persistently infected with EBV. Usually, this goes unnoticed or with transient infectious mononucleosis at young adult age. Nevertheless EBV is causally linked to two premalignant lymphoproliferative diseases and up to nine distinct types of human cancer, including Hodgkin's lymphoma, Burkitt lymphoma, B lymphoproliferative disease (all B-cell-derived) and nasopharyngeal carcinoma (NPC) and gastric carcinoma (both epithelial cell-derived). Together the worldwide disease burden due to EBV includes approximately 200,000 new cases of cancer per year.[1] Of these, NPC, gastric carcinomas, and Hodgkin lymphomas account for 80,000, 82,700, and 28,000 new cases per year, respectively.[1] EBV is also known to frequently cause B-cell-derived cancers during the temporary immunodeficiency associated with the installation of a newly emergent immune system following hematopoietic stem cell transplantation or in other T-cell immunodeficiency states.[1,2] Such B-cell lymphomas express a variety of EBV-associated antigens (EBNA1, EBER, LMP1, LMP2A&B, EBNA3A,B&C. EBNA6/LP), together representing latency state 3,[2] because there is lack of immune editing normally taking place under pressure of a robust antitumor immune response. Interestingly, such a wide array of EBV-associated antigens is also expressed on EBV + diffuse large B-cell lymphomas arising in immunocompetent persons, suggesting some intricate way of tumor-associated immune evasion that needs to be elucidated. Therapy of the stem cell transplantation-associated lymphomas is efficiently achieved by T lymphocyte infusion of stem cell donor origin, in particular of T lymphocytes educated ex vivo against EBV antigens.[3,4] More recently, infusion of allogeneic T cell lines specific for EBV antigens and matched for EBV antigen presenting HLA antigens with the patient has led to "off-the-shelf" therapy development.[5,6] EBV can also cause other cancers in immunocompetent hosts, in particular, Burkitt lymphoma, Hodgkin's lymphoma, NPC, and gastric cancer.[1,2] In all of these cancers, EBV antigen expression is more restricted. Hodgkin lymphomas and nasopharyngeal cancers are characterized as EBV latency II with the same antigen expression as in latency III tumors with exception of EBNA3A,B&C and EBNA6, which are lacking. In Burkitt lymphomas and gastric carcinomas, EBV antigen expression is even more restricted, remaining confined to expression of the latency I EBV antigens EBNA1 and EBER.[2,7] The LMP1 and LMP2 proteins are oncogenic proteins, as shown by induction of B-cell tumors in LMP1- or LPMP2 transgenic mice. More recently, the B-cell pro-proliferative activities of LMP1, LMP2A, and EBNA2 were shown to critically control B-cell lymphomagenesis in combination with EBNA3-mediated inhibition of terminal plasma cell differentiation.[8] EBV-induced NPC and gastric carcinomas are both epithelial cell-derived cancers as opposed to the B-cell lymphomas caused by EBV and are highly frequent cancers in Southeast Asia, including China, with yearly incidence rates of 80,000 and 82,500, respectively.[1] Several clinical therapeutic vaccination trials against established cancers expressing EBV antigens have been conducted. An early clinical trial treated 16 patients with advanced NPC with autologous DC pulsed with several LMP2 CTL epitope-containing peptides (A*1101, A*2402, and B*40011), using intranodal injection.[9] T-cell immune responses to these CTL epitopes were measured postvaccination and partial tumor regression coincided with such a CTL response in two patients. Another approach has been intradermal vaccination of NPC patients with a recombinant vaccinia virus (modified vaccinia Ankara, MVA) encoding the EBNA1 and LMP2 EBV proteins (MVA-EL). Of 14 vaccinated NPC patients, 8/18 showed vaccine-induced T-cell responses against known epitopes in these EBV proteins (7/14 EBNA1; 6/14 LMP), illustrating positivity against expected epitopes according to HLA types of patients of different ethnic origins.[10] However, the size of these T-cell responses was low, as expected with recombinant pox vectors, due to antigenic competition between the large pox vector and the inserted EBV proteins. Therefore alternative vaccination approaches were tried in two mouse lymphoma prevention models in which combined CD4+ and CD8+ T-cell responses against EBNA1 were induced by heterologous prime/boosting regimens.[11] In one of these EBNA1 expressed as a transgene in B cells was associated with spontaneous B-cell lymphoma and these lymphomas were prevented by recombinant adenovirus EBNA1 priming followed by MVA EBNA1 boosting. Lymphoma prevention was less efficient with anti-DEC205 antibody targeted EBNA1 priming followed by MVA/EBNA1 boosting in

DEC205/EBNA1 double transgenic mice. DEC 205 is a dendritic cell-specific lectin receptor on DC. Also prevention of outgrowth of EL4 tumor cells transduced with EBNA1 was shown by the same prime/boost approaches.[11]

Therapeutic vaccination against cancers caused by human T-cell lymphotropic virus type 1

HTLV-1 is the first human retrovirus discovered.[12] It causes persistent infection in approximately 10 million people worldwide across different continents.[13] HTLV-1 infects immune cells, especially T lymphocytes, which can lead to adult T-cell leukemia (ATLL) in around 5% of infected individuals, or HTLV-1-associated myelopathy–tropical spastic paraparesis in 3% (reviewed in Ref. 14). Other persistently infected persons, in comparison with noninfected individuals, more often develop a range of other diseases such as uveitis, bronchiectasis, infectious dermatitis and scabies, or coinfections with other infectious agents, indicating dysregulation of the immune system. Even asymptomatic HTLV-1 carriers experience premature mortality without clear pathogenesis (reviewed in Ref. 14). Thus far there is no satisfactory treatment of the malignant or inflammatory diseases cause by HTLV-1.[15] Also, no preventive or therapeutic vaccine has been developed. The tax and HBZ proteins of HTLV-1 together have a major role in ATLL causation by HTLV-1.[14] These two proteins do not act as classical oncogenic proteins but cause sustained proliferation of polyclonal T-cell populations with integration of viral genes by a mechanism that reflects retrovirus-mediated promotion insertion (reviewed in Ref. 15). A main natural viral target protein for CD8 + CTL is the tax protein, whereas a main natural target for CD4 + T cells against HTLV-1 is the envelope (env) protein next to the tax protein (reviewed in Ref. 15). Indeed expanded CD8 + CTL from ATLL patients kill HLA class I-matched ATTL cell lines in vitro.[16] The size of the natural CTL response against HTLV-1 shows an inverse relationship to the levels of FoxP3-positive CD4 + T cells.[17] A therapeutic vaccine consisting of autologous monocyte-derived DC pulsed with peptides containing HTLV-1 Tax epitopes presented by HLA-A2, HLA-A11, and HLA-A24 has induced CD8 + CTL responses in patients with ATLL, associated with long remission of disease in two out of three treated patients.[18,19] These patients also had a reduced proviral load compared with control patients. Although this small pilot clinical study shows some promise, larger clinical studies with many more patients and with additional coverage of HLA class I alleles and induction of HTLV-1-specific CD4 + T-cell responses for optimal CTL effector cells and CTL memory are necessary. Several teams of investigators have published HTLV-1 preventive and/or therapeutic vaccine designs based on B- and T-cell epitope predictions in various viral proteins.[20–23] In the paper of Kabiri et al., robust immunogenicity for induction of T-cell responses was shown in mice. Unfortunately, none of these designs has led to the production of clinical-grade vaccines and clinical trials. Because of the subversion of the normal physiological response of myeloid cells and T lymphocytes by persistent HTLV-1 infection, immunotherapy of ATLL, or other diseases caused by HTLV-1 may be complicated. This is also brought home by the observation that the treatment of three ATLL patients with anti-PD-1 checkpoint blocker nivolumab led to accelerated disease progression rather than disease reduction, suggesting that PD-1 can be a tumor suppressor molecule for T-cell lymphomas.[24] Nevertheless, induction of a sufficiently powerful combined CD4 + /CD8 + T-cell response against multiple HTLV-1 proteins in combination with non-T cell depleting chemotherapy and/or proper immunomodulator(s) must be considered for treating ATLL and other diseases caused by this very sneaky virus.

Therapeutic vaccination against hepatitis B virus

HBV is usually acquired perinatally by the newborn from a persistently infected mother. This frequently leads likewise to persistent infection by the child. Despite the fact that preventive vaccines against HBV have been available for decades, close to 300 million people are still persistently infected with HBV, particularly in Southeast Asia, with an annual mortality from liver cirrhosis or hepatocellular cancer [hepatocellular carcinoma (HCC)] of approximately 900,000. Antiviral agents are an important part of the armamentarium against chronic HBV infection. These agents include type I interferon, such as Pegintron and nucleoside analogues, and more recently, RNAi species (reviewed in Refs. 25,26) The Cornberg et al. review is a common guidance document from European and US experts entitled "*Guidance for design and endpoints of clinical trials in chronic hepatitis B*" and states in its summary: "*Functional but not sterilising cure is achievable and should be defined as sustained HBsAg loss in addition to undetectable HBV DNA 6 months post-treatment. The primary endpoint of phase III trials should be functional cure; HBsAg loss in >30% of patients was suggested as an acceptable rate of response in these trials. Sustained virologic suppression (undetectable serum HBV DNA) without HBsAg loss 6 months after discontinuation of treatment would be an intermediate goal. Demonstrated validity for the prediction of sustained HBsAg loss was considered the most appropriate criterion for the approval of new HBV assays to determine efficacy endpoints. Clinical trials aimed at HBV functional cure should initially focus on patients with HBeAg-positive or negative chronic hepatitis, who are treatment-naïve or virally suppressed on nucleos(t)ide analogues. A hepatitis flare associated with an increase in bilirubin or international normalised ratio should prompt temporary or permanent cessation of an investigational treatment. New treatments must be as safe as existing nucleos(t)ide analogues. In conclusion, regarding HBV 'functional cure', the primary goal is sustained HBsAg loss with undetectable HBV DNA after completion of treatment and the intermediate goal is sustained undetectable HBV DNA without HBsAg loss after stopping treatment.*"[26] So far, none of the antiviral agents has been capable of achieving eradication of chronic HBV infection as characterized by permanent HBsAg negativity and undetectable serum HBV DNA after termination of treatment. Therefore combination of treatment with antiviral agents and immunotherapy, including therapeutic vaccination is considered an attractive option. Therapeutic vaccination trials of chronic HBV infection were recently comprehensively reviewed.[27] Vaccination platforms have included HBV-derived peptides, proteins, nucleic acids encoding (segments) of HBV antigens, or monocyte-derived DC loaded ex vivo with HBV antigens. Initially, many vaccination attempts were directed at HBsAg. This is probably not a good idea, because HBsAg is abundantly present as a secreted protein at sites devoid of virus replication and as such serves as a decoy mechanism to distract the immune system. More attractive target molecules for therapeutic vaccination are probably the HBV core, polymerase, and X proteins.[27] Appropriate designs of therapeutic vaccines need to take into account sufficient coverage of HLA class I and II epitopes of these proteins presented by HLA types of diverse ethnic origins. In addition, the genetic diversity of different types of HBV itself needs to represented in the vaccine. So far therapeutic vaccination trials with different vaccines have not been successful in causing functional cure, let alone permanent eradication of HBV infection.[27,28] A proper population of chronic HBV carriers for therapeutic vaccine trials appears to be persons who spontaneously have low viral load or who experience low viral load upon treatment with antiviral nucleosides or RNAi. Because of the immunosuppressive conditions prevailing in the infected liver, additional combination of therapeutic vaccination with immunomodulatory drugs like PD-1 or PD-L1 blockers could be important for successful virus eradication[29] and/or treatment of HBV-induced HCC.[30] One striking feature in a detailed study of the specificity and phenotype of T cells that had infiltrated HBV-related HCC tissue was the presence of intratumoral HBV-specific CD103 + CD69 + resident memory (Trm) T cells. The presence of these HBV-specific CD8 + Trm cells correlated with relapse-free survival following surgery and hence a better prognosis of HBV-associated HCC.[31] This study indicates that therapeutic vaccination should not only be considered for the treatment of chronic HBV infection but also for immunotherapy of HBV-induced HCC, for example, to expand these HBV-specific Trm cells, also found in the healthy liver margin of the surgical specimens and thereby further reduce disease relapse.[31] In HBV-related HCC, HBV antigens are not the only targets for therapeutic vaccination, because also mutation-based neo-antigens and overexpressed antigens like cancer testis antigens elicit T-cell responses.[31]

Therapeutic vaccination against hepatitis C virus

Although more than 95% of persistent HCV infections can now be cleared by antiviral drugs, reinfections can occur and again establish HCV persistence.[32,33] Also, liver disease can still develop in some patients who have cleared the virus.[32] The WHO has developed a strategy to eradicate persistent HCV infection by 2030 by massive HSV screening programs in countries like Egypt.[34] In view of the great efficacy of antiviral HCV drugs, there is currently no great medical need of additional therapeutic vaccines against HCV. A preventive vaccine next to the antiviral drugs would, on the other hand, assist the WHO in its ambition of HCV eradication and would also be important to prevent reinfection following virus eradication by treatment with antiviral drugs.

Therapeutic vaccination against high-risk human papilloma virus

High-risk (hr) HPV consists of a family of closely related viruses that can all cause anogenital cancers and oral cancers, the most frequent of which are cervical cancer and oropharyngeal cancer (OPC). HPV16 is the hrHPV virus most often implicated in the causation of both cervical cancer and OPC[35] with HPV18 occupying second place. HPV16 is responsible for approximately 50% of cervical cancers and 85% of OPCs. Therefore most therapeutic vaccines against hrHPV-induced cancers are directed against HPV16 and HPV18. Virtually, all therapeutic vaccines against hr HPV are directed against the two oncogenic proteins E6 and E7. Since these proteins are required for the maintenance of the transformed cell state, they are particularly attractive as targets for therapy, because downregulation or loss of E6/E7 by the (pre-)malignant cells would inevitably lead to reduction or loss of that transformed state. Many different therapeutic vaccine platforms have been used against premalignant disease or cancers caused by HPV16 or HPV18, including DNA, mRNA, synthetic long peptide (SLP), and recombinant virus vaccines (reviewed in Ref. 36). Therapeutic vaccination appears to be effective against premalignant cervical intraepithelial neoplasia or vulvar intraepithelial neoplasia (VIN) as monotherapy with either DNA vaccines[37,38] or an SLP vaccine.[39,40] In late-stage HPV16 + cancer, combination of therapeutic vaccination with either anti-PD-1 monoclonal antibody (nivolumab)[41] or with the standard of care chemotherapy (carboplatin + paclitaxel)[42,43] is necessary for clinical benefit. Indeed, late-stage disease is characterized by immunosuppressive mechanisms counteracting effective T-cell immunity both in the tumor microenvironment and in the periphery (reviewed in Ref. 36). The cotreatment with anti-PD-1 addresses the checkpoint control in the TME and tumor draining lymph nodes,[41] whereas the chemotherapy deletes mononuclear immunosuppressive myeloid cells without depleting the tumor-specific T cells.[42,43] In three independent studies, the size of the T-cell response against HPV16 E6/E7 induced by SLP vaccination correlated positively with clinical benefit in both premalignant disease[39,40] and late-stage cervical cancer.[42] An important condition for success therefore is induction of a robust and sustained T-cell response as demonstrable in PBMC. An important observation made in monotherapy SLP vaccination of VIN lesions is that an inflammatory state of the original lesions before therapy is predictive of the response to therapeutic SLP vaccination.[44,45]

Therapeutic vaccination against Merkel carcinoma virus

Merkel cell carcinoma is a rare and aggressive skin disease caused by Merkel cell polyoma virus, a small DNA cancer virus. The virus encodes two oncoproteins, large T and small T.[46] First-line therapy with anti-PD-1 monoclonal antibody against established disease is quite effective. In a series of 50 patients with Merkel cell carcinoma, 64% of whom were polyoma virus positive, the overall response rate to treatment with pembrolizumab was 56% (59% in the virus positive tumors and 53% in the virus negative ones) with 24% complete responses.[47] There is also anecdotal evidence of a good clinical response to adoptive T-cell therapy with ex vivo expanded T cells.[48] Because slightly more than 50% of patients respond to monotherapy with anti-PD-(L)1 and many of these patients ultimately develop therapy resistance, further improvement of immunotherapeutic approaches is important. Combination with a potent therapeutic vaccine to induce T cells against the oncogenic proteins of the virus is attractive but has as yet not been accomplished.[49] This could lead to similar synergy of therapeutic vaccination and anti-PD-1 checkpoint blocking as observed in the therapy of late-stage HPV16-induced OPC.[41]

References

1. Shannon-Lowe C, Rickinson A. The global landscape of ebv-associated tumors. *Front Oncol.* 2019;9:713.
2. Jean-Pierre V, Lupo J, Buisson M, Morand P, Germi R. Main targets of interest for the development of a prophylactic or therapeutic Epstein-Barr virus vaccine. *Front Microbiol.* 2021;12:1543.
3. Leen AM, Heslop HE, Brenner MK. Antiviral T-cell therapy. *Immunological Rev.* 2014;258:12–29.
4. Tashiro H, Brenner MK. Immunotherapy against cancer-related viruses. *Cell Res.* 2017;27:59–73.
5. Prockop S, Doubrovina E, Suser S, et al. Off-the-shelf EBV-specific T cell immunotherapy for rituximab-refractory EBV-associated lymphoma following transplantation. *J Clin Invest.* 2020;130:733–747.
6. Tzannou I, Papadopoulou A, Naik S, et al. Off-the-shelf virus-specific t cells to treat BK virus, human herpesvirus 6, cytomegalovirus, Epstein-Barr virus, and adenovirus infections after allogeneic hematopoietic stem-cell transplantation. *J Clin Oncol.* 2017;35:3547–3557.
7. Taylor GS, Long HM, Brooks JM, Rickinson AB, Hislop AD. The immunology of Epstein-Barr virus-induced disease. *Annu Rev Immunology.* 2015;33:787–821.
8. Sommermann T, Yasuda T, Ronen J, et al. Functional interplay of Epstein-Barr virus oncoproteins in a mouse model of B cell lymphomagenesis. *PNAS.* 2020;117:14421–14432.
9. Lin C-L, Lo W-F, Lee T-H, et al. Immunization with Epstein-Barr Virus (EBV) peptide-pulsed dendritic cells induces functional CD8 + T-cell immunity and may lead to tumor regression in patients with EBV-positive nasopharyngeal carcinoma. *Cancer Res.* 2002;62.
10. Taylor GS, Jia H, Harrington K, et al. A recombinant modified vaccinia Ankara vaccine encoding Epstein-Barr Virus (EBV) target antigens: a phase I trial in UK patients with EBV-positive cancer. *Clin Cancer Res.* 2014;20:5009–5022.

11. Rühl J, Citterio C, Engelmann C, et al. Heterologous prime-boost vaccination protects against EBV antigen-expressing lymphomas. *J Clin Investigation*. 2019;129:2071–2087.
12. Poiesz BJ, Ruscetti FW, Gazdar AF, Bunn PA, Minna JD, Gallo RC. Detection and isolation of type C retrovirus particles from fresh and cultured lymphocytes of a patient with cutaneous T-cell lymphoma. *PNAS*. 1980;77:7415–7419.
13. Gessain A, Cassar O. Epidemiological aspects and world distribution of HTLV-1 infection. *Front Microbiology*. 2012;3.
14. Hirons A, Khoury G, Purcell DFJ. Human T-cell lymphotropic virus type-1: a lifelong persistent infection, yet never truly silent. *Lancet Infect Dis*. 2021;21:e2–e10.
15. Bangham CRM. Human T cell leukemia virus type 1: persistence and pathogenesis. *Annu Rev Immunol*. 2018;36:43–71.
16. Suzuki S, Masaki A, Ishida T, et al. Tax is a potential molecular target for immunotherapy of adult T-cell leukemia/lymphoma. *Cancer Sci*. 2012;103:1764–1773.
17. Toulza F, Heaps A, Tanaka Y, Taylor GP, Bangham CRM. High frequency of CD4 + FoxP3 + cells in HTLV-1 infection: inverse correlation with HTLV-1-specific CTL response. *Blood*. 2008;111:5047–5053.
18. Kannagi M, Hasegawa A, Nagano Y, Iino T, Okamura J, Suehiro Y. Maintenance of long remission in adult T-cell leukemia by Tax-targeted vaccine: a hope for disease-preventive therapy. *Cancer Sci*. 2019;110:849–857.
19. Suehiro Y, Hasegawa A, Iino T, et al. Clinical outcomes of a novel therapeutic vaccine with Tax peptide-pulsed dendritic cells for adult T cell leukaemia/lymphoma in a pilot study. *Br J Haematology*. 2015;169:356–367.
20. Kabiri M, Sankian M, Sadri K, Tafaghodi M. Robust mucosal and systemic responses against HTLV-1 by delivery of multi-epitope vaccine in PLGA nanoparticles. *Eur J Pharm Biopharm*. 2018;133:321–330.
21. Alam S, Hasan MK, Manjur OH, et al. Predicting and designing epitope ensemble vaccines against HTLV-1. *J Integr Bioinforma*. 2020;16.
22. Pandey RK, Ojha R, Chatterjee N, Upadhyay N, Mishra A, Prajapati VK. Combinatorial screening algorithm to engineer multiepitope subunit vaccine targeting human T-lymphotropic virus-1 infection. *J Cell Physiol*. 2019;234:8717–8726.
23. Raza MT, Mizan S, Yasmin F, Akash AS, Shahik SM. Epitope-based universal vaccine for Human T-lymphotropic virus-1 (HTLV-1). *PLoS one*. 2021;16.
24. Ratner L, Waldmann TA, Janakiram M, Brammer JE. Rapid progression of adult T-cell leukemia-lymphoma after PD-1 inhibitor therapy. *N Engl J Med*. 2018;378:1947–1948.
25. van den Berg F, Limani SW, Mnyandu N, Maepa MB, Ely A, Arbuthnot P. Advances with RNAi-based therapy for hepatitis B virus infection. *Viruses*. 2020;12.
26. Cornberg M, Lok ASF, Terrault NA, et al. Guidance for design and endpoints of clinical trials in chronic hepatitis B - Report from the 2019 EASL-AASLD HBV Treatment Endpoints Conference‡. *J Hepatology*. 2020;72:539–557.
27. Jansen DTSL, Dou Y, de Wilde JW, Woltman AM, Buschow SI. Designing the next-generation therapeutic vaccines to cure chronic hepatitis B: focus on antigen presentation, vaccine properties and effect measures. *Clin & Transl Immunol*. 2021;10.
28. Lim SG, Agcaoili J, de Souza NNA, Chan E. Therapeutic vaccination for chronic hepatitis B: a systematic review and *meta*-analysis. *J Viral Hepat*. 2019;26:803–817.
29. Tsounis EP, Tourkochristou E, Mouzaki A, Triantos C. Toward a new era of hepatitis B virus therapeutics: the pursuit of a functional cure. *World J Gastroenterology*. 2021;27:2727–2757.

30. Li B, Yan C, Zhu J, et al. Anti–PD-1/PD-L1 blockade immunotherapy employed in treating hepatitis B virus infection–related advanced hepatocellular carcinoma: a literature review. *Front Immunol.* 2020;11:1037.
31. Cheng Y, Gunasegaran B, Singh HD, et al. Non-terminally exhausted tumor-resident memory HBV-specific T cell responses correlate with relapse-free survival in hepatocellular carcinoma. *Immunity.* 2021;54:1825–1840. e7.
32. Negro F. Residual risk of liver disease after hepatitis C virus eradication. *J Hepatology.* 2021;74:952–963.
33. Stuart JD, Salinas E, Grakoui A. Immune system control of hepatitis C virus infection. *Curr OpVirology.* 2021;46:36–44.
34. Dore GJ, Bajis S. Hepatitis C virus elimination: laying the foundation for achieving 2030 targets. *Nat Rev Gastroenterology & Hepatology.* 2020;18(2):91–92.
35. de Martel C, Plummer M, Vignat J, Franceschi S. Worldwide burden of cancer attributable to HPV by site, country and HPV type. *Int J Cancer.* 2017;141:664–670.
36. Saxena M, van der Burg SH, Melief CJM, Bhardwaj N. Therapeutic cancer vaccines. *Nat Rev Cancer.* 2021;21:360–378.
37. Kim TJ, Jin HT, Hur SY, et al. Clearance of persistent HPV infection and cervical lesion by therapeutic DNA vaccine in CIN3 patients. *Nat Commun.* 2014;5.
38. Trimble CL, Morrow MP, Kraynyak KA, et al. Safety, efficacy, and immunogenicity of VGX-3100, a therapeutic synthetic DNA vaccine targeting human papillomavirus 16 and 18 E6 and E7 proteins for cervical intraepithelial neoplasia 2/3: a randomised, double-blind, placebo-controlled phase 2b trial. *Lancet.* 2015;386:2078–2088.
39. Kenter GG, Welters MJP, Valentijn ARPM, et al. Vaccination against HPV-16 oncoproteins for vulvar intraepithelial neoplasia. *N Engl J Med.* 2009;361:1838–1847.
40. van Poelgeest M, Welters M, Vermeij R, et al. Vaccination against oncoproteins of HPV16 for noninvasive vulvar/vaginal lesions: lesion clearance is related to the strength of the T-cell response. *Clin Cancer Res.* 2016;22:2342–2350.
41. Massarelli E, William W, Johnson F, et al. Combining immune checkpoint blockade and tumor-specific vaccine for patients with incurable human papillomavirus 16-related cancer: a phase 2 clinical trial. *JAMA Oncol.* 2019;5:67–73.
42. Melief CJM, Welters MJP, Vergote I, et al. Strong vaccine responses during chemotherapy are associated with prolonged cancer survival. *Sci Transl Med.* 2020;12.
43. Welters MJ, van der Sluis TC, van Meir H, et al. Vaccination during myeloid cell depletion by cancer chemotherapy fosters robust T cell responses. *Sci Transl Med.* 2016;8.
44. Abdulrahman Z, de Miranda N, van Esch EMG, et al. Pre-existing inflammatory immune microenvironment predicts the clinical response of vulvar high-grade squamous intraepithelial lesions to therapeutic HPV16 vaccination. *J Immunotherapy Cancer.* 2020;8.
45. Sousa L, Rajapakshe K, Canales J, et al., In Press. ISA101 and nivolumab for HPV-16 + cancer: updated clinical efficacy and immune correlates of response. Journal for Immunotherapy of Cancer, in press.
46. Dalianis T. Immunotherapy for polyomaviruses: opportunities and challenges. *Immunotherapy.* 2012;4:617–628.
47. Nghiem P, Bhatia S, Lipson EJ, et al. Durable tumor regression and overall survival in patients with advanced Merkel cell carcinoma receiving pembrolizumab as first-line therapy. *J Clin Oncol.* 2019;37:693–702.

48. Chapuis AG, Afanasiev OK, Iyer JG, et al. Regression of metastatic Merkel cell carcinoma following transfer of polyomavirus-specific T cells and therapies capable of re-inducing HLA class-I. *Cancer Immunol Res*. 2014;2:27–36.
49. Tabachnick-Cherny S, Pulliam T, Church C, Koelle DM, Nghiem P. Polyomavirus-driven Merkel cell carcinoma: prospects for therapeutic vaccine development. *Mol Carcinogenesis*. 2020;59:807.

Chapter 7

Novel dendritic cell vaccine strategies

Paul V. Munson, Lisa H. Butterfield and Juraj Adamik
Parker Institute for Cancer Immunotherapy, University of California San Francisco, Microbiology and Immunology, San Francisco, CA, United States

Introduction

Dendritic cells (DCs) are the most potent antigen-presenting cells (APCs) in vivo and represent a critical link between the innate and adaptive immune responses. As soon as methods were developed to culture these rare blood cells in numbers sufficient to study[1] in vitro and in vivo murine models, studies to harness their antigen presentation capabilities were begun. The ability to culture similar cells from human blood[2] soon enabled the first studies of DC as vaccines for cancer.[3] These trials have shown DC vaccines to be safe, immunogenic, and able to promote antitumor immunity that can induce tumor regression, but only in a small minority of patients with cancer. Recent approaches in immune profiling have enabled the study of DC vaccines and their in vivo impact in patients with human cancer in new ways that are yielding insights that will enable more potent clinical impact from this vaccination platform.

Dendritic cells biology

Ralph Steinman initially described DCs in 1973 due to their treelike morphology (dendritic originating from the Greek word for tree—dendreon) and their robust expression of MHC, work for which he later won the Nobel prize in 2011.[4] These cells originate from the common myeloid bone marrow progenitor, are negative for the lineage markers CD3/CD19/CD20/CD56, and exist primarily in peripheral nonlymphoid organs as immature DCs (iDCs).[5] iDCs patrol tissues and efficiently acquire and present antigens due to their robust endocytosis, phagocytosis, pinocytosis activities, and high expression of MHC class I and II molecules.[6] iDCs can be activated by inflammation or infection by danger signals, such as pattern recognition receptors[7] and Toll-like receptors (TLRs).[8] Upon activation, they mature and

Cancer Vaccines as Immunotherapy of Cancer. DOI: https://doi.org/10.1016/B978-0-12-823901-8.00003-0
© 2022 Elsevier Inc. All rights reserved.

migrate to lymphoid organs, such as the lymph node or the spleen, through the expression of chemokine coreceptors, such as CCR7.[9] During their maturation to mature DC (mDC), they upregulate costimulatory molecules critical for inducing T cellular immune responses, such as CD80 and CD86.[10] Successfully generating a de novo T-cell response requires three different signals between the DC and the T cell: (1) binding of the MHC: peptide complex and the T-cell receptor (TCR), (2) interaction of costimulatory molecules and their ligands, and (3) production of cytokines to skew the T-cell responses.[11,12] Once in the lymphoid tissue, DCs interact with naive $CD4^+$ and $CD8^+$ T cells in an antigen-specific way via binding of MHC-restricted peptide and the TCR, the so-called "signal 1." Signal 2 requires the interaction of costimulatory molecules with their T-cell ligands, such as B7−1 (CD80, on DC) with CD28 (on the T cell). Lastly, signal 3 refers to the cytokines secreted by the DC that can skew the nature of the T-cell response.[11,12] Generally, DC vaccines attempt to promote DCs to produce IL-12p70 to generate a Th1 response instead of interleukin (IL)-4 for Th2 or IL-17 for Th17s.[13]

Dendritic cells dysfunction in cancer

DCs play an essential role in regulating immunity and tolerance in the setting of cancer.[14] Tumors can evolve mechanisms for limiting DC function, such as excluding them from the tumor microenvironment (TME), induction of inhibitory ligands,[15] increasing metabolic stress,[16] and limiting differentiation.[17] For example, tumor production of oxidized lipid metabolites decreases the cross-presentation capacity of DCs.[18] It is becoming increasingly clear that alterations in tumor-infiltrating DC, associated with mitochondrial, glycolytic, and lipid metabolism, leading to their impaired capacity to activate T-cell responses.[18] Therefore targeting metabolic reprogramming, such as hypoxia and glycolysis using HIF-1α inhibitors, has proven an interesting strategy that increased tumor regression and survival in combination with DC immunotherapy in preclinical studies.[19]

These suppressive effects are not limited to DC in the TME but systemically modulate the immune macroenvironment.[20,21] One example is hepatocellular carcinoma (HCC) production and secretion of the fetal glycoprotein alpha-fetoprotein[22,23] or tumor-derived alpha-fetoprotein (tAFP) potently inhibits the differentiation of DCs[17] by limiting fatty acid metabolism and oxidative phosphorylation.[16] Developing a better understanding of how tAFP limits DC function—and how cancers interfere with DC function generally—could allow for more targeted approaches to improve vaccine efficacy.

Dendritic cells subsets

DCs are a heterogeneous population of cells with unique subsets that differ in their phenotype and immunological functions. With the emergence of single-cell RNA sequencing analysis of monocytes and DCs, the number of subsets, their characteristics, and functional plasticity undergo continual refinement.[24] Here, we will review the DC subsets, with an emphasis on subsets frequently used for DC vaccines, including $CD14^+$- and $CD34^+$-derived DCs, plasmacytoid DCs (pDCs), and Langerhans cells (LCs).[25]

DCs in the blood include pDCs versus myeloid or conventional DCs. Conventional DCs can be further divided into BDCA-1 ($CD1c^+$) DCs or BDCA-3 ($CD141^+$) myeloid DC. BDCA-1 DCs produce large amounts of IL-12 to stimulate Th1s. BDCA-1s express TLR1−8 and can therefore sense a wide variety of danger signals. In contrast, BDCA-3 (similar to murine cDC1) is a minor population in the blood, expresses high levels of TLR3[26] and TLR8, and synthesizes robust amounts of type-I interferon in response to dsRNA.[27] BDCA-3 cells are also more efficient at cross-presenting antigen than BDCA-1s.

pDCs are a third subset of blood DCs and are lineage negative and $CD303^+$ (BDCA-2), $CD304^+$ (BDCA-4), and MHC-II^+ cells.[28] In vivo, their tissue distribution is more limited than their conventional DC counterparts, residing primarily in lymphoid organs but can be detected in the liver and nasal mucosa. They express TLR7 and TLR9, allowing them to recognize viral ssRNA and unmethylated CpG motifs in bacterial DNA. Upon activation can generate massive amounts of type I interferons—although this function is impaired in the TME. Furthermore, tumor-associated pDCs produce indoleamine 2,3-dioxygenase (IDO), an enzyme that degrades tryptophan and can stimulate T regulatory (Treg) proliferation, which may shield the tumor from host immunity.[29] In addition to modulating immunity through the production of secreted factors [interferon (IFN) or IDO], they can directly process and present antigen, including cross-presentation.[30,31] Their ability to process antigen has made them the subject of newer DC vaccines discussed later in this chapter.

Skin is one of the largest organs in humans and is a significant physical barrier to the environment. As such, the skin is continuously exposed to microbes, allergens, chemicals, and physical damage. Therefore the cutaneous immune environment is critical for balancing wound healing, detecting pathogens, and limiting autoimmunity.[32] Anatomically the skin has two main physical compartments, the dermis (outer) and epidermis (inner), each with unique immune cell subsets. LCs are the primary DC subset of the epidermis and are positive for the C-type lectin Langerin (CD207), MHC-II, E-cadherin, and CD11c.[33] LCs also contain Birbeck granules, which are cytoplasmic organelles with an unusual racket or rod-shaped structures with Langerin and may be involved in receptor-mediated endocytosis. LCs can potently stimulate T-cell responses and are capable of cross-presentation[34] and of self-renewal.[35]

In the dermis, DCs are primarily $CD1a^+$ or $CD14^+$ DC. While less is known about CD1a, it is a transmembrane glycoprotein that has been implicated in antigen presentation by DCs. It has been shown that $CD1a^+$ and $CD14^-$ $CD207^-$ DCs were superior to dermal $CD14^+$ DC for the activation of $CD8^+$ T cells but inferior to LCs.[36] The dermal $CD14^+$ DCs induce humoral responses and regulate cellular immunity via the production of IL-10[36] that, in combination with vitamin D, induces Treg cells.[37] Furthermore, dermal $CD14^+$ DCs express the inhibitory ligands ILT2 and ILT4, which bind MHC-I and prevent peptides' binding.[38]

Making the dendritic cell vaccine

DCs are rare in the blood (~1%), making obtaining sufficient amounts (10^5-10^8) for vaccination difficult. Therefore many approaches focus on generating monocyte-derived DCs (moDCs) in vitro from blood-derived $CD14^+$ monocytes (Fig. 7.1). These approaches generally rely on peripheral blood mononuclear cells from the whole blood or leukapheresis. These monocytes are commonly isolated by adherence to tissue culture surfaces by incubating for 1–2 h and gently washing away the nonadherent cells. In 1994 Sallusto and Lanzavecchia[39] and Romani et al.[40] established a protocol using granulocyte-macrophage colony-stimulating factor (GM-CSF) and IL-4 to promote monocyte differentiation to DC and suppress macrophage proliferation, respectively. The GM-CSF/IL-4 expansion typically lasts around 5 days for the monocyte to become iDCs. Some DC vaccines relied on antigen-pulsed iDCs infused into the patient; however, it is generally considered that mDCs are superior generating immune responses in patients with cancer. The standard maturation cocktail to convert iDC to mDC was TNF-α, IL-1β, IL-6, and prostaglandin E_2 (PGE_2) for 1–2 days. Currently,

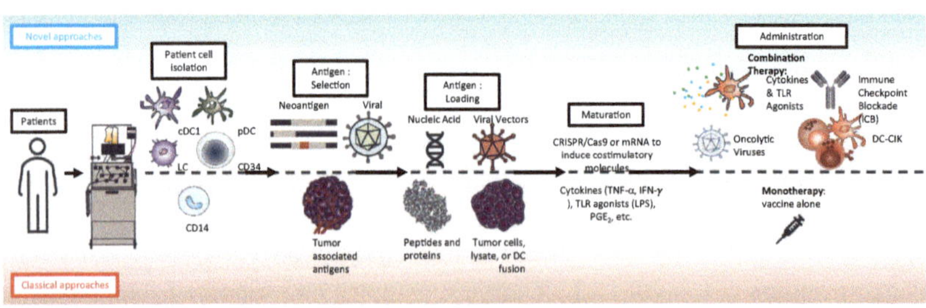

FIGURE 7.1 Development of dendritic cell vaccines: novel and classic approaches. Shown is a depiction of the development of a clinical dendritic cell vaccine. First, patient blood cells are obtained by obtaining a large volume of blood, typically by leukapheresis ± elutriation. The circulating dendritic cells are purified or, more commonly, progenitors are differentiated to dendritic cell. The cells are exposed to antigens (TAA, neoantigens, or viral) in the form of peptides/proteins, complex tumor products (cells, lysate, or dendritic cell tumor fusions), nucleic acids, or viral vectors. Cytokines, Toll-like receptor agonists, and PGE_2 typically mature the immature cells, but new approaches include using CRISPR/Cas9. Vaccines can be administered alone or in combination with other immunotherapeutic approaches.

the more preferred maturation cocktails are combinations of IFN-γ and lipopolysaccharides (LPS) or monophosphoryl lipid A (MPLA).[41] The final mDC (or in some cases iDC) product is then harvested, counted, aliquoted, and administered fresh or cryopreserved for later use.

While most protocols typically require 6–7 days to generate DCs, other approaches attempt to decrease differentiation time with 2-day fast-protocols. Koski et al.[42] established a 2-day protocol in a pilot trial for patients with HER-2/neu ductal carcinoma in situ breast cancer ($n = 27$). To achieve this, they cultured blood monocytes with GM-CSF + IL-4 and a HER-2/neu peptide and LPS and IFN-γ. These cells displayed high levels of DC costimulatory molecules (CD80/CD86), CCR7, and produced IL-12. A subset of patients with breast cancer treated with these fast-DC experienced an objective clinical response.[43] Functional differences remain between fast-DC and standard-DC, including the ability to phagocytose antigen, making fast-DC not suitable for tumor lysate-based vaccines.[44]

Alternatively, other groups have sought to simply use mRNA to engineer DC to express desired costimulatory molecules of interest. Van Nuffel et al.[45] made their vaccine by initially using a conventional approach of isolating blood monocytes by adherence and culturing with IL-4 + GM-CSF. However, instead of using a traditional maturation cocktail, they electroporated the iDC with TriMix mRNA and mRNA for tumor antigens. The TriMix mRNA is a cocktail of mRNAs for CD40L, CD70, and an activated form of TLR4. Of 35 patients enrolled, they report a single patient with melanoma (MA)[46] that successfully developed broad $CD8^+$ T-cell responses to multiple epitopes from each antigen and experienced a partial response (PR). Forty-three percentage of patients (9 of 21 patients tested) developed an increased $CD8^+$ T-cell responses to multiple vaccine antigens. The vaccine-specific $CD8^+$ T cells were at about fivefold greater frequencies when compared to baseline and directed primarily to tyrosinase and MAGE-C2. This suggests that approaches to engineer phenotypes associated with mDC genetically may be a viable approach compared to traditional maturation cocktails.

The previously described moDCs have been extensively tested in numerous clinical trials and be safe and immunogenic, and induce an objective response in up to 10% of vaccinated patients. However, they have some drawbacks, including several different *ex vivo* manipulations that add to the time and ultimately the vaccine's cost. moDC cannot be propagated *ex vivo* since they are terminally differentiated and do not proliferate. In addition, moDCs are not the most efficient DC type at cross-presentation, meaning that antigen must be produced endogenously to get a robust MHC-I presentation, and exogenous antigen from killed tumor cells cannot be utilized for spreading of the response from one antigen to another, which may be a key mechanism for improved clinical response.[47,48] These drawbacks have prompted researchers to explore other avenues for either generating DC in vitro or obtaining them directly from the patient.

Another approach is to use $CD34^+$ stem cells, which give rise to all DC progenitors, which can be isolated from donor bone marrow or increased in the blood through cytokine mobilization. These cells can propagate in vitro with the addition of GM-CSF, Flt3L, TNF-α, TNF-β, and stem cell factor (SCF) for up to 12 days.[49] Compared to moDCs, these cells are more similar to LCs and secrete IL-15 to stimulate more potent $CD8^+$ T-cell responses. In 2011 Romano et al.[50] directly compared LCs to moDCs in patients with MA. Importantly, they found that moDC required IL-15 to observe robust TA-specific memory T cells over several weeks in vitro, whereas LCs endogenously make IL-15. The authors speculate that moDCs will require exogenous IL-15 to achieve clinical efficacy. This could additionally boost T-cell responses because IL-15 has been shown to activate CTLs.[51] Other groups have also used retroviral vectors to generate a sustained expression of antigens in LCs; in this instance, retroviral transduction was more effective than peptide pulsing of DCs at inducing immune responses to a HLA-A*0201 restricted influenza matrix epitope.[52]

In addition to generating DCs from $CD14^+$ or $CD34^+$, another approach is to isolate DC directly from the patient. This has the advantage of selecting DC subtypes that may be difficult to generate in vitro; however, it can be challenging to obtain a sufficient yield, especially when patients undergo chemotherapy. While pDCs are mostly thought of for their ability to generate type I interferons, they are essential for T-cell immunity. Tel et al.[53] performed the first-in-human study of vaccinating metastatic MA patients with peptide-pulsed pDCs. pDCs were isolated using GMP-grade beads with anti-BDCA-4 antibodies pDCs were stimulated with IL-3 as well as inactivated tick-borne encephalitis (TBE) virus as a natural TLR agonist. Most of the patients developed TBE-specific IgG responses. However, there was no increase in TBE-specific T-cell responses. In contrast, approximately half of the patients tested ($N = 6$) developed a tenfold increase in the frequency of $gp100_{154}$-specific $CD8^+$ T cells. These data indicate that TBE does not substantially interfere with the development of tumor-specific T cells. While this study was not designed to determine clinical efficacy, pDC-vaccinated patients had significantly longer 1- and 2 -year survival rates of 60% and 45% compared to controls with 30% and less than 10%. These findings are particularly impressive given the relatively low dose ($3 \times 10^5 - 3 \times 10^6$) per vaccination.

In addition to directly isolated pDCs, isolation of BDCA-1^+ (CD1c) DCs has garnered more attention recently. Schreibelt et al.[54] used biotin-coated CD1c antibodies to isolate the BDCA-1^+ DCs from patients with stage IV MA. The cells were supplemented with GM-CSF and keyhole limpet hemocyanin (KLH, as a nonspecific xeno-helper antigen), giving rise to mDCs. The mDCs were loaded with MA HLA-A*0201-restricted tumor antigens from gp100 and tyrosinase. A subset of the vaccinated subjects (4/14) experienced longer-term progression-free survival (PFS). The long-term PFS phenotype was associated with polyfunctional CD8 T cells in three of these four patients. A similar vaccine was developed for patients with prostate cancer, and while the patients developed KLH-specific immunity, they did not detect responses to the vaccine antigen PSA. In addition, they did not observe any evidence of increased survival.[55]

Both in vitro differentiated DC and isolating naturally occurring DC rely upon developing unique autologous vaccines. The complexity of manufacturing these products has led to the development of "off-the-shelf" allogeneic DC-like cell lines. Ideally, these vaccines could be mass-produced and would not require a unique vaccine to be generated for every patient. A major roadblock for such an approach would be the lack of HLA compatibility between the vaccine and the patient. The novel vaccine, DCP-001, was generated from acute myeloid leukemia (AML)-derived cell line that expresses multiple AML-associated antigens.[56] Patients were advanced-stage AML that were at high risk for relapse. To assess immunogenicity, the authors developed an "Overall T cell Score" based on the T-cell proliferative capacity, ELISpot reactivity, and delayed-type hypersensitivity (DTH) responses. They found that long-term survivors (> 6 months) had higher overall T-cell scores (15 positive assays of 22) when compared to short-term survivors (<6 months, 5 positive assays of 15). While promising, additional work is needed to improve these vaccines' efficacy and make them more widely available for other malignancies.

Antigen selection

Compared to vaccines for pathogens (nonself), cancer vaccines are more challenging to develop in part because high-affinity responses to self-antigens are eliminated through central or peripheral tolerance mechanisms. T cells that do overcome these hurdles are typically reduced in frequency and affinity than viral-specific T cells. In many cases, vaccine-induced T cells are challenged with large established tumors with a myriad of immunosuppressive mechanisms. Historically, cancer vaccines have been composed of tumor-associated antigens (TAAs), comprising cancer-testes, differentiation, and overexpressed antigens. Cancer-testes antigens are only expressed by the tumor and male germ cells (e.g., NY-ESO-1 and MAGE). Differentiation antigens are expressed by specific lineages of tumors and a limited subset of normal tissues, which are precursors for those tumors (e.g., gp100 and tyrosinase are expressed by melanocytes and MA). Lastly, overexpressed antigens are present in both normal and tumor cells but are more highly expressed in cancer (e.g., HER2). Due to tumor heterogeneity, several antigens are typically selected to ensure that one or more vaccine antigens are present in the patient tumor without having to test each tumor for each antigen.

Multiple antigens are chosen to address cancer immunoediting, whereby the initial immune response can eliminate highly immunogenic tumor cells and select for escape mutations, such as downregulation or deletion of the antigen. Choosing multiple antigens also addresses naturally occurring antigen expression heterogeneity. For DC vaccines, these TAA-specific responses can be induced through several different approaches, including incubating killed tumor or tumor lysate, recombinant full-length proteins, or HLA-restricted peptides that will be discussed in more detail in the antigen loading section.

While TAAs are conserved across different patients, high-affinity TAA-specific TCRs are deleted during central tolerance, and the remaining TCRs are susceptible to peripheral tolerance.[57]

In contrast, neoantigens are more "nonself" and elicit robust frequencies of T cells with high affinity TCRs. Neoantigens have attracted a significant amount of attention recently because they are both tumor specific and resistant to central tolerance due to later (adult) expression. This led to the hypothesis that neoantigen-specific $CD8^+$ T cells could be of more strong avidity and could be superior against tumor cells than TAA-specific CD8s. However, private neoantigens are patient specific, meaning that the antigens must be identified for each patient, adding to the time and cost to generate a vaccine. This approach is complicated because many tumor mutations are not immunogenic, and biochemically determining which epitopes are presented has been challenging. In addition, most neoantigens are patients specific and require a personalized vaccine development. To overcome this, several groups have generated in silico methods to help narrow down the pool of mutations to candidates that are likely to be bound by HLA molecules and could potentially stimulate T-cell responses.[58]

In pursuit of this concept, Carreno et al.[59] developed a neoantigen-based DC vaccine in patients with MA and used gp100 as a TAA for a positive control for vaccination. In brief, three patients with MA that all previously received ipilimumab were subjected to total exome sequencing to identify candidate neoantigens. Mutations were analyzed in silico to predict binding to HLA-A*0201 and were later biochemically confirmed to be presented. Ultimately, seven peptide candidates were identified for each patient. Interestingly, while the vaccine only contained immunodominant neoantigens (short HLA*A02-restricted peptides), it induced epitope spreading to subdominant-neoantigens confirmed by TCR sequencing.[47] The authors speculate that developing a diverse repertoire of dominant and subdominant neoantigen-specific responses may be necessary to block all possible escape mutation pathways to provide a durable remission or cure. This study argues that private-neoantigen-based approaches are a feasible, safe, and effective approach for cancers with high mutational burdens.

While Carreno et al. demonstrated that neoantigen-based DC vaccines are feasible and immunogenic, developing these patient-specific approaches involves sophisticated computational and biochemical techniques. The drawbacks of both TAAs and private-neoantigens have led groups to develop shared or "public-neoantigen" vaccines, which are neoantigens arising from mutations shared in a significant proportion of patients a given tumor type.[60]

Recently, we developed a vaccine strategy aimed to promote antitumor innate immunity, determinant spreading by enhancing cross-presentation in vivo, and stimulate a polyclonal TA-specific T-cell response.[61] This approach used mDCs transduced with a replication-incompetent adenoviral vector to engineer the DCs to constitutively express three full-length TAs commonly expressed in patients with MA, and a subset of patients received 1 month of high-dose IFNα to promote innate and adaptive immune responses. IFNα boost did not result in significant clinical benefit in OS or PFS. However, analysis of vaccine-induced T-cell responses as measured by IFN-γ ELISPOT counts showed that CD8$^+$ and CD4$^+$ T-cell responses to any of the MA antigens associated with longer OS (20 months) and PFS (5 months) in patients with MA.

Interestingly, circulating immunoregulatory cells, including myeloid-derived suppressor cells (MDSCs) and T regulatory Treg cells, appeared to limit vaccine immunogenicity. In addition, we found that low expression of PD-1 on MA-specific T cells correlated with improved outcome. Since more than half of the enrolled patients were previously treated with checkpoint blockade, we further showed that CD8$^+$ IFN-γ responses against all TAs and frequencies specific for MART1 peptides were elevated in patients who had received previous checkpoint inhibitor therapy.[62]

These findings suggest that combining DC vaccines with novel combinations checkpoint inhibitors and agents that decrease Treg cells and MDSCs could be a strategy to increase vaccine efficacy. However, the precise timing of checkpoint inhibitor usage prior to, during, or postvaccination needs further considerations in future vaccine trial designs.

Certain cancers have a viral etiology, including hepatitis B and C viruses for liver cancer. In addition, human papillomavirus can cause a variety of cancers in men and women. While the mechanistic connection between cytomegalovirus (CMV) and glioblastoma (GBM) remains controversial,[63] multiple studies have detected CMV nucleic acid and proteins in more than 90% of GBM tumors.[64] Batich et al.[65] reported data from numerous clinical trials pulsed autologous in vitro generated DCs with pp65 mRNA and found about one-third of the 5-year survival rate. In support of this concept, CMV pp65 mRNA-based vaccines have been tested in a small number of patients with GBM and significantly improved survival.[66] Based on mouse models, the vaccine efficacy seems to depend upon pretreating the injection site with tetanus toxoid that increased expression of CCL3 and improving homing to the lymph node.[67] While viral-antigen-based vaccines would not be possible for all cancers, they may be an attractive option for virus-associated cancers.

Antigen loading

All DC-based vaccines include pulsing, or loading, the DCs with a source of cancer antigens that has tumor products, including (1) autologous tumor cells, (2) autologous lysate, and (3) allogeneic tumor cells, peptides (short and long), full-length proteins, nucleic acids (mRNA and DNA), and viral vectors.

Pulsing DCs with a variety of autologous or allogeneic tumor cells or lysates have been extensively studied. Lysate/homogenate preparation techniques aim to induce tumor necrosis and apoptosis for maximal release of an array of tumor antigens as well as undefined neoantigens. However, in some patients, it can be challenging to obtain sufficient tumor cells for vaccine formulation. Some of the strategies involving autologous tumor processing include DC/tumor fusions and hypochlorous acid-mediated (HOCl) tumor lysate oxidation.[68]

In early studies, fusing DC with multiple myeloma (MM)-engineered J558 cells expressing CD40L using polyethylene glycol exhibited effective cytotoxic T-cell responses (CD4/CD8) against J558 tumor cells in vivo.[69] Patient-derived MM cells chemically fused with autologous MUC1-pulsed DCs display a broad repertoire of MM-associated antigens. In a phase I clinical study, fusion vaccines promoted the expansion of myeloma-reactive circulating CD4 and CD8 lymphocytes in 11 out of 15 patients together with disease stabilization in the majority of enrolled patients with advanced disease.[70] While the DC–tumor fusion vaccines have significantly improved immune activation and survival of patients with certain classes of hematological malignancies, such as AML,[71] the clinical benefit of these vaccines had limited efficacy in solid tumors. Therefore numerous adjuvant therapies, including IL2, IL12, IL18, CD40, CpG oligodeoxynucleotides, poly(I:C) with IL-10 siRNA, and D-1-methyl-tryptophan (IDO inhibitor), have been implemented in preclinical studies.[72]

To improve the immunogenicity of the killed autologous tumor, another group killed cells through oxidation with HOCl and then freeze–thaw cycles.[73] The vaccine-induced tumor-specific T cells correlated with improved survival. Interestingly, this vaccine also generated T-cell responses to previously undescribed neoepitopes with markedly greater avidity than typical neoepitope-specific T cells. Additional innovative strategies aiming to enhance the immunogenicity of personalized DC vaccines, such as tumor lysate antigen loading into yeast cell wall particles described by Hardin et al.,[74] are being investigated in a current Stage III and Stage IV (resected) MA (NCT02301611) and colon cancer (NCT04147078) clinical trials.

Among the earliest vaccine approaches used small HLA-defined peptides. While short peptides effectively induce $CD8^+$ T-cell responses, they are too short to induce $CD4^+$ T cells that typically require a minimum of 12–15 amino acids. This can be problematic because inducing $CD8^+$ T-cell responses in the absence of a helper $CD4^+$ T-cell response can limit their functionality and durability. To address this issue, several groups have added xenogeneic helper antigens, which are typically full-length heterologous proteins that induce potent CD4 responses (KLH, tetanus toxoid). Another drawback to using tumor-specific short peptides is that they must be HLA defined, and many cancer epitopes remain undefined, as are the biological rules for identifying them.

Viral vectors provide an attractive approach to genetically engineer DC to express cancer antigens. Numerous viral vectors have been tested in cancer vaccines, including Poxviridae, Adenoviridae, and adeno-associated virus. For DC vaccines, most of the vaccines have focused on transducing DCs with adenovirus, genetically engineered to express cancer antigens. While theoretically, the immune system focusing on viral antigens instead of the cancer antigens because the immunodominance was a concern, this is not an issue. Approaches that involve the direct administration of viral vectors have considered using novel serotypes to avoid problems arising from preexisting immune responses; it is unclear to what extent these approaches would improve viruses applied to DCs ex vivo.

More novel approaches include the use of mRNA applied to DCs. mRNA has many desirable characteristics in that it cannot integrate into the genome, is inexpensive, and cannot be alternatively spliced. Since mRNA is translated in the cytosol, it can readily stimulate $CD8^+$ T-cell responses but is not as effective as expressing antigen on MHC-II to generate $CD4^+$ T-cell responses. Initially, to make DC vaccines, mRNA was applied to DC ex vivo and taken up through micropinocytosis.[75] However, this approach signals TLR-7 and reduces additional mRNA ingestion.[76] Approaches to improve delivery include electroporation,[77] sonoporation,[78] and nanofection.[79] One strategy to elicit $CD4^+$ T-cell immunity is to fuse the antigen of interest to lysosomal-associated membrane protein (LAMP). In a preclinical study using E7 expressing human papilloma oncogenic model and metastatic $HER2^+$ tumor breast cancer, the addition of LAMP enhanced vaccine potency by the intracellular routing of antigens to endosomal/lysosomal compartments, resulting in enhanced MHCI/II presentation driving polyfunctional T-cell responses.[80,81] Incorporation of LAMP in CMV-based targeting of pp65-expressing tumor cells combined with GM-CSF boost, following conventional chemotherapy is currently being explored in phase I clinical trials of newly diagnosed GBM (NCT02465268, NCT03927222).

In a single-arm phase II trial, DCs were electroporated with mRNA encoding gp100, tyrosinase, MAGE-A3, and MAGE-C2. To generate $CD4^+$ T-cell responses, the proteins were fused to DC.LAMP targets the antigens to the lysosome.[82] In addition, this vaccine was delivered with ipilimumab and had an overall response rate of 38% based on 8/39 CRs and 7/39 were PRs. Overall survival in all HLA-A*0201 patients ($N = 10$), T-cell responders ($N = 3$), and nonresponders ($N = 7$) was 25.6, 30.7, and 23.4 months, respectively. The T-cell responses induced by the vaccine were measured 12 weeks after vaccination. In comparison, other studies demonstrated ipilimumab alone or in combination with a gp100 vaccine have a median overall survival of 10.1 and 10.0 months, respectively, in HLA-A*0201 patients with metastatic MA ($N = 676$).[83]

A phase I study ($N = 9$) demonstrated that treating patients with MA with Langerhans cells electroporated with mRNA encoding TRP-2 was also safe and immunogenic.[84] All nine patients developed a mild DTH at the injection site during booster vaccinations, and no toxicities exceeded grade 1. One month postvaccination, $CD8^+$ T-cell secretion of IFN- γ, IL-2, and TNF-α increased by 1.71-, 2.05-, and 1.37-fold above prevaccination levels, respectively. Importantly, deep sequencing of the TCR-V-β CD3 genes revealed that the six patients who did not relapse had a greater increase in TCR clonality after vaccination (CD4: 2.31, CD8: 3.99) than the three patients who relapsed (CD4: 1.83, CD8: 1.54).

Notably, the Pfizer and BioNTech SARS-CoV-2 vaccine consists of nucleoside-modified mRNA formulated in lipid nanoparticle, which enhanced DC targeting in vivo, preventing coronavirus disease 2019 (Covid-19) in 95% of vaccinated subjects.[85] Future mRNA-based DC vaccines may include similar lipid nanoparticle formulations and nucleoside modifications to improve antigen expression.

Administration

Numerous groups have tested a variety of routes to identify the optimum delivery for vaccine efficacy. Studies in mice tested intravenous (i.v.), intraperitoneally (i.p.), subcutaneously (s.c.), and intradermally (i.d.)[86] delivered DC vaccines using OVA as a model Ag. The i.p. and i.v. routes performed poorly, whereas the i.d. and s.c. routes resulted in Ag-loaded DC trafficking to lymph nodes. Different DC routes were tested in patients with metastatic prostate cancer, and while all routes developed some level of anticancer immunity, the only routes that elicited IFN-γ responses were the i.d and intralymphatic routes.[87] Intralymphatic injection is achieved by cannulating lymphatic vessels in the feet so the DCs can enter the lymph nodes through afferent lymphatics. Similarly, DC vaccines were administered to patients with advanced MA using i.d. versus intranodal delivery and found that i.d resulted in superior immunogenicity.[88] Other approaches have attempted to develop in situ vaccination strategies by directly injecting DCs into a solid tumor.[89] The skin is a complex organ, and further work is needed to define the DC subsets and their unique functions in humans more clearly. In addition, given that many DC-vaccines are injected intradermally, a better understanding of this compartment could lead to the design of more effective DC vaccines, such as the delivery of immune-modulatory compounds at the time of vaccine injection.

Early successes and transition into novel combinatorial applications in clinical trials

The majority of DC vaccines with clinical testing have focused on prostate, MA, brain (GBM), and renal malignancies. In early trials, many DC vaccines focused on MA because it possessed characterized immunogenic tumor antigens, but advances in the field have led to the development of new vaccination strategies, which have been extended to numerous cancer types. The summary of 90 active/upcoming clinical trials is depicted in Fig. 7.2.

For prostate cancer, the FDA approved Provenge (Sipuleucel-T) in 2010,[90] consisting of a complex mixture of T, B, and NK cells, monocytes, and APCs stimulated with a prostatic acid phosphatase (PAP) antigen fused to GM-CSF. This product does not fall into a classically defined DC vaccine because it was a complex mixture of cells, including lymphocytes, and did not employ a maturation cocktail. Regardless, Provenge has been shown to induce humoral and cellular immunity with antigen spreading and increase T-cell infiltration into the TME. Additional studies are underway combining Provenge with checkpoint inhibitors (NCT03024216) and cytokines, such as IL-7 (NCT01881867), to improve its clinical benefits.

Notably, many patients participating in early-phase DC vaccine clinical trials have progressed despite the standard of care treatments and are in the

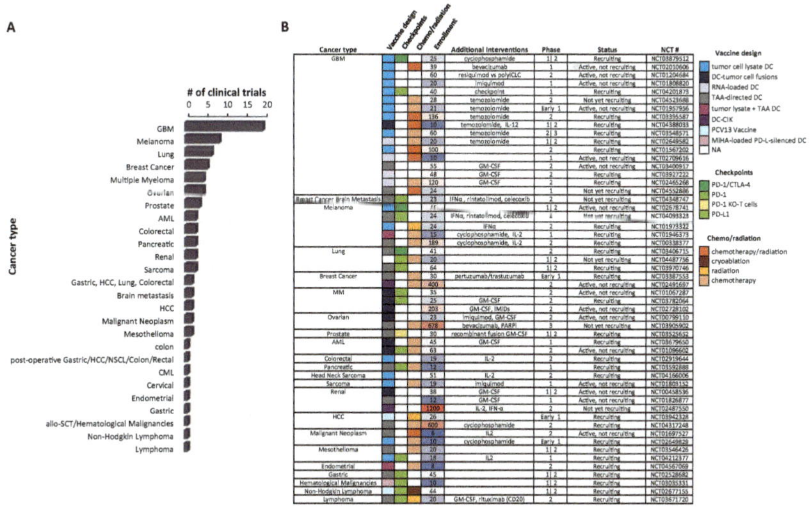

FIGURE 7.2 Overview of dendritic cell vaccine strategies and combinatorial treatment in current clinical trials. (A) Depicted is distribution of current (in status as active, recruiting or not yet recruiting) dendritic cell vaccine trials per cancer type. (B) Subsets of 54 dendritic cell vaccine trials, which use a variety of combination therapy approaches are further illustrated. Classification of dendritic cell vaccine preparation/antigen-loading methods using RNA, recombinant peptide pulsing, dendritic cell-tumor fusions, or tumor lysate preparations is also depicted.

later stages with metastatic disease. Typically, for patients to be eligible for immunotherapy, they must not have chronic viral infections (HIV and hepatitis B or C) or autoimmune diseases to ensure they could successfully respond to the vaccine. Even when DCs are generated from patients with cancer ex vivo, evidence of their dysfunction, such as defective NF-κB signaling with defective ICOSL presentation, is being uncovered.[10] Additional research is needed to overcome this immunosuppression, including studies vaccinating patients at earlier stages of the disease or after tumor debulking,[91] which may be a strategy to improve the immunogenicity and efficacy of these approaches.

While the full potential of DC vaccines as monotherapy in cancer has not delivered the expected clinical success, the future of this personalized cell therapy shows promise in combination approaches with small molecule inhibitors, checkpoint blockade, adoptive transfer cell therapies, immunomodulatory cytokines, and chemotherapy and/or radiation regiments. In addition to novel vaccine designs, new approaches in DC-based cancer vaccination primarily focus on improving DC functioning in the immunosuppressive TME by targeting DC migration, metabolic adaptation, antigen-presenting and costimulatory capabilities, deletion of Treg cells to collectively reverse DC-induced T-cell tolerance, and enhanced their antitumor responsiveness. In support of this, 54 out of 90 DC vaccine active/recruiting or upcoming clinical trial designs employ a combinatorial regimen that counteracts the hostile tumor environment to enhance the effectiveness of DC cancer vaccines (Fig. 7.2B).

Several classes of small molecule inhibitors, including MAPK, tyrosine kinase, and HSP90 inhibitors, that have been extensively used in oncology therapy show promising results in combination with DC vaccines in preclinical studies[92] (Table 7.1). A rationale for combination therapies was shown in murine subcutaneous M05 MA model using multikinase small molecule inhibitor dasatinib, which exhibited multiple antitumor properties, including eliminating MDSCs and Treg cell-attenuated hypoxic signaling axes to favor the influx of tumor-infiltrating $CD8^+$ T cells and $CD11^+$ DC. Most importantly, dasatinib use in conjunction with DC vaccination promoted superior antitumor responses with antigen spreading of CD8-specific responses to nonvaccine-loaded antigens to MA TAA EPhA2, gp100, and TRP2.[93]

Among recent studies, Guo et al. demonstrated that a drug cocktail of selected AKT, DNA-PK, and MEK (MKNUTRA cocktail) inhibitors effectively upregulated MHC class I, CCR7, and CD83 and decreased PD-L1 expression of moDCs as well as enhanced T-cell activation and proliferation by NY-ESO-1 peptide-pulsed moDCs. Furthermore, MKNUTRA treatment enhanced the immunogenicity of IL13Rα2 and TRP2-targeting ICT107 moDC vaccines by enhancing their maturation with a reduced tolerogenic protein surface profile in vitro, which resulted in the increased frequency of tumor-specific lymphocytes and reduced growth of human GBM tumors in NOD.Cg-$Prkdc^{scid}$ $Il2rg^{tm1Wjl}$/SzJ (NSG) mice.[94]

TABLE 7.1 Summary results from preclinical studies using dendritic cells vaccine combinatorial therapies.

Reference	93	94	95	96	97	98	99
Model	Subcutaneous M05 melanomas	Glioblastoma (U87 GBM cells)/xenogeneic (NSG) mouse model	Analysis of circulating DC subsets in patients with GBM	MO5 (B16.OVA) melanoma	Murine colon carcinoma (MC33-CEA) and breast cancer (4T1)	Orthotopic mesothelioma mouse models	• Renal cell carcinoma • SCID mice injected with R11 and R6 human cell line expressing GMCA-9 fusion protein
Combinatorial drug	Dasatinib (broad-spectrum multikinase inhibitor)	AKT, DNA-PK, and MEK inhibitors (MKNUTRA)	p38 MAPK inhibitor (p38i)	sunitinib (VEGFR/PDGFR inhibitor)	tyrosine kinase inhibitors (TKI), sunitinib and sorafenib	PLX3397 (pexidartinib)/CSF-1R kinase inhibitor	Recombinant GM-CSF-CA9 transgene
Vaccine	OVA257−264 peptide-pulsed DC overexpressing murine IL-12	GP100$_{(209-217)}$, TRP2$_{(180-188)}$, HER2$_{(773-782)}$, and IL13Rα2$_{(345-354)}$ peptide pulsed	cDC2 (nonvaccine study)	OVA 257−264, OVA 323−339 synthetic peptide pulsed	rMVA−CEA−TRICOM	Tumor lysate-pulsed DC	Recombinant adenovirus containing GM-CSF and CA9 antigen fusion gene
In vivo immune efficacy in combination with vaccine							
	• $CD8^+$ T cells (TDLN and TILLs) and • $CD11c^+$ DC	• Tumor-reactive lymphocytes within tumors and in peripheral tissues	• Increased IL-12 expression of TOL DCs • IL-12:IL-10 was skewed in favor of	• OVA-specific T-cell responses observed • Infiltration of tumor lesions	• $CD4^+$ and $CD8^+$ TILs, including TA-specific T cells • Expression of activation markers FAS-L, CXCL-9, CD31, and CD105	• $CD8^+$ T-cell numbers and functionality • Proliferation of CD4/CD8 double positive	

(*Continued*)

TABLE 7.1 (Continued)

Reference	93	94	95	96	97	98	99
	infiltration into TME	• Frequency of peptide-specific T cells	Th1 responses, similar to healthy donor cDC2s	by anti-OVA CD8 TILS • Enhanced type-1 T-cell infiltration due to increased VCAM-1/CXCR3 ligand chemokines in vascular/perivascular cells within the TME	in MDSCs and TAMs • $CD11^+$-scattered monocytes • Vascular normalization and increased tumor oxygenation • CD105 in MDSCs	• Superior T-cell memory responses • $CD8^+/CD4^+$ T-cell proliferation following tumor rechallenge	
	• $CD4^+$ T-cell infiltration • MDSC, Treg cells			MDSC, Treg cells	• $CD31^+$ tumor vascular area • Monocytic vasculature • Compactness • Tight junctions and intratumoral pressure • Percentage of TAMs	• TAMs • Circulating nonclassical PD-$L1^+$ monocytes • Treg cells • Numbers of exhausted $PD1^+LAG$-$3^+IFN\gamma^-$ TILs • Neoangiogenesis, ascites • Expression of PD-1 on tumor antigen-specific $CD8^+$ T cells	

Ex vivo analyses							
	• Type 1 $CD8^+$ T-cell response to nonvaccine tumor-associated antigens, such as EphA2, gp100, and TRP2	• moDC cell surface MHC class I, CCR7, CD83, • IL12p70 secretion • T-cell proliferation • IFNg and Granzyme B production	• T-cell stimulatory capacity of cDC2 • Proliferation and IFNγ release by matured cDC2				• GMCA-9 expressed inside the DCs induced DC maturation with HLA-Dr/CD83 upregulation with ability to stimulate CA9-specific CTLs
		• PD-L1, ILT3	• Reduced secretion of the regulatory cytokine IL-10				

Notes: Summarized immune profiling of in vivo and in vitro experiments from preclinical studies using DC vaccine combinatorial treatments are presented. Arrows indicate increased (UP) or decreased (DOWN) aspects of immune system altered by the combinatorial DC vaccine treatments. *DC*, dendritic cell; *MDSCs*, myeloid-derived suppressor cells; *OVA*, ovalbumin; *TAMs*, tumor-associated macrophages.

Inhibition of p38 in immune-compromised dexamethasone-induced tolerogenic DCs and circulating cDC2s obtained from patients with GBM restored their inflammatory gene repertoire, IL12:IL10 ratio, and CD86 expression, resulting in enhanced T-cell activation.[95] While these experiments were not conducted in the context of cDC2-based DC vaccines, these results suggest that targeting selected kinase activity such p38 may positively affect DC immunogenicity and potentially synergize with DC vaccination in patients with GBM.

Antiangiogenic kinase inhibitors sunitinib and sorafenib are another example of small molecule combination therapy, which showed superior vaccine efficacy in combatting the tumor immunosuppression in MA, colon carcinoma, and breast cancer murine models.[96,97] Lastly, blockade of tumor-associated macrophages (TAMs) using CSF1R kinase inhibitor also showed robust synergy in boosting durable antitumor immunity when used in conjunction with DC vaccination in preclinical murine mesothelioma study by Dammeijer et al.[98] Furthermore, CSF1R kinase inhibition reduced the levels of TAMs and exhausted T cells with decreased PD-1 and LAG3 expression and protected mice from tumor rechallenge.[98]

Frontline therapies battling immunosuppressive networks within TME include checkpoint blockade antagonistic monoclonal antibodies directed against CTLA-4 and PD-1/PD-L1 signaling axes. Overexpression of checkpoint molecules has been extensively studied in tumor immune evasion and dampening of antitumor T-cell effector responses. However, manipulation of the surface and soluble forms of the checkpoint expression in different subtypes of DCs is less well understood.[100] Therefore combinatorial DC vaccine-checkpoint therapy design and timing of anti-CTLA-4 and/or PD-1/PD-L1 antibodies pre-, during, or postvaccination require careful consideration in obtaining safe and desired synergistic effects and the different biology of each of these molecules. The addition of checkpoint blockade is currently explored by a number of current DC vaccine studies, with the majority focusing on the use of anti-PD-1 monoclonal antibody regiment in numerous cancer types: GBM, brain metastasis from multiple tumors, MA, lung, MM, pancreatic, gastric, AML, and non-Hodgkin's lymphoma. A combination of autologous DC vaccines with atezolizumab (anti-PD-L1) following standard chemotherapy care using carboplatin is being conducted in patients with advanced-stage small-cell lung malignancies (NCT04487756). Together with checkpoint monotherapies, a combination of ipilimumab (anti-CTLA-4) and nivolumab (anti-PD-1) plus a DC-based p53 vaccine, followed by maintenance immunotherapy and with potential retreatment regimen (NCT03406715), is being tested in relapsed small lung cancer. In addition to monoclonal antibodies, novel monocyte-derived MiHA-loaded PD-L1-silenced DC vaccines expressing siRNA targeting PD-L1 and PD-L2 expression[101] are being evaluated in preventing tumor-relapse following allogeneic stem cell transplantation in hematological malignancies (NCT02528682).

While checkpoint blockade is the new gold standard of immunotherapy, the key limitations are due to the lack of T-cell infiltrate in case of "cold" tumors and acquired/therapy-induced resistance.[102] Various combinations of immunomodulatory drugs are being used in an effort to overcome checkpoint resistance. For instance, a clinical trial of a type 1-polarized DC (aDC1) vaccine targeting tumor blood vessel antigens is being tested with a cytokine modulating regiment including IFNα and Toll-like receptor 3 (TLR3) agonist (rintatolimod) together with celecoxib (NSAID), in HLA-A2-positive (+) patients with advanced PD-1/PD-L1 refractory MA (NCT04093323). A similar combination strategy is being applied in patients with asymptomatic brain metastasis from triple-negative or $HER2^+$ breast cancer (NCT04348747).

Another combination approach using adjuvants, such as TLR7/8 agonist resiquimod, has been recently compared to polyICLC with tumor lysate-pulsed DC vaccination in GBM. Early reports (2020 SNO virtual meeting) indicate that the polyICLC-treated group of patients exhibited a larger increase in proinflammatory gene expression signatures with increased numbers of circulating lymphocytes, which likely attributed to enhanced overall survival benefit over resiquimod cohort (NCT01204684). The use of resiquimod in this setting may need further consideration.

Poly(ADP-ribose) polymerase inhibitors (PARPi) and antibodies directed against vascular endothelial growth factor (VEGF) have been tested in ovarian cancer as strategies to block DNA repair in tumor cells and battle tumor vasculature, respectively.[103,104]

These are now being tested clinically together with autologous DC vaccines. Following a successful phase II study (SOV02) using DCVAC/OvCa vaccine in combination with platinum-based chemotherapy in patients with recurrent ovarian cancer,[105] a multicenter phase III clinical study (VITALIA) encompassing 678 patients has been initiated (NCT03905902) to test the effectiveness of DCVAC/OvCa and chemotherapy with or without bevacizumab (VEGF-A inhibitor) during initial phase following by the combination of bevacizumab with PARPi with over placebo-control.[106]

Targeting tumor-infiltrating immunosuppressive cell types, such as Treg cells and MDSCs, is being explored as a novel combinatorial strategy to enhance the effectiveness of checkpoint blockades and DC vaccine therapy. Cyclophosphamide, implemented in combination with autologous DC vaccine, has been shown to reduce numbers of circulating Treg cells in patients with mesothelioma.[107] Currently, it is being used in combination with personalized tumor-lysate DC vaccines followed by cycles of combinatorial checkpoint blockade PD-1 and CTLA4 with subsequent PD-1 monotherapy maintenance in children with relapsed high-grade gliomas (NCT03879512) and with multiple antigen-loaded DC vaccine in conjunction with radiation in hepatitis B-related HCC (NCT04317248). Inhibition of MDSCs using gemcitabine in a preclinical mouse model enhanced in vitro expansion of splenic lymphocytes, IFN-γ secretion of antigen-specific T cells, and reduced tumor burden in vivo using 4T1 tumor mouse model.[108] While the use of Gemcitabine for MDSCs depletion has been successful in murine model, a caution needs to be exercised when applying murine findings to human biology, as a study by Santegoets at al. did not see effects of Gemcitabine in reducing the elevated MDSC to DC ratio in epithelial ovarian cancer patient cohort.[109] Despite this, Gemcitabine, together with TLR7 agonist imiquimod, is used in combination with DC vaccination and subsequent administration of tumor lysate boosts for experimental treatment of pediatric and adult Sarcoma (NCT01803152).

While administration of recombinant GM-CSF has been conventionally used in the past, new methods of GM-CSF delivery are being tested to augment DC vaccination in future trials.[110] A study by Hernandez et al. demonstrated that DC vaccines generated using adenoviral transduction of GMCA9 construct, which encodes partially processed GM-CSF-carbonic anhydrase IX (CA-9) fusion protein targeted to the cell surface, induced DC maturation and was capable of mounting CA9-specific T-cell responses and in vitro.[99] The effectiveness of this vaccination strategy was demonstrated using in vivo mouse model in this study. This technology is also being adopted in the treatment of patients with metastatic renal malignancy (NCT01826877).

Certain chemotherapy agents and/or radiation-induced immunogenic cell death that causes the release of tumor antigens and danger-associated molecular patterns, which promote DC-mediated antitumor immunity.[110] The immune effects of DC vaccines combined with chemotherapy, radiation, and in some cases, additional immunotherapies are explored in 31 recent clinical trials. In hand with these methods, cryoablation of tumor tissues demonstrated synergistic effects with immunotherapy in preclinical models,[111] and it is being evaluated in non-Hodgkin's lymphoma in conjunction with the intratumoral injection of autologous DCs and pembrolizumab (anti-PD-1) (NCT03035331).

Combination cell therapies designed to antagonize the immunosuppressive networks within the TME, with maximizing DC vaccine biology, offer great potential and represent a dynamic area of investigations. An approach in promoting synergistic activation between autologous DCs vaccines with cytokine-induced killer cells (DC-CIK), predominantly consisting of $CD3^+$ and $CD56^+$ T and NK lymphocytes, respectively, has shown significant clinical benefits in combination with chemotherapy and against poorly immunogenic tumors in numerous cancer types.[112] Autologous DC-CIK vaccines pulsed with tumor-lysates are being evaluated in comparison to capecitabine monotherapy in advanced breast cancer (NCT02491697) renal cell carcinoma (NCT02487550). Additional clinical trials in gastric (NCT04567069) and ovarian cancer (NCT03735589) propose the use of DC vaccines with autologous cytotoxic $CD3^+$ lymphocytes and natural killer cell-like cytotoxic T lymphocytes (CTLs), respectively. In addition to DC-educated/coculture generation of autologous tumor-directed CTL products, direct gene editing of T cells using CRISPR Cas9 technology is paving the way for the next generation of personalized cell therapy. The safety and efficacy of combination therapy using PD-1 KO T cells, which are infused together with PAP–GM-CSF fusion protein-activated DC vaccines, is under investigation in prostate cancer (NCT03525652). Rapid advancements in adoptive T-cell transfer therapies using chimeric antigen receptor (CAR) T cells are an exciting area in transfusion oncology medicine. Since the first use of CART cell therapy recognizing MART-1, gp100, and NY-ESO-1 MA tumor antigens,[113] new generations of modified TCRs are being developed to maximize specificity with limiting off-target toxicities of these transgenic lymphocytes.[114] Coadministration of CART cells with DC vaccines is being implemented in new clinical protocols along with checkpoint inhibitor regimens and cytokine therapy in patients with advanced malignancies (NCT01697527).

Conclusions

DCs have become an established platform for cancer vaccination. After decades of in vitro and in vivo studies, DCs have become a reliable approach to increase T-cell responses to cancers of many types. The challenge has been how to utilize these potent APCs to more reliably induce tumor regressions. A better understanding of DC biology and translating this knowledge to DC vaccine manufacture will lead to a new class of DC vaccines. The other challenge has been to identify the optimal clinical setting. The next-generation trials testing combinations of DC vaccines with not only checkpoint blockade but also other immune and tumor modulators will identify ways to improve the immune and outcomes and forge the building blocks of the next generation of trials.

Acknowledgments

This work was supported by the Parker Institute for Cancer Immunotherapy. L.H.B. declares scientific advisory relationships unrelated to this work: StemImmune/Calidi, SapVax, NextCure, Western Oncolytics, Torque Therapeutics, Khloris, Pyxis, Cytomix, Vir, DCprime, and RAPT.

References

1. Inaba K, et al. Generation of large numbers of dendritic cells from mouse bone marrow cultures supplemented with granulocyte/macrophage colony-stimulating factor. *J Exp Med.* 1992;176:1693–1702.
2. Markowicz S, Engleman EG. Granulocyte-macrophage colony-stimulating factor promotes differentiation and survival of human peripheral blood dendritic cells in vitro. *J Clin Invest.* 1990;85:955–961.

3. Mukherji B, et al. Induction of antigen-specific cytolytic T cells in situ in human melanoma by immunization with synthetic peptide-pulsed autologous antigen presenting cells. *Proc Natl Acad Sci USA*. 1995;92:8078–8082.
4. Steinman RM, Cohn ZA. Identification of a novel cell type in peripheral lymphoid organs of mice. I. Morphology, quantitation, tissue distribution. *J Exp Med*. 1973;137:1142–1162.
5. Dudek AM, Martin S, Garg AD, Agostinis P. Immature, semi-mature, and fully mature dendritic cells: toward a DC-cancer cells interface that augments anticancer immunity. *Front Immunol*. 2013;4.
6. Santambrogio L, et al. Extracellular antigen processing and presentation by immature dendritic cells. *Proc Natl Acad Sci USA*. 1999;96:15056–15061.
7. Pulendran B. Variegation of the immune response with dendritic cells and pathogen recognition receptors. *J Immunol*. 2005;174:2457–2465.
8. Hemmi H, Akira S. TLR signalling and the function of dendritic cells. *Chem Immunol Allergy*. 2005;86:120–135.
9. Alvarez D, Vollmann EH, von Andrian UH. Mechanisms and consequences of dendritic cell migration. *Immunity*. 2008;29:325.
10. Maurer DM, et al. Dysregulated NF-κB-dependent ICOSL expression in human dendritic cell vaccines impairs T-cell responses in patients with melanoma. *Cancer Immunol Res*. 2020;8:1554–1567.
11. Eisenbarth SC. Dendritic cell subsets in T cell programming: location dictates function. *Nat Rev Immunol*. 2019;19:89–103.
12. Benvenuti F. The dendritic cell synapse: a life dedicated to T cell activation. *Front Immunol*. 2016;7:70.
13. Driessens G, et al. Therapeutic efficacy of antitumor dendritic cell vaccinations correlates with persistent Th1 responses, high intratumor $CD8^+$ T cell recruitment and low relative regulatory T cell infiltration. *Cancer Immunol Immunother*. 2008;CII 57:1745–1756.
14. Wculek SK, et al. Dendritic cells in cancer immunology and immunotherapy. *Nat Rev Immunol*. 2020;20:7–24.
15. Oh SA, et al. PD-L1 expression by dendritic cells is a key regulator of T-cell immunity in cancer. *Nat Cancer*. 2020;1:681–691.
16. Santos PM, et al. Tumor-derived α-fetoprotein suppresses fatty acid metabolism and oxidative phosphorylation in dendritic cells. *Cancer Immunol Res*. 2019;7:1001–1012.
17. Pardee AD, Shi J, Butterfield LH. Tumor-derived α-fetoprotein impairs the differentiation and T cell stimulatory activity of human dendritic cells. *J Immunol*. 2014;193:5723–5732.
18. Giovanelli P, Sandoval TA, Cubillos-Ruiz JR. Dendritic cell metabolism and function in tumors. *Trends Immunol*. 2019;40:699–718.
19. Kheshtchin N, et al. Inhibition of HIF-1α enhances anti-tumor effects of dendritic cell-based vaccination in a mouse model of breast cancer. *Cancer Immunol Immunother CII*. 2016;65:1159–1167.
20. Allen BM, et al. Systemic dysfunction and plasticity of the immune macroenvironment in cancer models. *Nat Med*. 2020;26:1125–1134.
21. Pinzon-Charry A, Maxwell T, López JA. Dendritic cell dysfunction in cancer: a mechanism for immunosuppression. *Immunol Cell Biol*. 2005;83:451–461.
22. Abelev GI, Perova SD, Khramkova NI, Postnikova ZA, Irlin IS. Production of embryonal alpha-globulin by transplantable mouse hepatomas. *Transplantation*. 1963;1:174–180.
23. Tatarinov IS. Detection of embryo-specific alpha-globulin in the blood serum of a patient with primary liver cancer. *Vopr Med Khim*. 1964;10:90–91.

24. Villani A-C, et al. Single-cell RNA-seq reveals new types of human blood dendritic cells, monocytes, and progenitors. *Science*. 2017;356.
25. Del Prete A, et al. Functional role of dendritic cell subsets in cancer progression and clinical implications. *Int J Mol Sci*. 2020;21.
26. Colletti NJ, et al. TLR3 signaling promotes the induction of unique human BDCA-3 dendritic cell populations. *Front Immunol*. 2016;7.
27. Cancel J-C, Crozat K, Dalod M, Mattiuz R. Are conventional type 1 dendritic cells critical for protective antitumor immunity and how? *Front Immunol*. 2019;10.
28. Reizis B. Plasmacytoid dendritic cells: development, regulation, and function. *Immunity*. 2019;50:37–50.
29. Platten M, Wick W, Eynde den BJV. Tryptophan catabolism in cancer: beyond IDO and tryptophan depletion. *Cancer Res*. 2012;72:5435–5440.
30. Villadangos JA, Young L. Antigen-presentation properties of plasmacytoid dendritic cells. *Immunity*. 2008;29:352–361.
31. Mouriès J, et al. Plasmacytoid dendritic cells efficiently cross-prime naive T cells in vivo after TLR activation. *Blood*. 2008;112:3713–3722.
32. Kabashima K, Honda T, Ginhoux F, Egawa G. The immunological anatomy of the skin. *Nat Rev Immunol*. 2019;19:19–30.
33. Clausen BE, Kel JM. Langerhans cells: critical regulators of skin immunity? *Immunol Cell Biol*. 2010;88:351–360.
34. Stoitzner P, et al. Langerhans cells cross-present antigen derived from skin. *Proc Natl Acad Sci USA*. 2006;103:7783–7788.
35. Kanitakis J, Morelon E, Petruzzo P, Badet L, Dubernard J-M. Self-renewal capacity of human epidermal Langerhans cells: observations made on a composite tissue allograft. *Exp Dermatol*. 2011;20:145–146.
36. Klechevsky E, et al. Functional specializations of human epidermal Langerhans cells and CD14 + dermal dendritic cells. *Immunity*. 2008;29:497–510.
37. Chu C-C, et al. Resident CD141 (BDCA3) + dendritic cells in human skin produce IL-10 and induce regulatory T cells that suppress skin inflammation. *J Exp Med*. 2012;209:935–945.
38. Banchereau J, et al. Immunoglobulin-like transcript receptors on human dermal CD14 + dendritic cells act as a CD8-antagonist to control cytotoxic T cell priming. *Proc Natl Acad Sci USA*. 2012,109.18885–18890.
39. Sallusto F, Lanzavecchia A. Efficient presentation of soluble antigen by cultured human dendritic cells is maintained by granulocyte/macrophage colony-stimulating factor plus interleukin 4 and downregulated by tumor necrosis factor alpha. *J Exp Med*. 1994;179:1109–1118.
40. Romani N, et al. Proliferating dendritic cell progenitors in human blood. *J Exp Med*. 1994;180:83–93.
41. Massa C, Thomas C, Wang E, Marincola F, Seliger B. Different maturation cocktails provide dendritic cells with different chemoattractive properties. *J Transl Med*. 2015;13.
42. Koski GK, et al. A novel dendritic cell-based immunization approach for the induction of durable Th1-polarized anti-HER-2/neu responses in women with early breast cancer. *J Immunother*. 2012;35:54–65.
43. Kvistborg P, Boegh M, Pedersen AW, Claesson MH, Zocca MB. Fast generation of dendritic cells. *Cell Immunol*. 2009;260:56–62.
44. Chiang CL-L, et al. Day-4 myeloid dendritic cells pulsed with whole tumor lysate are highly immunogenic and elicit potent anti-tumor responses. *PLoS One*. 2011;6:e28732.

45. Van Nuffel AMT, et al. Intravenous and intradermal TriMix-dendritic cell therapy results in a broad T-cell response and durable tumor response in a chemorefractory stage IV-M1c melanoma patient. *Cancer Immunol Immunother CII*. 2012;61:1033–1043.
46. Wilgenhof S, et al. Therapeutic vaccination with an autologous mRNA electroporated dendritic cell vaccine in patients with advanced melanoma. *J Immunother Hagerstown Md*. 2011;1997(34):448–456.
47. Butterfield LH, et al. Determinant spreading associated with clinical response in dendritic cell-based immunotherapy for malignant melanoma. *Clin Cancer Res*. 2003;9:998–1008.
48. Ribas A, Timmerman JM, Butterfield LH, Economou JS. Determinant spreading and tumor responses after peptide-based cancer immunotherapy. *Trends Immunol*. 2003;24:58–61.
49. Ferlazzo G, Klein J, Paliard X, Wei WZ, Galy A. Dendritic cells generated from CD34 + progenitor cells with flt3 ligand, c-kit ligand, GM-CSF, IL-4, and TNF-alpha are functional antigen-presenting cells resembling mature monocyte-derived dendritic cells. *J Immunother*. 2000;23:48–58.
50. Romano E, et al. Peptide-loaded Langerhans cells, despite increased IL15 secretion and T-cell activation in vitro, elicit antitumor T-cell responses comparable to peptide-loaded monocyte-derived dendritic cells in vivo. *Clin Cancer Res Off J Am Assoc Cancer Res*. 2011;17:1984–1997.
51. Banchereau J, et al. The differential production of cytokines by human Langerhans cells and dermal CD14(+) DCs controls CTL priming. *Blood*. 2012;119:5742–5749.
52. Yuan J, et al. Langerhans cells derived from genetically modified human $CD34^+$ hemopoietic progenitors are more potent than peptide-pulsed Langerhans cells for inducing antigen-specific $CD8^+$ cytolytic T lymphocyte responses. *J Immunol*. 2005;174:758–766.
53. Tel J, et al. Natural human plasmacytoid dendritic cells induce antigen-specific T-cell responses in melanoma patients. *Cancer Res*. 2013;73:1063–1075.
54. Schreibelt G, et al. Effective clinical responses in metastatic melanoma patients after vaccination with primary myeloid dendritic cells. *Clin Cancer Res Off J Am Assoc Cancer Res*. 2016;22:2155–2166.
55. Prue RL, et al. A phase I clinical trial of CD1c (BDCA-1) + dendritic cells pulsed with HLA-A*0201 peptides for immunotherapy of metastatic hormone refractory prostate cancer. *J Immunother*. 2015;38:71–76.
56. van de Loosdrecht AA, et al. A novel allogeneic off-the-shelf dendritic cell vaccine for post-remission treatment of elderly patients with acute myeloid leukemia. *Cancer Immunol Immunother*. 2018;67:1505–1518.
57. Aleksic M, et al. Different affinity windows for virus and cancer-specific T-cell receptors: implications for therapeutic strategies. *Eur J Immunol*. 2012;42:3174–3179.
58. Wells DK, et al. Key parameters of tumor epitope immunogenicity revealed through a consortium approach improve neoantigen prediction. *Cell*. 2020;183:818–834.e13.
59. Carreno BM, et al. A dendritic cell vaccine increases the breadth and diversity of melanoma neoantigen-specific T cells. *Science*. 2015;348:803–808.
60. Klebanoff CA, Wolchok JD. Shared cancer neoantigens: Making private matters public. *J Exp Med*. 2018;215:5–7.
61. Butterfield LH, et al. Multiple antigen-engineered DC vaccines with or without IFNα to promote antitumor immunity in melanoma. *J Immunother Cancer*. 2019;7:113.
62. Santos PM, et al. Impact of checkpoint blockade on cancer vaccine-activated $CD8^+$ T cell responses. *J Exp Med*. 2020;217.
63. Rahman M, Dastmalchi F, Karachi A, Mitchell D. The role of CMV in glioblastoma and implications for immunotherapeutic strategies. *Oncoimmunology*. 2018;8.

64. Batich KA, et al. Long-term survival in glioblastoma with cytomegalovirus pp65-targeted vaccination. *Clin Cancer Res Off J Am Assoc Cancer Res*. 2017;23:1898–1909.
65. Batich KA, Mitchell DA, Healy P, Herndon JE, Sampson JH. Once, twice, three times a finding: reproducibility of dendritic cell vaccine trials targeting cytomegalovirus in glioblastoma. *Clin Cancer Res*. 2020;26:5297–5303.
66. Mitchell DA, et al. Tetanus toxoid and CCL3 improve DC vaccines in mice and glioblastoma patients. *Nature*. 2015;519:366–369.
67. Mitchell DA, et al. Tetanus toxoid and CCL3 improve dendritic cell vaccines in mice and glioblastoma patients. *Nature*. 2015;519:366–369.
68. Chiang CL-L, et al. A dendritic cell vaccine pulsed with autologous hypochlorous acid-oxidized ovarian cancer lysate primes effective broad antitumor immunity: from bench to bedside. *Clin Cancer Res Off J Am Assoc Cancer Res*. 2013;19:4801–4815.
69. Hao S, et al. Enhanced antitumor immunity derived from a novel vaccine of fusion hybrid between dendritic and engineered myeloma cells. *Exp Oncol*. 2004;26:300–306.
70. Rosenblatt J, et al. Vaccination with dendritic cell/tumor fusion cells results in cellular and humoral antitumor immune responses in patients with multiple myeloma. *Blood*. 2011;117:393–402.
71. Rosenblatt J, et al. Individualized vaccination of AML patients in remission is associated with induction of antileukemia immunity and prolonged remissions. *Sci Transl Med*. 2016;8. 368ra171.
72. Koido S. Dendritic-tumor fusion cell-based cancer vaccines. *Int J Mol Sci*. 2016;17.
73. Tanyi JL, et al. Personalized cancer vaccine effectively mobilizes antitumor T cell immunity in ovarian cancer. *Sci Transl Med*. 2018;10.
74. Hardin MO, et al. Tumor lysate particle loaded dendritic cell vaccine: preclinical testing of a novel personalized cancer vaccine. *Immunotherapy*. 2018;10:373–382.
75. Boczkowski D, Nair SK, Snyder D, Gilboa E. Dendritic cells pulsed with RNA are potent antigen-presenting cells in vitro and in vivo. *J Exp Med*. 1996;184:465–472.
76. Wadhwa A, Aljabbari A, Lokras A, Foged C, Thakur A. Opportunities and challenges in the delivery of mRNA-based vaccines. *Pharmaceutics*. 2020;12.
77. Van Meirvenne S, et al. Efficient genetic modification of murine dendritic cells by electroporation with mRNA. *Cancer Gene Ther*. 2002;9:787–797.
78. Dewitte H, et al. The potential of antigen and TriMix sonoporation using mRNA-loaded microbubbles for ultrasound-triggered cancer immunotherapy. *J Controlled Release*. 2014;194:28–36.
79. McCullough KC, et al. Self-replicating replicon-RNA delivery to dendritic cells by chitosan-nanoparticles for translation in vitro and in vivo. *Mol Ther Nucleic Acids*. 2014;3: e173.
80. Kang TH, et al. Enhancement of dendritic cell-based vaccine potency by targeting antigen to endosomal/lysosomal compartments. *Immunol Lett*. 2006;106:126–134.
81. Chen AC, et al. HER2-LAMP vaccines effectively traffic to endolysosomal compartments and generate enhanced polyfunctional T cell responses that induce complete tumor regression. *J Immunother Cancer*. 2020;8.
82. Wilgenhof S, et al. Phase II study of autologous monocyte-derived mRNA electroporated dendritic cells (TriMixDC-MEL) plus ipilimumab in patients with pretreated advanced melanoma. *J Clin Oncol Off J Am Soc Clin Oncol*. 2016;34:1330–1338.
83. Hodi FS, et al. Improved survival with ipilimumab in patients with metastatic melanoma. *N Engl J Med*. 2010;363:711–723.

84. Chung DJ, et al. Langerhans-type dendritic cells electroporated with TRP-2 mRNA stimulate cellular immunity against melanoma: results of a phase I vaccine trial. *Oncoimmunology*. 2017;7:e1372081.
85. Polack FP, et al. Safety and efficacy of the BNT162b2 mRNA Covid-19 vaccine. *N Engl J Med*. 2020;383:2603–2615.
86. Okada N, et al. Administration route-dependent vaccine efficiency of murine dendritic cells pulsed with antigens. *Br J Cancer*. 2001;84:1564–1570.
87. Fong L, Brockstedt D, Benike C, Wu L, Engleman EG. Dendritic cells injected via different routes induce immunity in cancer patients. *J Immunol*. 2001;166:4254–4259.
88. Lesterhuis WJ, et al. Route of administration modulates the induction of dendritic cell vaccine-induced antigen-specific T cells in advanced melanoma patients. *Clin Cancer Res Off J Am Assoc Cancer Res*. 2011;17:5725–5735.
89. Castiello L, Aricò E, D'Agostino G, Santodonato L, Belardelli F. In situ vaccination by direct dendritic cell inoculation: the coming of age of an old idea? *Front Immunol*. 2019;10.
90. Kantoff PW, et al. Sipuleucel-T immunotherapy for castration-resistant prostate cancer. *N Engl J Med*. 2010;363:411–422.
91. Bm A, et al. Systemic dysfunction and plasticity of the immune macroenvironment in cancer models. *Nat. Med*. 2020;26. Available from: https://pubmed.ncbi.nlm.nih.gov/32451499/.
92. Fecek RJ, Storkus WJ. Combination strategies to enhance the potency of monocyte-derived dendritic cell-based cancer vaccines. *Immunotherapy*. 2016;8:1205–1218.
93. Lowe DB, et al. Dasatinib promotes the expansion of a therapeutically superior T-cell repertoire in response to dendritic cell vaccination against melanoma. *Oncoimmunology*. 2014;3:e27589.
94. Guo J, Muse E, Christians AJ, Swanson SJ, Davila E. An anticancer drug cocktail of three kinase inhibitors improved response to a dendritic cell-based cancer vaccine. *Cancer Immunol Res*. 2019;7:1523–1534.
95. Adhikaree J, et al. Impaired circulating myeloid CD1c + dendritic cell function in human glioblastoma is restored by p38 inhibition - implications for the next generation of DC vaccines. *Oncoimmunology*. 2019;8:1593803.
96. Bose A, et al. Sunitinib facilitates the activation and recruitment of therapeutic anti-tumor immunity in concert with specific vaccination. *Int J Cancer*. 2011;129:2158–2170.
97. Farsaci B, et al. Immune consequences of decreasing tumor vasculature with antiangiogenic tyrosine kinase inhibitors in combination with therapeutic vaccines. *Cancer Immunol Res*. 2014;2:1090–1102.
98. Dammeijer F, et al. Depletion of tumor-associated macrophages with a CSF-1R kinase inhibitor enhances antitumor immunity and survival induced by DC immunotherapy. *Cancer Immunol Res*. 2017;5:535–546.
99. Hernández JM, et al. Novel kidney cancer immunotherapy based on the granulocyte-macrophage colony-stimulating factor and carbonic anhydrase IX fusion gene. *Clin Cancer Res Off J Am Assoc Cancer Res*. 2003;9:1906–1916.
100. Saxena M, Balan S, Roudko V, Bhardwaj N. Towards superior dendritic-cell vaccines for cancer therapy. *Nat Biomed Eng*. 2018;2:341–346.
101. Hobo W, et al. siRNA silencing of PD-L1 and PD-L2 on dendritic cells augments expansion and function of minor histocompatibility antigen-specific CD8 + T cells. *Blood*. 2010;116:4501–4511.

102. Barrueto L, et al. Resistance to checkpoint inhibition in cancer immunotherapy. *Transl Oncol*. 2020;13:100738.
103. Onstad M, Coleman RL, Westin SN. Movement of poly-ADP ribose (PARP) inhibition into frontline treatment of ovarian cancer. *Drugs*. 2020;80:1525–1535.
104. Choi H-J, et al. Anti-vascular therapies in ovarian cancer: moving beyond anti-VEGF approaches. *Cancer Metastasis Rev*. 2015;34:19–40.
105. Cibula D, et al. Dendritic cell-based immunotherapy (DCVAC/OvCa) with chemotherapy in patients with platinum-sensitive, relapsed, epithelial ovarian carcinoma: survival analysis of a phase II, open-label, randomized, multicenter trial (study SOV02). *Gynecol Oncol*. 2019;154:18.
106. Guo Q, et al. Advanced clinical trials of dendritic cell vaccines in ovarian cancer. *J Investig Med Off Publ Am Fed Clin Res*. 2020;68:1223–1227.
107. Noordam L, et al. Low-dose cyclophosphamide depletes circulating naïve and activated regulatory T cells in malignant pleural mesothelioma patients synergistically treated with dendritic cell-based immunotherapy. *Oncoimmunology*. 2018;7:e1474318.
108. Le HK, et al. Gemcitabine directly inhibits myeloid derived suppressor cells in BALB/c mice bearing 4T1 mammary carcinoma and augments expansion of T cells from tumor-bearing mice. *Int Immunopharmacol*. 2009;9:900–909.
109. Santegoets S J a M, et al. The blood mMDSC to DC ratio is a sensitive and easy to assess independent predictive factor for epithelial ovarian cancer survival. *Oncoimmunology*. 2018;7:e1465166.
110. Wculek SK, Khouili SC, Priego E, Heras-Murillo I, Sancho D. Metabolic control of dendritic cell functions: digesting information. *Front Immunol*. 2019;10:775.
111. Aarts BM, et al. Cryoablation and immunotherapy: an overview of evidence on its synergy. *Insights Imaging*. 2019;10:53.
112. Wang S, et al. DC-CIK as a widely applicable cancer immunotherapy. *Expert Opin Biol Ther*. 2020;20:601–607.
113. Morgan RA, et al. Cancer regression in patients after transfer of genetically engineered lymphocytes. *Science*. 2006;314:126–129.
114. June CH, O'Connor RS, Kawalekar OU, Ghassemi S, Milone MC. CAR T cell immunotherapy for human cancer. *Science*. 2018;359:1361–1365.

Chapter 8

Combinatorial immunotherapy strategies for cancer vaccines

Maria Tagliamonte[1] and Gennaro Ciliberto[2]
[1]Laboratory of Innovative Immunological Models, National Cancer Institute "Fond. Pascale" - IRCCS, Naples, Italy, [2]Scientific Direction, IRCCS Regina Elena National Cancer Institute, Rome, Italy

Introduction

Cancer immunotherapies have significantly changed cancer treatment in the last decade.[1] They can be divided in three major classes, according to their mechanism of action: (1) immune checkpoint inhibitors (ICIs) that unleash the "natural" antitumor immunity; (2) chimeric antigen receptor (CAR)-T cells, consisting in the passive transfer of in vitro engineered antitumor T cells; and (3) cancer vaccines (including oncolytic viruses) aiming to induce antitumor immunity against tumor antigens.

ICIs have revolutionized the clinical management of patients with cancer, inducing a near to complete and definitive cure of selected histotypes of advanced neoplasia.[2] CAR-T cell therapy has shown impressive results mostly in nonsolid tumors.[3] On the contrary, cancer vaccines have not held the promise to date and the only strategies approved by FDA are represented by Provenge (sipuleucel-T), for the treatment of asymptomatic or minimally symptomatic metastatic castration-resistant prostate cancer (mCRPC),[4] and T-VEC (Imlygic), for the treatment of advanced melanoma.[5] However, both of them provide a limited benefit in the progression-free survival and overall survival (OS).

The limited efficacy of cancer vaccines is consequent to a series of "auto-protective" and immune-evasive mechanisms developing in the so-called tumor microenvironment (TME). Tumor cells may be characterized by the downregulation of MHC class I molecules on their surface, affecting the expression of target tumor antigens, and/or upregulation of surface molecules (e.g., PD-L1 or PD-L2) [immune checkpoints (ICs)] that engage receptors on the surfaces of activated T cells (PD-1), causing T-cell anergy or exhaustion.[6,7] Moreover, tumors may trigger the accumulation or expansion

Cancer Vaccines as Immunotherapy of Cancer. DOI: https://doi.org/10.1016/B978-0-12-823901-8.00004-2
© 2022 Elsevier Inc. All rights reserved.

of regulatory T (Treg) cells as well as of myeloid-derived suppressor cells (MDSCs), leading to the accumulation in the TME of several suppressors immune cell types, such as tumor-associated neutrophils and cancer-associated fibroblasts.[8–10] For these reasons, a successful immunotherapy must be able to counteract all these critical points to evoke durable immune effector and cytotoxic T cells to control the tumor growth.[11,12]

In several preclinical and clinical studies, it has been proven that immune escape mechanisms and the suppressive TME can be effectively targeted by combining different strategies based on immunotherapy and conventional anticancer therapies, such as chemotherapy, radiation, and epigenetic modulators.

Here, we focus on combinatorial strategies directed at improving the efficacy of cancer vaccines that were evaluated in preclinical and in clinical studies in recent years.

Immune modulatory approaches to increase tumor immunogenicity and potentiate cancer vaccines

Limitations and pitfalls of therapeutic cancer vaccines

Therapeutic cancer vaccines aim at eliciting a specific de novo immune response against tumor antigens or at amplifying an existing antitumor immune response. This mechanism involves professional antigen-presenting cells (APCs) to trigger T-helper cell responses to support antibody production and/or expansion of cytotoxic effector T cells, as well as to sustain the antitumor effector mechanisms of macrophages and neutrophils. Several vaccine strategies in different tumor settings have been evaluated in the last years, however, with limited results.

The main reason of this failure, among others, is the immunosuppressive TME leading to a limited infiltration of cancer-specific cytotoxic T lymphocytes and/or their functional anergy/exhaustion. Such a so-called "cold" non-immunogenic tumor needs to be converted into a "hot" immunogenic tumor, to improve the efficacy of cancer vaccines. Indeed, combination strategies based on cancer vaccines in association with other immunotherapies, such as ICIs, standard or metronomic chemotherapy (MCT), and radiotherapy (RT) and epigenetic therapy, have been shown to be more effective than cancer vaccines as monotherapy[13,14] (Fig. 8.1).

Standard and metronomic chemotherapy and combination with cancer vaccine

Chemotherapy has been shown to enhance antitumor immunity, both at standard and at low-dose metronomic schedule. In particular, chemotherapy can induce several effects: (1) modulate the immunogenicity of tumor cells by inducing immunogenic cell death (ICD) in cancer cells with the release of

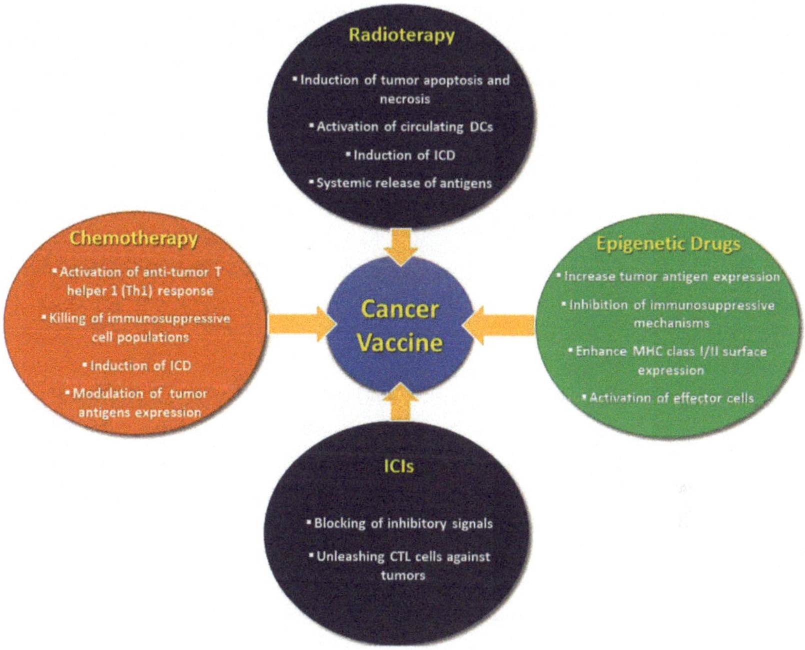

FIGURE 8.1 Schematic representation of different combinatorial strategies to potentiate cancer vaccines. Positive effects of tumor microenvironment on each standard therapies or immune checkpoint inhibitors are described.

danger signals able to polarize dendritic cells (DCs) and activate an antitumor T helper 1 (Th1) responses; (2) increase the tumor mutational burden and the tumor-associated neoantigen load; (3) upregulate costimulatory molecules (B7−1) and downregulate coinhibitory molecules (B7-H1/PDL1) expressed on the tumor cell surface, increasing effector T-cell function; and (4) inhibit the immunosuppressive cell populations (e.g., MDSCs and Tregs), thus modifying the TME and improving the efficacy of immunotherapy treatments.[15−18]

Preclinical and clinical studies have shown that a low dose of cyclophosphamide (CTX) or doxorubicin or paclitaxel can enhance the antitumor immune responses of immunotherapies, including cancer vaccines, by inhibiting Tregs and increasing interferon gamma (IFN-γ) secreting $CD8^+$/ $CD44^+$ effector memory T cells. Moreover, anthracycline-based neoadjuvant chemotherapy induces ICD and an increased effector/Tregs ratio in the tumor immune infiltrate, leading to favorable outcomes in multiple solid cancers.[19−22] Similarly, gemcitabine has been shown to kill either myeloid-derived suppressors in vitro and in mouse model or Tregs in humans[23] cells (MDSCs) and docetaxel has been reported to modulate different cell subsets, enhancing $CD8^+$ T cells function and deleting Tregs.[24,25]

Other studies have demonstrated that, much more than standard chemotherapy, low doses of chemotherapy drugs administered at short intervals (MCT) induce positive immunomodulatory effects in the TME. In particular, low-dose CTX can induce an ICD, which generates the release from tumor cells of danger signals polarizing DCs toward a proinflammatory phenotype, which drives a Th1 response reviewed in,[26] as well as an inhibition of immunosuppressive Treg cell population.[27,28]

The use of cisplatin chemotherapy has been shown to increase tumor immune infiltration[29] and tumor cell killing by tumor necrosis factor alpha (TNFα)-producing T cells.[30]

In the last years our group has conducted several studies in a preclinical model to show immunomodulatory effects of MCT on TME. For this purpose, we have combined taxanes (docetaxel and paclitaxel) and alkylating (CTX) agents, to hit different targets of the immune suppressive environment. In particular, we demonstrated that the MCT approach was able to significantly delay tumor growth and prolong animal survival but was not sufficient to provide full protection. Such antitumor biological effects directly correlated with the induction of immunological cell death of cancer cells, the enhancement of effector T-cell response, and the reduction of the immune suppressive Treg cell population.[28] Overall, these studies demonstrated that chemotherapeutic regimens, in standard and metronomic fashion, can induce a double effect, killing cancer cells while eliciting complementary immune-stimulatory effects.[31] Ultimately, immunological "cold" tumors are converted into "hot" tumors with a strong intratumor invasion of $CD8^+$ and $CD4^+$ Th1 cells as well as a durable control of established lesions.[32] All these immunological effects make chemotherapy the optimal complementary strategy for cancer vaccines to improve their efficacy. In this regard, several preclinical studies have provided encouraging results supporting the design of clinical trials.[33–37]

Unfortunately, transferring these concepts to the clinic resulted in inconsistent results as reported below.

The clinical effects of ISA101 vaccination were assessed during chemotherapy in patients with advanced, recurrent, or metastatic cervical cancer. Tumor regressions were observed in 43% of evaluable patients and the depletion of myeloid suppressive cells by carboplatin/paclitaxel was associated with the detection of low frequency of spontaneous HPV16-specific immunity in 34% of tested patients.[38]

A modified vaccinia Ankara expressing the oncofetal antigen 5T4 (MVA-5T4) was evaluated in patients with metastatic colorectal cancer in combination with CTX. Transient depletion of regulatory T cells by CTX did not increase immune responses generated by MVA-5T4 vaccination. However, CTX and MVA-5T4 each independently induced beneficial antitumor immune responses, resulting in prolonged survival without toxic effects.[39]

A phase 2 clinical study was conducted on 68 patients with advanced renal cell carcinoma assessing the efficacy of a multipeptide vaccine,

IMA901 + GMCSF adjuvant, with or without a low dose of CTX 3 days before vaccination. Patients who received CTX showed longer OS as compared to the ones who received IMA901 only.[40]

Moreover, in some trials, these drugs failed to efficiently deplete the immunosuppressive cells.[41–43] In particular, low-dose CTX, when associated with a p53-synthetic long peptide vaccine in patients with recurrent ovarian cancer, neither induced a quantitative reduction of Tregs nor a demonstrable qualitative difference in Treg function tested in vitro.[41] Similarly, six cycles of CTX combined with a vaccine based on DCs transfected with mRNA in patients with metastatic malignant melanoma (MM) did not have any measurable effect on the percentage of Tregs.[42] Low-dose CTX associated with tumor peptides versus observation induced only a transient modulation in the frequencies of Tregs in PBMCs collected after low-dose CTX administration[43] (Table 8.1).

Overall, the results of clinical trials assessing the efficacy of chemotherapy and vaccine combinatorial strategies do not hold the promise of preclinical data. Among the possible reasons for such failure are: (1) the selection of chemotherapy drugs; (2) the choice of chemotherapy dose and schedule; and (3) the selection of patients' cohort. Therefore additional work is required to fully decode the pharmacological and immunological interactions between peptide-based anticancer vaccines and chemotherapy.

Radiotherapy and cancer vaccine

Another option to improve cancer vaccines' efficacy is to design combination protocols including RT. Indeed, the latter has been shown to stimulate tumor-reactive T cells by inducing the release of antigens from tumor cells as well as proinflammatory signals that trigger the innate immune system.[55] Ultimately, irradiated tumors might be considered as an "in situ vaccine" releasing tumor antigens and inducing an ICD with production of cytokines. Overall, these effects are able to recruit and activate tumor-specific T cells as well as promote DCs maturation into effective APCs.[56] Interestingly, localized irradiation has been shown to generate a so-called "abscopal effect," consisting in the eradication of distant metastasis though the activation of systemic antitumor immunity.[57–60] Moreover, Gameiro et al. demonstrated in vitro and in vivo that lower doses of RT can induce significant changes in TME, reshaping tumors to be more susceptible to T-cell killing through increasing type I IFN secretion and inducing ICD, as reflected by calreticulin expression.[61]

Consequently, combinations of RT and immunotherapies, including cancer vaccines, have been evaluated in the treatment of local and metastatic cancers. Preclinical studies have demonstrated that the combination of vaccines and RT is synergistic, leading to the destruction of tumor cells via upregulation of MHC, Fas, ICAM-1, and TAAs, as well as by enhancing

TABLE 8.1 Clinical trials completed combining cancer vaccines and standard therapies or ICIs.

Combination therapy	Drug used	Cancer vaccine	Phase	Tumor type	Reference
Chemotherapy	Full dose of carboplatin with paclitaxel (CarboTaxol)	HPV16-SLP vaccine (ISA101)	Phase 1/2	HPV16-positive advanced cervical cancer	[37]
	Cyclophosphamide	Oncofetal antigen 5T4 (MVA-5T4)	Phase 1/2	Metastatic colorecteal cancer	[38]
	Low dose of cyclophosphamide	Multipeptide (IMA901)	Phase 2	Advanced renal cell carcinoma	[39]
	Low dose of cyclophosphamide	p53-SLP	Phase 1/2	Epithelial ovarian cancer	[40]
Radiotherapy	Radiotherapy	NY-ESO-1, MAGEC1, MAGEC2, 5T4, and MUC1 (CV9202)	Phase 1/2	Stage IV NSCLC	[44]
	External beam radiation therapy	rV/F-PSA/B7.1	Phase 2	Localized prostate cancer	[45]
	153-Samarium (Quadramet)	PROSTVAC	Phase 2	mCRPC	NCT00450619[46]
Epigenetic therapy	5-AZA-CdR	NY-ESO-1 vaccine	Phase 1	Epithelial ovarian cancer	[47]
	Decitabine (DNMT inhibitor)	NYESO-1 vaccine (CDX-1401)	Phase 1	Myelodysplastic syndrome	[48]
	Decitabine (DNMT inhibitor)	DC vaccine pulsed with MAGE-A1, MAGE-A3, and NY-ESO-1 peptides	Phase 1	Childhood neuroblastoma and sarcoma	[49]

ICIs	Ipilimumab	T-VEC	Phase Ib	Unresectable stage IIIB/ IV	50
	Ipilimumab	Sipuleucel-T	Phase 1	mCRPC	51
	Ipilimumab	GVAX	Phase 1	Pancreatic cancer	52
	Pembrolizumab	GVAX	Phase 1	Advanced MMRp CRC	53
	Nivolumab	Multipeptide vaccines	Phase 1/2	Ipilimumab-refractory and -naive melanoma	54

5-AZA, 5-azacytidine; *mCRPC*, metastatic castration-resistant prostate cancer; *NSCLC*, nonsmall-cell lung cancer; *PSA*, prostate-specific antigen; *SLP*, synthetic long peptide; *ICIs*, immune checkpoint inhibitors.

vaccine-mediated tumor lysis in mouse models.[62,63] Several studies have shown augmented antitumor immune responses when radiation is combined with tumor-specific vaccines[64,65] and this effect can be achieved by administering the cancer vaccine either after or before RT.[44,64]

In very recent years these highly promising preclinical data have inspired a handful number of clinical trials evaluating the efficacy of combinatorial therapies based on RT and cancer vaccines, again with contrasting results.

An mRNA cancer vaccine (RNActive), called CV9202, targeting NY-ESO-1, MAGEC1, MAGEC2, 5 T4, survivin, and MUC1, has been tested in combination with RT in patients with stage IV nonsmall-cell lung cancer.[45] The vaccine and RT were coadministered, and then, the vaccine was continued until disease progression. The majority of patients showed antigen-specific immune responses and these were targeting multiple vaccine antigens in about half of the patients. Immune responses against tumor antigens not covered by the RNA vaccine (antigen spread) were observed in a subgroup of patients, which could be an effect of the radiation inducing an ICD of cancer cells.

A phase 2 clinical trial assessed the use of a recombinant poxviral-based vaccine expressing prostate-specific antigen (PSA) combined with standard definitive RT in patients with localized prostate cancer.[66] Results from this clinical trial indicated that the combination was safe, well tolerated, and, more importantly, effective at generating a PSA-specific immune response. Approximately 76.5% of patients in the combination therapy arm showed a threefold or greater increase in PSA-specific T cells versus 0% in the radiation-alone arm ($P < 0.0005$).

PSA-TRICOM (PROSTVAC, rilimogene galvacirepvec/rilimogene glafolivec) is an active immunotherapeutic cancer vaccine designed to induce the activation of T cells specific against PSA.[46] In a randomized placebo-controlled phase 2 study, PSA-TRICOM vaccine was evaluated in patients with metastatic mCRPC in combination with Samarium-153-ethylene diamine tetramethylene phosphonate (Sm-153-EDTMP; Quadramet), a radiopharmaceutical that binds osteoblastic bone lesions and emits beta particles causing local tumor cell destruction. Decline in PSA was achieved in the combination arm but there was no statistical difference in radiographic disease progression at 4 months compared to Sm-153-EDTMP alone[67] (Table 8.1).

The limited number of clinical studies based on the combination of RT and cancer vaccines shows promising results. However, the current combination strategies are not yet fully optimized, with many unknowns, including the appropriate sequence of vaccine-radiation administration. Furthermore, the observed differences in efficacy between preclinical and clinical studies are likely due to the peculiar tumor characteristics and immune microenvironment of a given patient. Hence, a deeper understanding of the inter- and intrapatient heterogeneities of the TME is required. This may enable the logical design of personalized approaches to complex combined-modality treatment.[68]

Epigenetic drugs and cancer vaccine

Cancer development and progression is regulated by the accumulation of genetic alterations and mutations as well as epigenetic modifications. In particular, epigenetic modifications induce an overall silencing of gene expression, impairing immunogenicity and immune recognition of cancer cells. Tumors frequently show an alteration in DNA methylation,[69] such as hypermethylation, mediated by DNA methyltransferase (DNMTs) enzymes, and histone deacetylation, mediated by histone deacetylase (HDACs) enzymes. Both mechanisms silence and inactivate tumor suppressor genes promoting cancer progression.[70]

Because of the relevant role of epigenetic alterations in tumorigenesis, specific anticancer therapies have been developed and tested in preclinical and clinical settings. Consequently, the Food and Drug Administration approved the DNA methyltransferase inhibitors [5-azacytidine (5-AZA)] and the histone deacetylase inhibitor (MS-275) in the treatments of several solid and hematological malignancies.[71,72]

It has been shown that epigenetic drugs have not only a direct antitumor effect by reducing cell proliferation and viability and inducing cell-cycle arrest and apoptosis, but they can simultaneously modify antitumor immunity, playing an important role at enhancing the response to immunotherapy. Indeed, they increase tumor antigen expression and enhance MHC class I/II surface expression for their presentation and recognition by T cells.[73–78] Moreover, they may directly act on the immune system, activating effector cells and inhibiting immunosuppressive mechanisms.[79–81] In particular, they are able to inhibit tumor-associated de novo methylation, which plays an important role in the promotion of $CD8^+$ T-cell exhaustion occurring during checkpoint inhibitor therapy.[82] Consequently, epigenetic treatments may potentiate immunotherapy strategies and several preclinical/clinical studies have reported such positive effects.[47,82,83]

The immunomodulatory and the tumor immune activation abilities of epigenetic drugs point to the fact that they can work synergistically not only with ICIs immunotherapy but also with therapeutic cancer vaccines. In particular, the ability of these drugs to enhance gene expression can lead to an increased presentation of new antigens on tumor cells inducing a strong and effective cytotoxic T-cell response. This effect may enable the discovery and consequently build new cancer vaccine platforms. However, few studies have investigated the combination of epigenetic therapies with vaccination strategies.

Odunsi et al. performed a phase I dose escalation of 5-AZA-CdR, in addition to NY-ESO-1 vaccine and doxorubicin liposome chemotherapy, in patients with relapsed epithelial ovarian cancer. Increased NY-ESO-1 serum antibodies and T-cell responses were observed in most patients, encouraging further evaluation in other tumor types.[48] A phase I study in myelodysplastic syndrome tested the combination of NY-ESO-1 vaccination (CDX-1401) and decitabine, resulting in the induction of NY-ESO-1 expression by decitabine- and NYESO-1-specific $CD4^+$ and $CD8^+$ T-cell responses, which was associated with increased frequency of $CD141^+$ DC in peripheral blood.[49] In childhood neuroblastoma and sarcoma, decitabine has been tested together with a DC vaccine pulsed with MAGE-A1, MAGE-A3, and NY-ESO-1 peptides. The majority of patients developed a response to vaccine antigens and an objective clinical response was documented in 1/10 patients, showing a complete response[84] (Table 8.1).

In the reported clinical studies, clinical responses were heterogeneous and short-term; however, the combination of vaccines with epigenetic drugs was effective in inducing a vaccine-specific T-cell response. Several challenges remain, such as designing selective epigenetic drugs with optimal pharmacokinetics, target inhibition, safety, biodistribution, and efficacy. Moreover, a full understanding of their activity in the TME is still missing. In addition, the identification of biomarkers to both select patients eligible for the therapy and monitor therapy response is required.

Immune checkpoint inhibitors and cancer vaccines

The description of the biological functions of PD-1 and CTLA-4 molecules expressed on the surface of T cells, able to inhibit their functional immune effector activity (so-called IC),[85,86] represented the grounding discovery for the development of monoclonal antibodies, which have revolutionized the anticancer treatment. Cancer cells and DCs may express molecules (PD-L1 and B27, respectively), which bind these ICs and activate intracellular inhibitory signal cascades in T cells, resulting in the loss of anticancer immune response.[87] In 2011 the anti-CTLA-4 mAb ipilimumab was the first FDA-approved checkpoint inhibitor for unresectable or metastatic melanoma, followed by the approval of anti-PD1 Nivolumab in 2014 for the same indication. Since then several other ICIs have been approved by FDA for the treatment of an increasing number of indications (https://www.cancerresearch.org/scientists/immuno-oncology-landscape/fda-approval-timeline-of-active-immunotherapies). Although ICIs have completely reshaped the landscape of cancer treatment, resulting in the tumor growth control and prolongation of both tumor-free survival and OS of patients with cancer, approximately two-third of patients do not benefit from treatment with single ICI. Therefore full clinical potential of ICIs is being evaluated in combination with other drugs.[88–90]

In this context, unleashing of antitumor T cells by ICIs represents the optimal strategy to complement anticancer vaccines. Indeed, T cells induced by the cancer vaccines would find a favorable TME generated by the ICIs that may allow them to exert their antitumor effect.

Comparative studies in preclinical settings have shown superior efficacy of vaccines in combination with a checkpoint inhibitor in controlling tumor

growth.[33,91–94] Several clinical trials have assessed cancer vaccines combined with a checkpoint inhibitor, such as anti-CTLA-4, PD-1, or PD-L1[50]; the list of those currently recruiting patients is available at https://clinicaltrials.gov/ct2/results?term = cancer + vaccines + %
26 + immune + checkpoint + inhibitors&
Search = Apply&recrs = a&age_v = &gndr = &type = &rslt.

The two cancer vaccines approved by FDA; talimogene laherparepvec (T-VEC), for metastatic melanoma, and sipuleucel-T, for metastatic castrate resistant (hormone refractory) prostate cancer, have been evaluated in clinical trials in combination with ICIs.

T-VEC is an oncolytic virus that has been evaluated in phase Ib clinical trials in combination with either ipilimumab or pembrolizumab in patients with unresectable stage IIIB to IV melanoma.[51] In both cases, the combination-induced enhanced antitumor activity and the objective response rate (ORR) were significantly higher in the combination therapies than in the ICIs alone. However, the subsequent phase III KEYNOTE-034 trial was stopped for futility after an interim analysis by the Data Monitoring Committee (NCT02263508).

Sipuleucel-T was evaluated in a phase I clinical trial in combination with dose-escalation ipilimumab in patients with docetaxel-naive progressive mCRPC. Tumor-specific antibodies directed at PAP and PA2024 were induced by the combination above the level achieved with SIP-T alone.[95] A prolonged follow-up of patients included in the trial showed a significantly improved median survival in 6/9 patients with a sustained low median PSA level.[96]

GVAX is an engineered cellular vaccine derived from allogeneic cancer cells transfected with GM-CSF, which has been shown to induce durable and specific antitumor immune responses.[52] It has been evaluated in combination with ICIs in two clinical trials with contrasting and limited clinical outcomes. The first study showed an improved prolonged disease stabilization as well as a trend to longer OS in few patients with pancreatic cancer treated with the GVAX combined with ipilimumab.[53] The second study was conducted in patients with advanced MMRp CRC combining GVAX with the PD1 inhibitor pembrolizumab. No improvement in ORR was observed nor in the secondary objectives.[54]

Finally, the therapeutic efficacy of nivolumab with or without multipeptide vaccines was assessed in ipilimumab-refractory and -naive melanoma. No increased responses in patients' PBMC to multipeptide were observed, showing that peptide vaccines failed to improve the clinical efficacy of nivolumab[97] (Table 8.1).

Also, this kind of combination shows discordant results between preclinical and clinical studies. Indeed, the latter show heterogeneous efficacy data. Tumor characteristics and immune microenvironment (cold vs. hot immunological TME) of a given patient may play a key role in such contrasting results. Additional studies are required to identify the most appropriate clinical settings for optimal efficacy.

Conclusions and perspectives

Although therapeutic cancer vaccines have failed in the past, the era of combinatorial strategies in the treatment of cancer prompts their reconsideration. Strategies have been optimized and immunological enhancement due to vaccines is now accepted. The overwhelming immunosuppressive TME that reduces the clinical efficacy of vaccines can now be modified by different approaches. Combinations of cancer vaccines and standard therapies or ICIs have emerged. However, to date, the highly encouraging results observed in preclinical models have not translated into promising clinical trial data. Possible reasons for such failures could be the different immunogenicity of vaccine antigens in mice and humans. Indeed, the TAAs are overexpressed self-antigens in humans but are mostly nonself antigens in mice and can elicit a much stronger immune response in the preclinical model. Furthermore, the efficacy in clinical trials may be hampered by the enrollment of patients in a late clinical stage, when the simultaneous activation of a plethora of immunosuppressive mechanisms characterizes the TME. Furthermore, given the heterogeneity of the TME, the etiology of the tumor, and the environmental conditions, it is expected that patients, even with the same type of cancer, show highly variable responses to treatments. The provided examples of therapy resistance indicate that the mechanisms underlying primary and secondary immune evasion can be versatile. Importantly, the complex system of immune regulation in the TME instinctively predicts that in some cases the low rate of efficacy results from the high tumor heterogeneity and the interaction of multiple genetic and immune factors in cancer. In this context, a more in-depth tumor characterization, using high-throughput technologies (i.e., digital spatial technology, single cell sequencing, assessment of the tumor mutational burden) could be critical for understanding the tumor-immune contexture and identifying biomarkers to stratify patients and predict response to immunotherapies. Finally, a window of opportunity should be sought designing clinical trials to hit cancer at a much earlier stage, or in an adjuvant phase when disease burden is low and immune-evasive mechanisms are more reversible.

References

1. Mellman I, Coukos G, Dranoff G. Cancer immunotherapy comes of age. *Nature*. 2011;480:480–489.
2. Kroemer G, Zitvogel L. Immune checkpoint inhibitors. *J Exp Med*. 2021;218.
3. Sterner RC, Sterner RM. CAR-T cell therapy: current limitations and potential strategies. *Blood Cancer J*. 2021;11:69.
4. Kantoff PW, Higano CS, Shore ND, Berger ER, Small EJ, Penson DF, et al. Sipuleucel-T immunotherapy for castration-resistant prostate cancer. *N Engl J Med*. 2010;363:411–422.
5. Andtbacka RH, Kaufman HL, Collichio F, Amatruda T, Senzer N, Chesney J, et al. Talimogene laherparepvec improves durable response rate in patients with advanced melanoma. *J Clin Oncol*. 2015;33:2780–2788.

6. Kooi S, Zhang HZ, Patenia R, Edwards CL, Platsoucas CD, Freedman RS. HLA class I expression on human ovarian carcinoma cells correlates with T-cell infiltration in vivo and T-cell expansion in vitro in low concentrations of recombinant interleukin-2. *Cell Immunol.* 1996;174:116–128.
7. Hamanishi J, Mandai M, Iwasaki M, Okazaki T, Tanaka Y, Yamaguchi K, et al. Programmed cell death 1 ligand 1 and tumor-infiltrating CD8 + T lymphocytes are prognostic factors of human ovarian cancer. *Proc Natl Acad Sci U S A.* 2007;104:3360–3365.
8. Marigo I, Dolcetti L, Serafini P, Zanovello P, Bronte V. Tumor-induced tolerance and immune suppression by myeloid derived suppressor cells. *Immunol Rev.* 2008;222:162–179.
9. An Y, Liu F, Chen Y, Yang Q. Crosstalk between cancer-associated fibroblasts and immune cells in cancer. *J Cell Mol Med.* 2020;24:13–24.
10. Jaillon S, Ponzetta A, Di Mitri D, Santoni A, Bonecchi R, Mantovani A. Neutrophil diversity and plasticity in tumour progression and therapy. *Nat Rev Cancer.* 2020;20:485–503.
11. Galon J, Angell HK, Bedognetti D, Marincola FM. The continuum of cancer immunosurveillance: prognostic, predictive, and mechanistic signatures. *Immunity.* 2013;39:11–26.
12. DeNardo DG, Andreu P, Coussens LM. Interactions between lymphocytes and myeloid cells regulate pro- vs anti-tumor immunity. *Cancer Metastasis Rev.* 2010;29:309–316.
13. van der Burg SH, Arens R, Ossendorp F, van HT, Melief CJ. Vaccines for established cancer: overcoming the challenges posed by immune evasion. *Nat Rev Cancer.* 2016;16:219–233.
14. Saxena M, van der Burg SH, Melief CJM, Bhardwaj N. Therapeutic cancer vaccines. *Nat Rev Cancer.* 2021;21:360–378.
15. Galluzzi L, BuquÃ© A, Kepp O, Zitvogel L, Kroemer G. Immunological effects of conventional chemotherapy and targeted anticancer agents. *Cancer Cell.* 2015;28:690–714.
16. Ghiringhelli F, Menard C, Puig PE, Ladoire S, Roux S, Martin F, et al. Metronomic cyclophosphamide regimen selectively depletes $CD4^{+}CD25^{+}$ regulatory T cells and restores T and NK effector functions in end stage cancer patients. *Cancer Immunol Immunother.* 2007;56:641–648.
17. Ghiringhelli F, Apetoh L, Tesniere A, Aymeric L, Ma Y, Ortiz C, et al. Activation of the NLRP3 inflammasome in dendritic cells induces IL-1beta-dependent adaptive immunity against tumors. *Nat Med.* 2009;15:1170–1178.
18. Apetoh L, Ghiringhelli F, Tesniere A, Obeid M, Ortiz C, Criollo A, et al. Toll-like receptor 4-dependent contribution of the immune system to anticancer chemotherapy and radiotherapy. *Nat Med.* 2007;13:1050–1059.
19. Fridman WH, PagÃ̈s F, SautÃ̈s-Fridman C, Galon J. The immune contexture in human tumours: impact on clinical outcome. *Nat Rev Cancer.* 2012;12:298–306.
20. Shang B, Liu Y, Jiang SJ, Liu Y. Prognostic value of tumor-infiltrating FoxP3 + regulatory T cells in cancers: a systematic review and *meta*-analysis. *Sci Rep.* 2015;5:15179.
21. Ladoire S, Mignot G, Dabakuyo S, Arnould L, Apetoh L, Rébé C, et al. In situ immune response after neoadjuvant chemotherapy for breast cancer predicts survival. *J Pathol.* 2011;224:389–400.
22. Kroemer G, Galluzzi L, Kepp O, Zitvogel L. Immunogenic cell death in cancer therapy. *Annu Rev Immunol.* 2013;31:51–72.
23. Santegoets SJAM, de Groot AF, Dijkgraaf EM, Carnaz Simões AM, van der Noord VE, van Ham JJ, et al. The blood mMDSC to DC ratio is a sensitive and easy to assess independent predictive factor for epithelial ovarian cancer survival. *Oncoimmunology.* 2018;7:e1465166.

24. Suzuki E, Kapoor V, Jassar AS, Kaiser LR, Albelda SM. Gemcitabine selectively eliminates splenic Gr-1 + /CD11b + myeloid suppressor cells in tumor-bearing animals and enhances antitumor immune activity. *Clin Cancer Res*. 2005;11:6713–6721.
25. Garnett CT, Schlom J, Hodge JW. Combination of docetaxel and recombinant vaccine enhances T-cell responses and antitumor activity: effects of docetaxel on immune enhancement. *Clin Cancer Res*. 2008;14:3536–3544.
26. Zitvogel L, Apetoh L, Ghiringhelli F, AndrÃ© F, Tesniere A, Kroemer G. The anticancer immune response: indispensable for therapeutic success? *J Clin Invest*. 2008;118:1991–2001.
27. Hermans IF, Chong TW, Palmowski MJ, Harris AL, Cerundolo V. Synergistic effect of metronomic dosing of cyclophosphamide combined with specific antitumor immunotherapy in a murine melanoma model. *Cancer Res*. 2003;63:8408–8413.
28. Tagliamonte M, Petrizzo A, Napolitano M, Luciano A, Rea D, Barbieri A, et al. A novel multi-drug metronomic chemotherapy significantly delays tumor growth in mice. *J Transl Med*. 2016;14:58.
29. Beyranvand NE, van der Sluis TC, van DS, Yagita H, Janssen GM, van Veelen PA, et al. Tumor eradication by cisplatin is sustained by CD80/86-mediated costimulation of CD8 + T cells. *Cancer Res*. 2016;76:6017–6029.
30. van der Sluis TC, van DS, Huppelschoten S, Jordanova ES, Beyranvand NE, Sloots A, et al. Vaccine-induced tumor necrosis factor-producing T cells synergize with cisplatin to promote tumor cell death. *Clin Cancer Res*. 2015;21:781–794.
31. Galluzzi L, Bravo-San Pedro JM, Vitale I, Aaronson SA, Abrams JM, Adam D, et al. Essential vs accessory aspects of cell death: recommendations of the NCCD 2015. *Cell Death Differ*. 2015;22:58–73.
32. Hanoteau A, Newton JM, Krupar R, Huang C, Liu HC, Gaspero A, et al. Tumor microenvironment modulation enhances immunologic benefit of chemoradiotherapy. *J Immunother Cancer*. 2019;7:10.
33. Petrizzo A, Mauriello A, Luciano A, Rea D, Barbieri A, Arra C, et al. Inhibition of tumor growth by cancer vaccine combined with metronomic chemotherapy and anti-PD-1 in a pre-clinical setting. *Oncotarget*. 2018;9:3576–3589.
34. Tagliamonte M, Petrizzo A, Napolitano M, Luciano A, Arra C, Maiolino P, et al. Novel metronomic chemotherapy and cancer vaccine combinatorial strategy for hepatocellular carcinoma in a mouse model. *Cancer Immunol Immunother*. 2015;64:1305–1314.
35. Denies S, Cicchelero L, Van A I, Sanders NN. Combination of interleukin-12 gene therapy, metronomic cyclophosphamide and DNA cancer vaccination directs all arms of the immune system towards tumor eradication. *J Control Release*. 2014;187:175–182.
36. Weir GM, Hrytsenko O, Stanford MM, Berinstein NL, Karkada M, Liwski RS, et al. Metronomic cyclophosphamide enhances HPV16E7 peptide vaccine induced antigen-specific and cytotoxic T-cell mediated antitumor immune response. *Oncoimmunology*. 2014;3:e953407.
37. Welters MJ, van der Sluis TC, van MH, Loof NM, van H, V, van DS, et al. Vaccination during myeloid cell depletion by cancer chemotherapy fosters robust T cell responses. *Sci Transl Med*. 2016;8. 334ra52.
38. Melief CJM, Welters MJP, Vergote I, Kroep JR, Kenter GG, Ottevanger PB, et al. Strong vaccine responses during chemotherapy are associated with prolonged cancer survival. *Sci Transl Med*. 2020;12.
39. Scurr M, Pembroke T, Bloom A, Roberts D, Thomson A, Smart K, et al. Effect of modified vaccinia ankara-5T4 and low-dose cyclophosphamide on antitumor immunity in metastatic colorectal cancer: a randomized clinical trial. *JAMA Oncol*. 2017;3:e172579.

40. Walter S, Weinschenk T, Stenzl A, Zdrojowy R, Pluzanska A, Szczylik C, et al. Multipeptide immune response to cancer vaccine IMA901 after single-dose cyclophosphamide associates with longer patient survival. *Nat Med.* 2012;18:1254–1261.
41. Vermeij R, Leffers N, Hoogeboom BN, Hamming IL, Wolf R, Reyners AK, et al. Potentiation of a p53-SLP vaccine by cyclophosphamide in ovarian cancer: a single-arm phase II study. *Int J Cancer.* 2012;131:E670–E680.
42. Borch TH, Engell-Noerregaard L, Zeeberg IT, Ellebaek E, Met Ã, Hansen M, et al. mRNA-transfected dendritic cell vaccine in combination with metronomic cyclophosphamide as treatment for patients with advanced malignant melanoma. *Oncoimmunology.* 2016;5:e1207842.
43. Camisaschi C, Filipazzi P, Tazzari M, Casati C, Beretta V, Pilla L, et al. Effects of cyclophosphamide and IL-2 on regulatory CD4 + T cell frequency and function in melanoma patients vaccinated with HLA-class I peptides: impact on the antigen-specific T cell response. *Cancer Immunol Immunother.* 2013;62:897–908.
44. Mondini M, Nizard M, Tran T, Mauge L, Loi M, ClÃ©menson C, et al. Synergy of radiotherapy and a cancer vaccine for the treatment of HPV-associated head and neck cancer. *Mol Cancer Ther.* 2015;14:1336–1345.
45. Papachristofilou A, Hipp MM, Klinkhardt U, FrÃ¼h M, Sebastian M, Weiss C, et al. Phase Ib evaluation of a self-adjuvanted protamine formulated mRNA-based active cancer immunotherapy, BI1361849 (CV9202), combined with local radiation treatment in patients with stage IV non-small cell lung cancer. *J Immunother Cancer.* 2019;7:38.
46. Madan RA, Arlen PM, Mohebtash M, Hodge JW, Gulley JL. Prostvac-VF: a vector-based vaccine targeting PSA in prostate cancer. *Expert Opin Investig Drugs.* 2009;18:1001–1011.
47. Juergens RA, Wrangle J, Vendetti FP, Murphy SC, Zhao M, Coleman B, et al. Combination epigenetic therapy has efficacy in patients with refractory advanced non-small cell lung cancer. *Cancer Discov.* 2011;1:598–607.
48. Odunsi K, Matsuzaki J, James SR, Mhawech-Fauceglia P, Tsuji T, Miller A, et al. Epigenetic potentiation of NY-ESO-1 vaccine therapy in human ovarian cancer. *Cancer Immunol Res.* 2014;2:37–49.
49. Griffiths EA, Srivastava P, Matsuzaki J, Brumberger Z, Wang ES, Kocent J, et al. NY-ESO-1 vaccination in combination with decitabine induces antigen-specific T-lymphocyte responses in patients with myelodysplastic syndrome. *Clin Cancer Res.* 2018;24:1019–1029.
50. Collins JM, Redman JM, Gulley JL. Combining vaccines and immune checkpoint inhibitors to prime, expand, and facilitate effective tumor immunotherapy. *Expert Rev Vaccines.* 2018;17:697–705.
51. Puzanov I, Milhem MM, Minor D, Hamid O, Li A, Chen L, et al. Talimogene laherparepvec in combination with ipilimumab in previously untreated, unresectable stage IIIB-IV melanoma. *J Clin Oncol.* 2016;34:2619–2626.
52. Lutz E, Yeo CJ, Lillemoe KD, Biedrzycki B, Kobrin B, Herman J, et al. A lethally irradiated allogeneic granulocyte-macrophage colony stimulating factor-secreting tumor vaccine for pancreatic adenocarcinoma. A phase II trial of safety, efficacy, and immune activation. *Ann Surg.* 2011;253:328–335.
53. Le DT, Lutz E, Uram JN, Sugar EA, Onners B, Solt S, et al. Evaluation of ipilimumab in combination with allogeneic pancreatic tumor cells transfected with a GM-CSF gene in previously treated pancreatic cancer. *J Immunother.* 2013;36:382–389.
54. Yarchoan M, Huang CY, Zhu Q, Ferguson AK, Durham JN, Anders RA, et al. A phase 2 study of GVAX colon vaccine with cyclophosphamide and pembrolizumab in patients with mismatch repair proficient advanced colorectal cancer. *Cancer Med.* 2020;9:1485–1494.

55. Demaria S, Golden EB, Formenti SC. Role of local radiation therapy in cancer immunotherapy. *JAMA Oncol.* 2015;1:1325–1332.
56. Kulzer L, Rubner Y, Deloch L, AllgÃ¤uer A, Frey B, Fietkau R, et al. Norm- and hypofractionated radiotherapy is capable of activating human dendritic cells. *J Immunotoxicol.* 2014;11:328–336.
57. Weichselbaum RR, Liang H, Deng L, Fu YX. Radiotherapy and immunotherapy: a beneficial liaison? *Nat Rev Clin Oncol.* 2017;14:365–379.
58. Stamell EF, Wolchok JD, Gnjatic S, Lee NY, Brownell I. The abscopal effect associated with a systemic anti-melanoma immune response. *Int J Radiat Oncol Biol Phys.* 2013;85:293–295.
59. Golden EB, Demaria S, Schiff PB, Chachoua A, Formenti SC. An abscopal response to radiation and ipilimumab in a patient with metastatic non-small cell lung cancer. *Cancer Immunol Res.* 2013;1:365–372.
60. Chandra RA, Wilhite TJ, Balboni TA, Alexander BM, Spektor A, Ott PA, et al. A systematic evaluation of abscopal responses following radiotherapy in patients with metastatic melanoma treated with ipilimumab. *Oncoimmunology.* 2015;4:e1046028.
61. Gameiro SR, Jammeh ML, Wattenberg MM, Tsang KY, Ferrone S, Hodge JW. Radiation-induced immunogenic modulation of tumor enhances antigen processing and calreticulin exposure, resulting in enhanced T-cell killing. *Oncotarget.* 2014;5:403–416.
62. Chakraborty M, Abrams SI, Camphausen K, Liu K, Scott T, Coleman CN, et al. Irradiation of tumor cells up-regulates Fas and enhances CTL lytic activity and CTL adoptive immunotherapy. *J Immunol.* 2003;170:6338–6347.
63. Chakraborty M, Abrams SI, Coleman CN, Camphausen K, Schlom J, Hodge JW. External beam radiation of tumors alters phenotype of tumor cells to render them susceptible to vaccine-mediated T-cell killing. *Cancer Res.* 2004;64:4328–4337.
64. Hillman GG, Reich LA, Rothstein SE, Abernathy LM, Fountain MD, Hankerd K, et al. Radiotherapy and MVA-MUC1-IL-2 vaccine act synergistically for inducing specific immunity to MUC-1 tumor antigen. *J Immunother Cancer.* 2017;5:4.
65. Chang MC, Chen YL, Lin HW, Chiang YC, Chang CF, Hsieh SF, et al. Irradiation enhances abscopal anti-tumor effects of antigen-specific immunotherapy through regulating tumor microenvironment. *Mol Ther.* 2018;26:404–419.
66. Gulley JL, Arlen PM, Bastian A, Morin S, Marte J, Beetham P, et al. Combining a recombinant cancer vaccine with standard definitive radiotherapy in patients with localized prostate cancer. *Clin Cancer Res.* 2005;11:3353–3362.
67. Heery CR, Madan RA, Stein MN, Stadler WM, Di Paola RS, Rauckhorst M, et al. Samarium-153-EDTMP (QuadrametÂ®) with or without vaccine in metastatic castration-resistant prostate cancer: a randomized Phase 2 trial. *Oncotarget.* 2016;7:69014–69023.
68. Bristow RG, Alexander B, Baumann M, Bratman SV, Brown JM, Camphausen K, et al. Combining precision radiotherapy with molecular targeting and immunomodulatory agents: a guideline by the American Society for Radiation Oncology. *Lancet Oncol.* 2018;19:e240–e251.
69. Baylin SB, Jones PA. A decade of exploring the cancer epigenome—biological and translational implications. *Nat Rev Cancer.* 2011;11:726–734.
70. Nebbioso A, Tambaro FP, Dell'Aversana C, Altucci L. Cancer epigenetics: moving forward. *PLoS Genet.* 2018;14:e1007362.
71. Kaminskas E, Farrell AT, Wang YC, Sridhara R, Pazdur R. FDA drug approval summary: azacitidine (5-azacytidine, Vidaza) for injectable suspension. *Oncologist.* 2005;10:176–182.
72. Mann BS, Johnson JR, Cohen MH, Justice R, Pazdur R. FDA approval summary: vorinostat for treatment of advanced primary cutaneous T-cell lymphoma. *Oncologist.* 2007;12:1247–1252.

73. Saleh MH, Wang L, Goldberg MS. Improving cancer immunotherapy with DNA methyltransferase inhibitors. *Cancer Immunol Immunother*. 2016;65:787–796.
74. Khan AN, Tomasi TB. Histone deacetylase regulation of immune gene expression in tumor cells. *Immunol Res*. 2008;40:164–178.
75. Petrizzo A, Ragone C, Cavalluzzo B, Mauriello A, Manolio C, Tagliamonte M, et al. Human endogenous retrovirus reactivation: implications for cancer immunotherapy. *Cancers (Basel)*. 2021;13.
76. Zahnow CA, Topper M, Stone M, Murray-Stewart T, Li H, Baylin SB, et al. Inhibitors of DNA methylation, histone deacetylation, and histone demethylation: a perfect combination for cancer therapy. *Adv Cancer Res*. 2016;130:55–111.
77. Chiappinelli KB, Strissel PL, Desrichard A, Li H, Henke C, Akman B, et al. Inhibiting DNA methylation causes an interferon response in cancer via dsRNA including endogenous retroviruses. *Cell*. 2015;162:974–986.
78. Roulois D, Loo YH, Singhania R, Wang Y, Danesh A, Shen SY, et al. DNA-demethylating agents target colorectal cancer cells by inducing viral mimicry by endogenous transcripts. *Cell*. 2015;162:961–973.
79. Zhu S, Denman CJ, Cobanoglu ZS, Kiany S, Lau CC, Gottschalk SM, et al. The narrow-spectrum HDAC inhibitor entinostat enhances NKG2D expression without NK cell toxicity, leading to enhanced recognition of cancer cells. *Pharm Res*. 2015;32:779–792.
80. Costantini B, Kordasti SY, Kulasekararaj AG, Jiang J, Seidl T, Abellan PP, et al. The effects of 5-azacytidine on the function and number of regulatory T cells and T-effectors in myelodysplastic syndrome. *Haematologica*. 2013;98:1196–1205.
81. Kim K, Skora AD, Li Z, Liu Q, Tam AJ, Blosser RL, et al. Eradication of metastatic mouse cancers resistant to immune checkpoint blockade by suppression of myeloid-derived cells. *Proc Natl Acad Sci USA*. 2014;111:11774–11779.
82. Ghoneim HE, Fan Y, Moustaki A, Abdelsamed HA, Dash P, Dogra P, et al. De novo epigenetic programs inhibit PD-1 blockade-mediated T cell rejuvenation. *Cell*. 2017;170:142–157.
83. Chiappinelli KB, Zahnow CA, Ahuja N, Baylin SB. Combining epigenetic and immunotherapy to combat cancer. *Cancer Res*. 2016;76:1683–1689.
84. Krishnadas DK, Shusterman S, Bai F, Diller L, Sullivan JE, Cheerva AC, et al. A phase I trial combining decitabine/dendritic cell vaccine targeting MAGE-A1, MAGE-A3 and NY-ESO-1 for children with relapsed or therapy-refractory neuroblastoma and sarcoma. *Cancer Immunol Immunother*. 2015;64:1251–1260.
85. Nishimura H, Minato N, Nakano T, Honjo T. Immunological studies on PD-1 deficient mice: implication of PD-1 as a negative regulator for B cell responses. *Int Immunol*. 1998;10:1563–1572.
86. Krummel MF, Allison JP. CD28 and CTLA-4 have opposing effects on the response of T cells to stimulation. *J Exp Med*. 1995;182:459–465.
87. Pardoll DM. The blockade of immune checkpoints in cancer immunotherapy. *Nat Rev Cancer*. 2012;12:252–264.
88. Curran MA, Montalvo W, Yagita H, Allison JP. PD-1 and CTLA-4 combination blockade expands infiltrating T cells and reduces regulatory T and myeloid cells within B16 melanoma tumors. *Proc Natl Acad Sci USA*. 2010;107:4275–4280.
89. Wolchok JD, Kluger H, Callahan MK, Postow MA, Rizvi NA, Lesokhin AM, et al. Nivolumab plus ipilimumab in advanced melanoma. *N Engl J Med*. 2013;369:122–133.
90. Emens LA, Ascierto PA, Darcy PK, Demaria S, Eggermont AMM, Redmond WL, et al. Cancer immunotherapy: opportunities and challenges in the rapidly evolving clinical landscape. *Eur J Cancer*. 2017;81:116–129.

91. Ali OA, Lewin SA, Dranoff G, Mooney DJ. Vaccines combined with immune checkpoint antibodies promote cytotoxic T-cell activity and tumor eradication. *Cancer Immunol Res.* 2016;4:95–100.
92. Karyampudi L, Lamichhane P, Scheid AD, Kalli KR, Shreeder B, Krempski JW, et al. Accumulation of memory precursor CD8 T cells in regressing tumors following combination therapy with vaccine and anti-PD-1 antibody. *Cancer Res.* 2014;74:2974–2985.
93. Xue W, Brentville VA, Symonds P, Cook KW, Yagita H, Metheringham RL, et al. SCIB1, a huIgG1 antibody DNA vaccination, combined with PD-1 blockade induced efficient therapy of poorly immunogenic tumors. *Oncotarget.* 2016;7:83088–83100.
94. Xue W, Metheringham RL, Brentville VA, Gunn B, Symonds P, Yagita H, et al. SCIB2, an antibody DNA vaccine encoding NY-ESO-1 epitopes, induces potent antitumor immunity which is further enhanced by checkpoint blockade. *Oncoimmunology.* 2016;5: e1169353.
95. Scholz M, Yep S, Chancey M, Kelly C, Chau K, Turner J, et al. Phase I clinical trial of sipuleucel-T combined with escalating doses of ipilimumab in progressive metastatic castrate-resistant prostate cancer. *Immunotargets Ther.* 2017;6:11–16.
96. Ku J, Wilenius K, Larsen C, De Guzman K, Yoshinaga S, Turner JS, et al. Survival after Sipuleucel-T (SIP-T) and low dose Ipilimumab (IPI) in men with metastatic, progressive, castrate-resistant prostate cancer (M-CRPC). *J.Clin.Oncol.* 2018;36(S6):368. Ref Type: Abstract.
97. Weber JS, Kudchadkar RR, Yu B, Gallenstein D, Horak CE, Inzunza HD, et al. Safety, efficacy, and biomarkers of nivolumab with vaccine in ipilimumab-refractory or -naive melanoma. *J Clin Oncol.* 2013;31:4311–4318.

Chapter 9

Peptide-based vaccines

Duane H. Hamilton, Jeffrey Schlom and Caroline Jochems
Laboratory of Tumor Immunology and Biology, Center for Cancer Research, National Cancer Institute, National Institutes of Health, Bethesda, MD, United States

Introduction

The potential of the immune system to control the outgrowth of tumor cells has been demonstrated clinically by the remarkable responses observed in a subset of patients following treatment with therapies targeting immune checkpoints. These therapies are thought to promote the unique ability of the immune system to target and specifically eliminate tumor cells. Conventional $CD4^+$ and $CD8^+$ T cells are considered the arms of the adaptive immune system primarily responsible for the specific recognition and elimination of cancer cells. The T-cell receptors (TCRs) expressed by these tumor-reactive T cells recognize short peptide fragments derived from proteins expressed by the tumor, which have been processed and presented on the tumor cells' surface within the binding groove of the major histocompatibility complex (MHC).[1] There are two main classes of MHC molecules, with class I presenting bound 8–11-mer peptides to $CD8^+$ T cells, while class II MHC peptide complexes present larger 11–13-mer peptide fragments to $CD4^+$ T-helper cells.[2–5] Class I MHC molecules are expressed on the surface of the majority of nucleated cells within the body, while class II MHC expression is typically restricted to professional antigen-presenting cells, such as dendritic cells (DCs), macrophages, and B cells. These MHC-bound peptides allow the immune system to survey proteins being produced by cells in the whole body, allowing reactive T cells to identify and kill tumor cells expressing anomalous proteins.

In an effort to generate tumor-specific immunity capable of controlling the growth of carcinoma cells, there is a diverse array of therapeutic cancer vaccines currently undergoing clinical development. This chapter focuses primarily on peptide-based vaccines employing differential strategies to enhance the immune response, using as examples studies ongoing within our cancer immunotherapy program at the National Cancer Institute, National Institutes of Health, Bethesda, Maryland, USA. These trials demonstrate the

Cancer Vaccines as Immunotherapy of Cancer. DOI: https://doi.org/10.1016/B978-0-12-823901-8.00008-X
© 2022 Elsevier Inc. All rights reserved.

benefit of combining multiple, rationally selected immune-oncology agents in the development of effective immunotherapies. Combining vaccines with therapeutic agents targeting multiple mechanisms by which tumors can evade the immune system is essential in the development of the next generation of immunotherapies.

Vaccine targets

Therapeutic cancer vaccines target either tumor-associated antigens (TAAs) or tumor-specific antigens (TSAs). TAAs, such as mucin 1 (MUC1),[6–8] carcinogenic embryonic antigen cell adhesion molecule 5 (CEA-CAM5),[9–11] prostate-specific antigen (PSA),[12] and others, are normal proteins aberrantly expressed at high levels by multiple tumor types. However, these antigens are also expressed in healthy tissues; inducing a strong immune response directed against them can therefore potentially lead to off-target effects with highly activated T cells engaging with healthy tissues, giving rise to immune-related adverse events. TAAs employed as antigens for vaccine development have been extensively reviewed elsewhere.[13] Over the past decade whole exome sequencing of tumors has become common, which has allowed researchers to identify genetic alterations including frameshifts, insertions/deletions, and nonsynonymous mutations uniquely present within the tumor. Some of these mutations will result in the expression of altered proteins that can be targeted by the immune system.[14] Many oncovirus-associated malignancies constitutively express viral proteins on the tumor cell surface, such as the E6 and E7 proteins of the human papillomavirus (HPV). Expression of E6 and E7 is required to drive uncontrolled cell growth observed with HPV-associated malignancies.[15] The tumor-restricted expression of these viral proteins makes them ideal targets for therapeutic cancer vaccines, since there is no risk of off-target effects.

Peptide vaccines targeting tumor-associated antigens

There are several challenges when developing peptide vaccines targeting TAAs. T cells in the periphery capable of recognizing TAAs have undergone the process of negative selection in the thymus and are thought to be present at low frequencies with a relatively low affinity for their target antigens. For these reasons, such peptide vaccines may have limited applicability in targeting TAAs.[16,17] One strategy for enhancing the immunogenicity of self-antigen peptide vaccines is by employing "agonist epitopes," which are peptides that have been modified by exchanging certain amino acids to enhance their binding to either the TCR or MHC molecule. This may result in increased efficacy of T-cell stimulation, with higher interferon gamma (IFNγ) production upon activation, and higher cytotoxic activity.[18] One such agonist epitope of the TAA brachyury was recently identified and showed better binding to MHC class I, increased IFNγ production after the stimulation of T cells, generation of T cells that displayed higher lysis of carcinoma cells expressing the native epitope, and stronger brachyury-specific T-cell responses in vivo.[19] This agonist epitope has now been included in therapeutic cancer vaccines employing several different vaccine platforms.

Peptide-based vaccines targeting onco-viral proteins

There are over 200 strains of HPV, and the two most common types that confer a high risk of cancer are HPV16 and 18.[20] HPV-associated malignancies include cancer of the cervix, oropharynx, anus, rectum, penis, vagina, and vulva. Two oncoproteins, the early proteins E6 and E7, which induce uncontrolled cellular proliferation, are constitutively overexpressed in HPV-induced tumor cells but are not found anywhere else in the human body.[15,21] This makes them ideal targets for vaccine development, since it decreases the risk of harming healthy tissues. In addition, the E6 and E7 proteins are responsible for driving the malignant transformation, and thus antigen loss as a potential escape mechanism is not likely.[22] These oncoproteins are inherently very immunogenic. The human body can mount a response to the peptides, and clinical studies have shown that patients with cervical cancer who displayed HPV-specific T cells against E6 or E7 had a better prognosis than those who did not.[23]

The development of therapeutic cancer vaccines against HPV-associated malignancies is an active area of research, and the current status of ongoing trials was recently reviewed by Smalley Rumfield et al.[24] Here, we focus on the peptide-based therapeutic vaccines. There are currently two short peptide vaccines being tested in clinical trials: PDS0101 and DPX-E7. PDS0101 contains non-MHC-restricted peptides in R-DOTAP, which is a liposomal carrier that activates toll-like receptor (TLR) 7. PDS0101 was shown to be safe in a phase I trial.[25,26] It is currently in phase II trials in combination with bintrafusp alfa (a novel bifunctional agent consisting of antiprogrammed death ligand 1 (anti-PDL1) linked to the extracellular domain of the transforming growth factor beta receptor II (TGF-βRII) and NHS-IL-12 [an immunocytokine composed of two interleukin (IL)-12 heterodimers fused to an NHS76 antibody targeting histones in necrotic tissue with IL-12] (NCT04287868), as well as with pembrolizumab for patients with head and neck squamous cell carcinoma (NCT04260126). DPX-E7 consists of HPV16 E7 peptide emulsified in DepoVax adjuvant and is being evaluated in a phase Ib/II trial (NCT02865135). ISA101 is another therapeutic HPV vaccine that consists of a pool of synthetic long peptides targeting the E6 and E7 proteins of HPV16. This vaccine will be discussed in the combination treatment sections.[27]

Epstein Barr virus (EBV, human herpesvirus-4) can cause the development of malignancies of the nasopharynx. These tumors express the latent membrane proteins (LMP) 1, 2A, and 2B, as well as EBNA1–3, which contribute to cellular transformation and promote immune escape in EBV^+

malignant cells.[28] One phase I clinical study evaluating two different LMP2 peptide vaccines was recently completed (NCT00078494), and another peptide vaccine containing six different EBV peptides was found to induce specific CD4 T-cell cytotoxic activity, which was granzyme B mediated.[29] Peptide vaccines targeting EBV could potentially be used for the treatment of nasopharyngeal carcinoma and posttransplant lymphoproliferative disorders, which have been found to be induced by latent EBV infections.

Peptide-based vaccines targeting neoepitopes

Neoepitopes are uniquely expressed by a patient's tumor, representing a pool of antigens that could potentially be targeted by therapeutic cancer vaccines. Similar to the onco-viral proteins discussed in the previous section, these neoepitopes are perceived to be nonself by the immune system. As such, therapeutic vaccines targeting these neoepitopes have the potential to activate a pool of high-affinity T cells capable of recognizing these unique tumor antigens, without the risk of inadvertently targeting healthy tissues.[30] However, unlike vaccines targeting onco-viral proteins, which allow for the development of vaccines to a specific cancer type, neoepitope vaccines must be tailored to the unique mutations expressed by each patient's tumor. These personalized cancer vaccines present numerous unique challenges in their development and production at the scale to be a viable treatment regimen. Mutations must be identified and prioritized based upon their potential to bind either class I or II human leukocyte antigen (HLA),[14,31] at which point the selected peptides must rapidly be chemically synthesized based upon selected neoepitope sequences in a good manufacturing practice–compliant manner and formulated into a vaccine capable of inducing a robust antitumor immune response. While not insurmountable, each of these challenges represents significant barriers that must be overcome to allow for these neoepitope-targeted therapies to become a viable therapeutic strategy.

While the majority of neoepitopes are patient specific, there are some mutations in cancer driver genes, such as epidermal growth factor receptor (EGFR)[32] and TP53,[33] that are common in multiple cancer types. These shared neoepitopes could potentially be targeted by therapeutic vaccines; however, it is to be expected that their ability to expand T cells would vary greatly among patients. Furthermore, it is likely that many MHC alleles would be unable to efficiently bind the appropriate neoepitope for presentation to T cells. For these reasons, public neoepitopes may have only very limited clinical applicability.

Short versus long synthetic peptides

Peptide-based vaccines have the advantage over the more traditional vector-based vaccines of being well tolerated and relatively cheap and easy to produce at scale. These peptide-based vaccines are typically divided into two categories based upon their length. Short peptides consist of peptides of

9–11 amino acids representing the nominal epitope capable of binding the class I HLA without having to first be processed by the antigen-presentation machinery. The main downside with short peptide vaccines is that they are dependent on MHC restriction and have to be matched to the patient's HLA type. The issue of MHC restriction can potentially be overcome by employing a mixture of peptides restricted to HLA A1, A2, A3, A11, and A24, which would cover 85% of the population. One such peptide vaccine was successfully used in patients with melanoma, and no antigenic competition was observed.[34] The use of peptides is an attractive method to target oncoviral proteins, such as HPV gene products, because it is possible to remove potentially oncogenic sequences when producing these vaccines. These short peptides are appealing for use in vaccines as they are generally easy to synthesize and relatively cheap to produce; however, the difficulty is accurately identifying which peptides will efficiently bind to specific HLA alleles. This prediction is typically performed using software packages.[14] Although these programs are continually improving,[35,36] the accuracy varies significantly among particular class I HLA alleles[37,38] and continues to be relatively poor for the prediction of class II binding peptides.[39] To overcome some of the issues associated with short peptides, many vaccines now use synthetic long peptides, typically between 20 and 30 amino acids in length. These longer peptides make use of the patient's own antigen-presentation machinery to process the peptides for presentation on diverse class I and class II HLA molecules. These long peptides can facilitate the expansion of both specific $CD8^+$ and $CD4^+$ T cells. This was recently shown in a clinical study employing a 20-mer peptide vaccine targeting the TAA NY-ESO1.[40] Although $CD4^+$ T cells may have the ability to directly target and kill tumor cells, they primarily help facilitate the expansion and longevity of specific $CD8^+$ cytotoxic T cells,[41] which are presumably the main effector cells responsible for the direct killing of tumor cells.

Formulation of peptide-based vaccines

Peptide vaccines must be properly formulated to ensure the generation of robust immunity. In preclinical models, it has been demonstrated that vaccination with peptides can result in epitope-specific tolerization rather than activation.[42,43] In fact, the same peptide can either activate cytotoxic T lymphocytes (CTLs) or induce specific anergy depending on how the peptide is administered.[43] In addition, free peptides are susceptible to rapid cleavage by peptidases[44]; thus considerable care must be taken in formulating these peptide-based vaccines with adjuvants to increase their stability and promote their immunogenicity. One such method is emulsifying the peptide vaccine in a water-in-oil formulation, such as Montanide ISA 51 VG, which generates an antigen depot that slowly releases the peptide, resulting in the generation of strong T-cell immune responses. This approach has been used widely in clinical studies of peptide-based vaccines.[45,46] One caveat to using these depot-based adjuvants is that they have the potential to sequester T cells at the site of vaccination.[47] Another method to increase the stability of peptide vaccines is to incorporate them onto/into nanoparticles, some of which have the ability to preferentially target their peptide cargo to a particular cell type, which can drastically improve the magnitude of the immune response generated against the peptide.[48]

Another family of commonly used adjuvants is TLR agonists, which can be used to increase the immunogenicity of the peptide vaccine. TLRs comprise a family of 10 receptors,[49,50] each of which is capable of recognizing unique pathogen-associated molecular patterns. These are conserved structures uniquely expressed by a wide variety of microbes.[51–53] Upon binding their respective ligands, TLRs will promote the downstream activation of a panel of genes which are critical to promoting innate immune responses and enhancing immunity against the microbe.[51,54] TLRs can be incorporated into peptide-based therapeutic cancer vaccines to promote the efficient activation and specific expansion of T cells. To date, numerous TLRs have been used clinically as adjuvants for vaccines and have been extensively reviewed by Hennessy et al.[55]

DCs are very efficient at processing and presenting antigens to T cells. Several groups have tried to exploit them therapeutically by infusing patients with autologous DCs that have been pulsed in vitro with peptides. These types of vaccines are promising because they allow for efficient trafficking of peptide loaded DCs to draining lymph nodes, where they can efficiently activate T cells. Clinical studies have demonstrated that these pulsed DCs are efficient at inducing immune responses[34,56]; however, it has not been clearly demonstrated that peptide-pulsed DC vaccination strategies offer a clear therapeutic advantage over the more traditional peptide in adjuvant therapies.

Peptide-based vaccines in combination with additional immune modulators

It is becoming increasingly evident that while therapeutic cancer vaccines have been shown to induce the robust expansion of tumor-specific T cells in many patients,[57] this immunity is seldom capable of controlling the growth of large, established tumors.[58] The expansion of regulatory immune cells,[59] aberrant expression of immunological checkpoints,[60,61] accumulation of immune-suppressive myeloid cells,[62] and the acquisition of mesenchymal features by tumor cells[63,64] are only some of the potential mechanisms working to impede the ability of the immune system to effectively control the neoplastic growth. Combining immunotherapeutic cancer vaccines with rationally selected immune-modulating agents has the potential to enhance their efficacy in the generation of effective antitumor immunity (Fig. 9.1).[65]

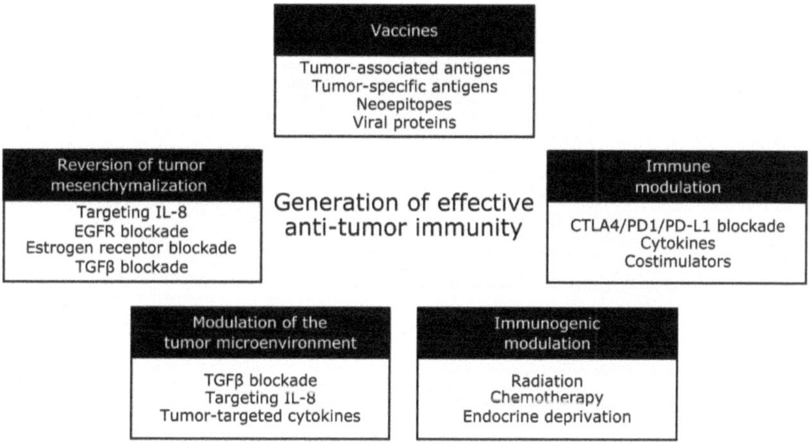

FIGURE 9.1 Therapeutic agents targeting mechanisms of immune resistance.

Vaccines in combination with cytokines and immunocytokines

One approach to enhance the antitumor activity of peptide vaccines is by combinations with stimulatory cytokines. IL-2 has been widely used as an immunotherapeutic agent, but systemic treatment is associated with dose-limiting toxicities, such as hypotension, flu-like symptoms, nausea, and capillary leak syndrome, as well as the induction of activation-induced cell death (AICD) of T cells, and proliferation of regulatory T (Treg) cells.[66] In contrast, IL-15 has been shown to have similar stimulatory effects on T cells and natural killer (NK) cells, but does not have a stimulatory effect on Treg cells or induce AICD.[67] Thus recombinant (r)IL-15 therapy has been successfully evaluated in clinical trials.[68,69] However, the half-life of rIL-15 is short in vivo, which limits its efficacy. To overcome this, several IL-15 superagonists have been developed. N803 is one such agent and consists of a mutated superagonist IL-15 bound to an IL-15Rα sushi domain linked to IgG1-Fc. Multiple clinical trials are ongoing and have shown promising activity for patients with diverse types of cancer.[70] The combination of N803 and a neoepitope peptide vaccine was used in a preclinical study with promising results,[71] and combinations with therapeutic cancer vaccines are ongoing. Another promising cytokine for combination therapies is IL-12, which regulates the transition from innate to adaptive immunity and directly acts on NK cells, NKT cells, and T cells to increase their cytotoxic activity.[72] Clinical trials evaluating rIL-12 revealed a very narrow therapeutic window[73,74]; therefore, IL-12 has now been formulated into an immunocytokine, NHS-IL-12.[75] NHS-IL-12 has a longer half-life than rIL-12 and has been shown to target the tumor in vivo. A phase I study showed safety, induction of IFNγ, and the influx of T cells into the tumor.[76] This agent is currently being evaluated

TABLE 9.1 Combination immunotherapy for HPV⁺ tumors.

Treatment	# mice with tumor volume <300 mm^3	T-cell clones per 25% of TCR repertoire (avg)
PBS control	0/16	18
R-DOTAP	0/8	ND
PDS0101	3/16	12
Bintrafusp alfa	0/16	22
NHS-IL-12	6/16	25
PDS0101 + bintrafusp alfa	5/16	8
PDS0101 + NHS-IL-12	10/16	6
NHS-IL-12 + bintrafusp alfa	8/16	18
PDS0101 + NHS-IL-12 + bintrafusp alfa	13/17	3

Notes: PDS0101 monotherapy reduced tumor volume in the TC-1 HPV$^+$ syngeneic tumor model, and the combination of PDS0101, bintrafusp alfa, and NHS-IL-12 resulted in further tumor control. TC-1 tumor bearing female C57BL/6J mice (n = 8–16 per group) were treated with PBS control (100 μL s.c.), R-DOTAP control (100 μL s.c.), PDS0101 (s.c., 3 weekly doses starting on day 7), bintrafusp alfa (250 μg, i.p., days 7, 9, and 11), and/or NHS-IL-12 (50 μg, s.c., day 7). The table shows "tumor control," the number of mice with tumors below 300 mm^3 at the end of study, and the average number of clones from three mice per group that make up the top 25% of the TCR productive rearrangements. A *meta*-analysis of two independent experiments is shown. TILs were purified from whole tumor. DNA isolated from TILs was analyzed by Adaptive Biotechnology (Seattle, WA, USA) for TCR repertoire.
IL-12, interleukin-12; *i.p.*, intraperitoneally; *ND*, not done; *s.c.*, subcutaneously; *TCR*, T-cell receptor; *TIL*, tumor infiltrating lymphocyte.
Source: Adapted from Smalley Rumfield C, Pellom ST, Morillon II YM, Schlom J, Jochems C. Immunomodulation to enhance the efficacy of an HPV therapeutic vaccine. *J Immunother Cancer*. 2020;8(1):e000612.[77]

(NCT0428768) in combination with the HPV16 peptide vaccine PDS0101 and bintrafusp alfa, based on the results of a preclinical study.[77] In the preclinical study, syngeneic mice bearing TC-1, an HPV$^+$ murine tumor, were treated with single agents or combinations of PDS0101, NHS-IL-12, and/or bintrafusp alfa. As seen in Table 9.1, vaccine monotherapy and monotherapy with NHS-IL-12 reduced tumor volume; however, the greatest tumor reduction was seen in mice treated with the triple combination of PDS0101, NHS-IL-12, and bintrafusp alfa. Furthermore, the table shows TCR clonality, as measured by the number of clones that make up the top 25% of the TCR repertoire. PDS0101 significantly increased TCR clonality in all treated groups, and the combination therapy further increased the clonality.

Vaccines in combination with radiation and chemotherapy

Studies have demonstrated that rationally selected chemotherapy and radiotherapy treatment regimens can enhance the development of effective antitumor immunity. Chemotherapies, such as oxaliplatin and doxorubicin, have been shown to induce immunogenic cell death, a process that promotes the uptake and presentation of dying tumor cells by DCs and facilitates the activation of reactive T cells.[78–81] In cases where tumor cells are not directly killed following their exposure to either chemotherapy or radiotherapy, they can undergo a process termed immunogenic modulation. This phenotypic modulation is associated with increased antigen presentation and expression of death receptors, rendering the tumor cells more susceptible to T-cell-mediated lysis.[82–87] In several preclinical studies, researchers have been able to exploit the processes of immunogenic modulation and antigenic cell death to improve the efficacy of therapeutic cancer vaccines.[82,88,89] Furthermore, specific chemotherapeutic agents and small-molecule inhibitors display differential effects on certain immune cell subsets, which can be exploited for combination therapy strategies. For example, Treg cells were shown to decrease both in numbers and in suppressive function after treatment with docetaxel in patients with metastatic breast cancer and prostate cancer, and cisplatin plus vinorelbine had a similar effect in patients with non-small-cell lung cancer.[90] In these trials, an increase in the CD4:Treg ratio was also observed, providing a rationale for further studies with combination therapies. Furthermore, combining the HPV16 targeting synthetic long peptide vaccine ISA101 with standard-of-care carboplatin/paclitaxel chemotherapy reduced the number of immunosuppressive myeloid cells in patients with advanced, recurrent, or metastatic cervical cancer, leading to tumor regression in 43% of the patients.[91,92] The decreased immune suppression also allowed for stronger HPV16-specific T-cell responses in patients treated with the combination, and the most advantageous time to vaccinate appeared to be at 2 weeks after the second cycle of chemotherapy. Moreover, the strong antigen-specific responses were retained after multiple cycles of chemotherapy.[92]

Vaccines in combination with checkpoint blockade

Immunotherapies targeting the programmed cell death protein-1 (PD-1)/L1 and CTL-associated protein 4 (CTLA-4) immune checkpoints have proven to be remarkably effective in a subset of patients, particularly those with a high tumor mutation burden.[93] These therapies are thought to facilitate the expansion and effector function of tumor-reactive T cells within the tumor microenvironment.[94] There is an obvious potential synergy in the combination of therapeutic cancer vaccines with checkpoint blockade to abrogate tumor-derived immunosuppression.[95] One example of how a vaccine can benefit from the addition of a checkpoint-targeted therapy is the HPV16 targeting vaccine ISA101. This vaccine consists of a pool of synthetic long peptides targeting the E6 and E7 proteins of HPV16. When used as a monotherapy, vaccination with ISA101 was associated with remarkable response rates in high-grade vulvar intraepithelial neoplasia. However, vaccine monotherapy did not show clinical activity in advanced cervical cancer.[27] As seen above, the combination of vaccine with chemotherapy led to tumor regression in 43%

of patients.[91] It is possible that unlike in hyperproliferative disease, advanced carcinomas may employ multiple mechanisms to actively dampen the efficacy of the immune system to successfully control the tumor. In a subsequent study, investigators combined ISA101 with PD-L1 blockade and observed a 36% response rate in patients with incurable $HPV16^{+}$ head and neck cancer.[96] Although this was a significant increase compared with the 16%–22% response rate observed in similar studies in which anti-PD-L1 treatment was used as a monotherapy,[97–99] only a minority of patients received any clinical benefit from the combination. It is clear that other mechanisms of immune failure may be playing a role in patients who did not respond to therapy. The inclusion of additional immune-oncology agents targeting multiple mechanisms of resistance may be required to develop therapeutic treatment regimens that could begin to offer clinical benefit to the majority of patients.

Reversion of mesenchymal-like features

Tumor cells can activate a differentiation program called epithelial–mesenchymal transition.[100,101] Carcinoma cells undergoing this process lose the expression of epithelial characteristics and gain mesenchymal characteristics, which promotes metastasis and resistance to chemotherapy and radiotherapy.[102,103] It has also been demonstrated that this process associates with an increased resistance to immune-mediated killing of tumor cells.[63] Soluble factors, such as TGF-β, IL-8, and EGFR and estrogen receptor alpha signaling, have been associated with promoting the acquisition of mesenchymal features by carcinoma cells, and the specific targeting of these molecules can improve responsiveness to vaccine.[63,104–111] Clinical studies are underway to investigate the effect of combining some of these agents with vaccine (NCT04432597, NCT04574583).

Multicombination therapies

As suggested by the discussion above, advanced tumors may utilize several mechanisms to inhibit the generation of effective antitumor immunity. Thus it is likely therapeutic cancer vaccines may require combining several rationally selected agents, each targeting discrete mechanisms of tumor resistance, to generate effective antitumor immunity (Fig. 9.1). A recent in vivo study in three different murine cancer models demonstrated that a therapeutic cancer vaccine combined with the IL-15 superagonist N803 to activate antigen-specific T cells, OX40, and glucocorticoid-induced tumor necrosis factor receptor agonists to expand activated immune cell populations, and an indoleamine 2,3-dioxygenase inhibitor to enable antitumor activity in the tumor microenvironment inhibited tumor growth and induced protection from tumor rechallenge.[112] Another study demonstrated the synergy in combining a neoepitope-targeted peptide vaccine with N803, PD-L1 blockade, and a tumor-targeted IL-12 agent (NHS-IL-12), in which all four agents were required to induce effective clearance of established tumors (Fig. 9.2).[71] These studies

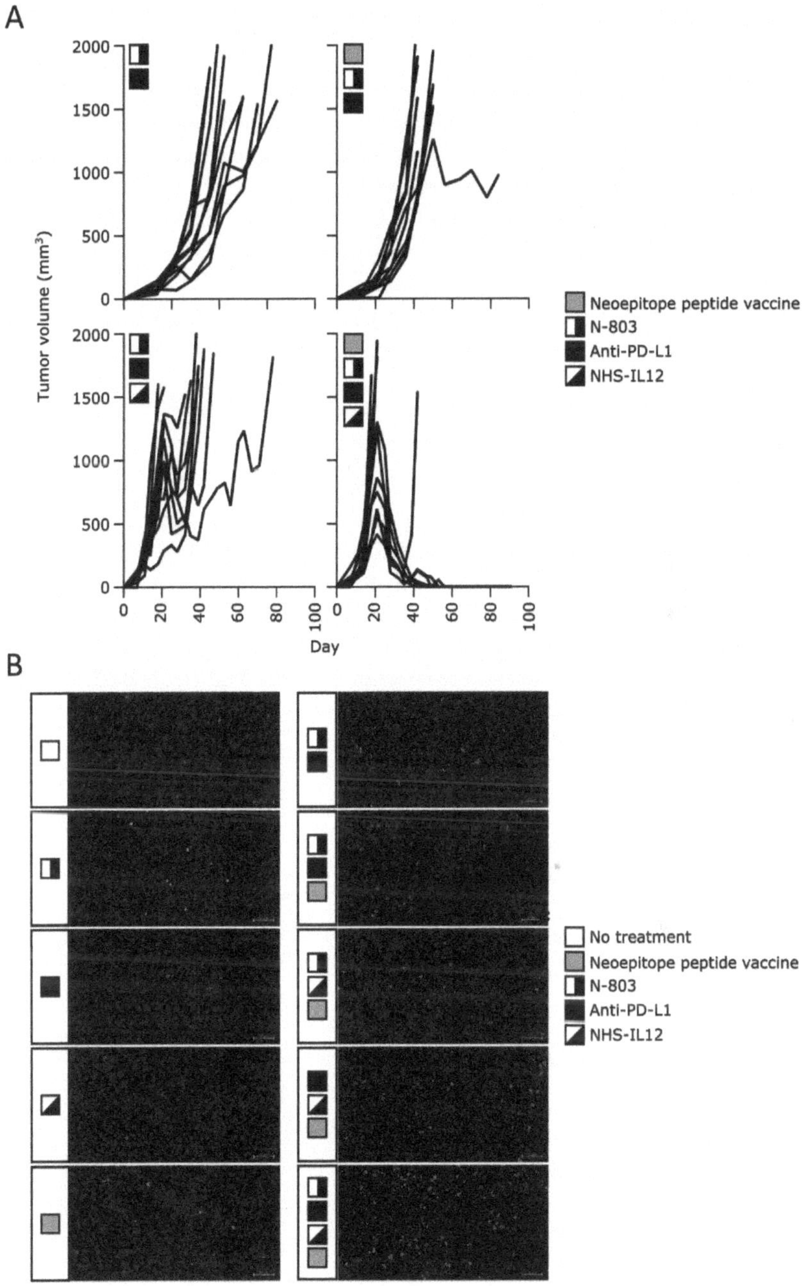

FIGURE 9.2 Combination of peptide vaccines with multiple immune-oncology agents is required for the generation of effective antitumor immunity. (A) Tumor growth curves and (B) Immunohistochemical analysis of tumor-infiltrating $CD8^+$ T cells (*red* signal) and DAPI (*blue* signal) of MC38 tumors. Animals were treated as indicated with a neoepitope-targeted peptide vaccine (days 4, 11, and 18), N803 (days 8, 11, and 15), and NHS-IL-12 (day 18). *Adapted from Lee KL, Benz SC, Hicks KC, Nguyen A, Gameiro SR, Palena C, et al. Efficient tumor clearance and diversified immunity through neoepitope vaccines and combinatorial immunotherapy.* Cancer Immunol Res. *2019;7(8):1359–1370.*[71]

provide the preclinical rationale for the combination of multiple agents in the treatment of cancer. Several studies are currently ongoing at the National Cancer Institute using a trial design where the safety and efficacy of several different combinations of immunotherapy agents are being evaluated simultaneously and quickly, termed "QuEST": Quick Efficacy Seeking Trials.[113] At this time, three such trials are open and recruiting patients: NCT03493945 in prostate cancer, NCT04296942 in breast cancer, and NCT04708470 in HPV-associated malignancies and colon cancer; additional studies in other malignancies are under review.

Concluding remarks

The development of therapeutic cancer vaccines is an active and important area of research to decrease the global cancer burden. A careful selection of vaccines with complementary treatments has the potential to orchestrate an immune response that is strong enough to effectively combat malignancies. It is also likely that the timing of the different treatment modalities is important to obtain the maximum benefits of a less suppressive tumor microenvironment, induction of antigen-specific T cells, and decreased immune checkpoint brakes on the activated immune cells.

Abbreviations

AICD, activation-induced cell death
CEA-CAM5, carcinogenic embryonic antigen cell adhesion molecule 5
CTL, cytotoxic T lymphocyte
CTLA-4, CTL-associated antigen 4
DC, dendritic cell
EBV, Epstein Barr virus
EGFR, epidermal growth factor receptor
HLA, human leukocyte antigen
IFN, interferon
IL, interleukin
LMP, latent membrane protein
MHC, major histocompatibility complex
MUC1, mucin 1
NK, natural killer
PD-1, programmed cell death protein-1
PD-L1, programmed death ligand 1
PSA, prostate-specific antigen
QuEST, Quick Efficacy Seeking Trial
r, recombinant
TAA, tumor-associated antigen
TGF-βRII, transforming growth factor beta receptor II
TLR, toll-like receptor

Treg, regulatory T
TSA, tumor-specific antigen

Acknowledgment

The authors thank Debra Weingarten for her editorial assistance in the preparation of this manuscript.

References

1. Zinkernagel RM, Doherty PC. Restriction of in vitro T cell-mediated cytotoxicity in lymphocytic choriomeningitis within a syngeneic or semiallogeneic system. *Nature*. 1974;248 (5450):701–702.
2. Brown JH, Jardetzky TS, Gorga JC, Stern LJ, Urban RG, Strominger JL, et al. Three-dimensional structure of the human class II histocompatibility antigen HLA-DR1. *Nature*. 1993;364(6432):33–39.
3. Chicz RM, Urban RG, Lane WS, Gorga JC, Stern LJ, Vignali DA, et al. Predominant naturally processed peptides bound to HLA-DR1 are derived from MHC-related molecules and are heterogeneous in size. *Nature*. 1992;358(6389):764–768.
4. Jones EY. MHC class I and class II structures. *Curr Opin Immunol*. 1997;9(1):75–79.
5. Sercarz EE, Maverakis E. Mhc-guided processing: binding of large antigen fragments. *Nat Rev Immunol*. 2003;3(8):621–629.
6. Gendler SJ, Lancaster CA, Taylor-Papadimitriou J, Duhig T, Peat N, Burchell J, et al. Molecular cloning and expression of human tumor-associated polymorphic epithelial mucin. *J Biol Chem*. 1990;265(25):15286–15293.
7. Hollingsworth MA, Swanson BJ. Mucins in cancer: protection and control of the cell surface. *Nat Rev Cancer*. 2004;4(1):45–60.
8. Kufe D, Inghirami G, Abe M, Hayes D, Justi-Wheeler H, Schlom J. Differential reactivity of a novel monoclonal antibody (DF3) with human malignant vs benign breast tumors. *Hybridoma*. 1984;3(3):223–232.
9. Benchimol S, Fuks A, Jothy S, Beauchemin N, Shirota K, Stanners CP. Carcinoembryonic antigen, a human tumor marker, functions as an intercellular adhesion molecule. *Cell*. 1989;57(2):327–334.
10. Robbins PF, Eggensperger D, Qi CF, Schlom J. Definition of the expression of the human carcinoembryonic antigen and non-specific cross-reacting antigen in human breast and lung carcinomas. *Int J Cancer*. 1993;53(6):892–897.
11. Tendler A, Kaufman HL, Kadish AS. Increased carcinoembryonic antigen expression in cervical intraepithelial neoplasia grade 3 and in cervical squamous cell carcinoma. *Hum Pathol*. 2000;31(11):1357–1362.
12. Balk SP, Ko YJ, Bubley GJ. Biology of prostate-specific antigen. *J Clin Oncol*. 2003;21 (2):383–391.
13. Schlom J. Therapeutic cancer vaccines: current status and moving forward. *J Natl Cancer Inst*. 2012;104(8):599–613.
14. Sahin U, Tureci O. Personalized vaccines for cancer immunotherapy. *Science*. 2018;359 (6382):1355–1360.
15. Lehoux M, D'Abramo CM, Archambault J. Molecular mechanisms of human papillomavirus-induced carcinogenesis. *Public Health Genomics*. 2009;12(5–6):268–280.

16. Marchand M, Weynants P, Rankin E, Arienti F, Belli F, Parmiani G, et al. Tumor regression responses in melanoma patients treated with a peptide encoded by gene MAGE-3. *Int J Cancer*. 1995;63(6):883–885.
17. Weber JS, Hua FL, Spears L, Marty V, Kuniyoshi C, Celis E. A phase I trial of an HLA-A1 restricted MAGE-3 epitope peptide with incomplete Freund's adjuvant in patients with resected high-risk melanoma. *J Immunother*. 1999;22(5):431–440.
18. Sette A, Vitiello A, Reherman B, Fowler P, Nayersina R, Kast WM, et al. The relationship between class I binding affinity and immunogenicity of potential cytotoxic T cell epitopes. *J Immunol*. 1994;153(12):5586–5592.
19. Tucker JA, Jochems C, Boyerinas B, Fallon J, Greiner JW, Palena C, et al. Identification and characterization of a cytotoxic T-lymphocyte agonist epitope of brachyury, a transcription factor involved in epithelial to mesenchymal transition and metastasis. *Cancer Immunol Immunother*. 2014;63(12):1307–1317.
20. Egawa N, Egawa K, Griffin H, Doorbar J. Human papillomaviruses; epithelial tropisms, and the development of neoplasia. *Viruses*. 2015;7(7):3863–3890.
21. Yeo-Teh NSL, Ito Y, Jha S. High-risk human papillomaviral oncogenes E6 and E7 target key cellular pathways to achieve oncogenesis. *Int J Mol Sci*. 2018;19(6).
22. Araldi RP, Assaf SMR, Carvalho RF, Carvalho M, Souza JM, Magnelli RF, et al. Papillomaviruses: a systematic review. *Genet Mol Biol*. 2017;40(1):1–21.
23. Nakagawa M, Gupta SK, Coleman HN, Sellers MA, Banken JA, Greenfield WW. A favorable clinical trend is associated with CD8 T-cell immune responses to the human papillomavirus type 16 e6 antigens in women being studied for abnormal pap smear results. *J Low Genit Tract Dis*. 2010;14(2):124–129.
24. Smalley Rumfield C, Roller N, Pellom ST, Schlom J, Jochems C. Therapeutic vaccines for HPV-associated malignancies. *Immunotargets Ther*. 2020;9:167–200.
25. Gandhapudi SK, Ward M, Bush JPC, Bedu-Addo F, Conn G, Woodward JG. Antigen priming with enantiospecific cationic lipid nanoparticles induces potent antitumor CTL responses through novel induction of a type I IFN response. *J Immunol*. 2019;202(12): 3524–3536.
26. Wood LV, Edwards LH, Ferris DG, Nevadunsky NS, Isani S, Conn G, et al. A novel enantio-specific cationic lipid R-DOTAP + HPV16 E6 & E7 antigens induces potent antigen-specific CD8 + T cell responses in-vivo in subjects with CIN and high-risk human papillomavirus infection (abstr). In: *The Society for Immunotherapy of Cancer (SITC) 34th Annual Meeting*, November 6–10, 2019, National Harbor, MD. *J ImmunoTher Cancer*. 2019;7(suppl 1):017.
27. Kenter GG, Welters MJ, Valentijn AR, Lowik MJ, Berends-van der Meer DM, Vloon AP, et al. Vaccination against HPV-16 oncoproteins for vulvar intraepithelial neoplasia. *N Engl J Med*. 2009;361(19):1838–1847.
28. Renaud S, Lefebvre A, Mordon S, Morales O, Delhem N. Novel therapies boosting T cell immunity in Epstein Barr virus-associated nasopharyngeal carcinoma. *Int J Mol Sci*. 2020;21(12).
29. Morales O, Depil S, Mrizak D, Martin N, Ndour PA, Dufosse F, et al. EBV Latency II-derived peptides induce a specific CD4 + cytotoxic T-cell activity and not a CD4 + regulatory T-cell response. *J Immunother*. 2012;35(3):254–266.
30. Tureci O, Vormehr M, Diken M, Kreiter S, Huber C, Sahin U. Targeting the heterogeneity of cancer with individualized neoepitope vaccines. *Clin Cancer Res*. 2016;22(8): 1885–1896.

31. Yadav M, Jhunjhunwala S, Phung QT, Lupardus P, Tanguay J, Bumbaca S, et al. Predicting immunogenic tumour mutations by combining mass spectrometry and exome sequencing. *Nature*. 2014;515(7528):572–576.
32. Nagane M, Coufal F, Lin H, Bogler O, Cavenee WK, Huang HJ. A common mutant epidermal growth factor receptor confers enhanced tumorigenicity on human glioblastoma cells by increasing proliferation and reducing apoptosis. *Cancer Res*. 1996;56(21):5079–5086.
33. Malekzadeh P, Pasetto A, Robbins PF, Parkhurst MR, Paria BC, Jia L, et al. Neoantigen screening identifies broad TP53 mutant immunogenicity in patients with epithelial cancers. *J Clin Invest*. 2019;129(3):1109–1114.
34. Slingluff Jr. CL, Petroni GR, Chianese-Bullock KA, Smolkin ME, Ross MI, Haas NB, et al. Randomized multicenter trial of the effects of melanoma-associated helper peptides and cyclophosphamide on the immunogenicity of a multipeptide melanoma vaccine. *J Clin Oncol*. 2011;29(21):2924–2932.
35. Sarkizova S, Klaeger S, Le PM, Li LW, Oliveira G, Keshishian H, et al. A large peptidome dataset improves HLA class I epitope prediction across most of the human population. *Nat Biotechnol*. 2020;38(2):199–209.
36. Shao XM, Bhattacharya R, Huang J, Sivakumar IKA, Tokheim C, Zheng L, et al. High-throughput prediction of MHC class I and II neoantigens with MHCnuggets. *Cancer Immunol Res*. 2020;8(3):396–408.
37. Gfeller D, Bassani-Sternberg M, Schmidt J, Luescher IF. Current tools for predicting cancer-specific T cell immunity. *Oncoimmunology*. 2016;5(7):e1177691.
38. Liu XS, Mardis ER. Applications of immunogenomics to cancer. *Cell*. 2017;168(4):600–612.
39. The problem with neoantigen prediction. *Nat Biotechnol*. 2017;35(2):97.
40. Kakimi K, Isobe M, Uenaka A, Wada H, Sato E, Doki Y, et al. A phase I study of vaccination with NY-ESO-1f peptide mixed with Picibanil OK-432 and Montanide ISA-51 in patients with cancers expressing the NY-ESO-1 antigen. *Int J Cancer*. 2011;129(12):2836–2846.
41. Kalams SA, Walker BD. The critical need for CD4 help in maintaining effective cytotoxic T lymphocyte responses. *J Exp Med*. 1998;188(12):2199–2204.
42. Aichele P, Brduscha-Riem K, Zinkernagel RM, Hengartner H, Pircher H. T cell priming vs T cell tolerance induced by synthetic peptides. *J Exp Med*. 1995;182(1):261–266.
43. Toes RE, Blom RJ, Offringa R, Kast WM, Melief CJ. Enhanced tumor outgrowth after peptide vaccination. Functional deletion of tumor-specific CTL induced by peptide vaccination can lead to the inability to reject tumors. *J Immunol*. 1996;156(10):3911–3918.
44. Tuettenberg A, Becker C, Huter E, Knop J, Enk AH, Jonuleit H. Induction of strong and persistent MelanA/MART-1-specific immune responses by adjuvant dendritic cell-based vaccination of stage II melanoma patients. *Int J Cancer*. 2006;118(10):2617–2627.
45. Bijker MS, van den Eeden SJ, Franken KL, Melief CJ, van der Burg SH, Offringa R. Superior induction of anti-tumor CTL immunity by extended peptide vaccines involves prolonged, DC-focused antigen presentation. *Eur J Immunol*. 2008;38(4):1033–1042.
46. Melief CJ, van der Burg SH. Immunotherapy of established (pre)malignant disease by synthetic long peptide vaccines. *Nat Rev Cancer*. 2008;8(5):351–360.
47. Hailemichael Y, Dai Z, Jaffarzad N, Ye Y, Medina MA, Huang XF, et al. Persistent antigen at vaccination sites induces tumor-specific CD8(+) T cell sequestration, dysfunction and deletion. *Nat Med*. 2013;19(4):465–472.

48. Reddy ST, Rehor A, Schmoekel HG, Hubbell JA, Swartz MA. In vivo targeting of dendritic cells in lymph nodes with poly(propylene sulfide) nanoparticles. *J Control Release*. 2006;112(1):26–34.
49. Gay NJ, Gangloff M. Structure and function of Toll receptors and their ligands. *Annu Rev Biochem*. 2007;76:141–165.
50. Takeda K, Kaisho T, Akira S. Toll-like receptors. *Annu Rev Immunol*. 2003;21:335–376.
51. Akira S, Takeda K, Kaisho T. Toll-like receptors: critical proteins linking innate and acquired immunity. *Nat Immunol*. 2001;2(8):675–680.
52. Takeda K, Akira S. Toll-like receptors in innate immunity. *Int Immunol*. 2005;17(1): 1–14.
53. Theofilopoulos AN, Baccala R, Beutler B, Kono DH. Type I interferons (alpha/beta) in immunity and autoimmunity. *Annu Rev Immunol*. 2005;23:307–336.
54. Medzhitov R. Toll-like receptors and innate immunity. *Nat Rev Immunol*. 2001;1 (2):135–145.
55. Hennessy EJ, Parker AE, O'Neill LA. Targeting Toll-like receptors: emerging therapeutics? *Nat Rev Drug Discov*. 2010;9(4):293–307.
56. Lesterhuis WJ, Schreibelt G, Scharenborg NM, Brouwer HM, Gerritsen MJ, Croockewit S, et al. Wild-type and modified gp100 peptide-pulsed dendritic cell vaccination of advanced melanoma patients can lead to long-term clinical responses independent of the peptide used. *Cancer Immunol Immunother*. 2011;60(2):249–260.
57. Gulley JL, Madan RA, Tsang KY, Jochems C, Marte JL, Farsaci B, et al. Immune impact induced by PROSTVAC (PSA-TRICOM), a therapeutic vaccine for prostate cancer. *Cancer Immunol Res*. 2014;2(2):133–141.
58. Lindau D, Gielen P, Kroesen M, Wesseling P, Adema GJ. The immunosuppressive tumour network: myeloid-derived suppressor cells, regulatory T cells and natural killer T cells. *Immunology*. 2013;138(2):105–115.
59. Zou W. Regulatory T cells, tumour immunity and immunotherapy. *Nat Rev Immunol*. 2006;6(4):295–307.
60. Kinkead HL, Hopkins A, Lutz E, Wu AA, Yarchoan M, Cruz K, et al. Combining STING-based neoantigen-targeted vaccine with checkpoint modulators enhances antitumor immunity in murine pancreatic cancer. *JCI Insight*. 2018;3(20).
61. Zhu G, Lynn GM, Jacobson O, Chen K, Liu Y, Zhang H, et al. Albumin/vaccine nanocomplexes that assemble in vivo for combination cancer immunotherapy. *Nat Commun*. 2017;8 (1):1954.
62. Talmadge JE, Gabrilovich DI. History of myeloid-derived suppressor cells. *Nat Rev Cancer*. 2013;13(10):739–752.
63. Hamilton DH, Huang B, Fernando RI, Tsang KY, Palena C. WEE1 inhibition alleviates resistance to immune attack of tumor cells undergoing epithelial-mesenchymal transition. *Cancer Res*. 2014;74(9):2510–2519.
64. Hamilton DH, Litzinger MT, Fernando RI, Huang B, Palena C. Cancer vaccines targeting the epithelial-mesenchymal transition: tissue distribution of brachyury and other drivers of the mesenchymal-like phenotype of carcinomas. *Semin Oncol*. 2012;39(3):358–366.
65. Gatti-Mays ME, Redman JM, Collins JM, Bilusic M. Cancer vaccines: Enhanced immunogenic modulation through therapeutic combinations. *Hum Vaccin Immunother*. 2017;13 (11):2561–2574.
66. Neri D, Sondel PM. Immunocytokines for cancer treatment: past, present and future. *Curr Opin Immunol*. 2016;40:96–102.

67. Waldmann TA. The shared and contrasting roles of IL2 and IL15 in the life and death of normal and neoplastic lymphocytes: implications for cancer therapy. *Cancer Immunol Res.* 2015;3(3):219–227.
68. Conlon KC, Lugli E, Welles HC, Rosenberg SA, Fojo AT, Morris JC, et al. Redistribution, hyperproliferation, activation of natural killer cells and CD8 T cells, and cytokine production during first-in-human clinical trial of recombinant human interleukin-15 in patients with cancer. *J Clin Oncol.* 2015;33(1):74–82.
69. Miller JS, Morishima C, McNeel DG, Patel MR, Kohrt HEK, Thompson JA, et al. A first-in-human phase I study of subcutaneous outpatient recombinant human IL15 (rhIL15) in adults with advanced solid tumors. *Clin Cancer Res.* 2018;24(7):1525–1535.
70. Romee R, Cooley S, Berrien-Elliott MM, Westervelt P, Verneris MR, Wagner JE, et al. First-in-human phase 1 clinical study of the IL-15 superagonist complex ALT-803 to treat relapse after transplantation. *Blood.* 2018;131(23):2515–2527.
71. Lee KL, Benz SC, Hicks KC, Nguyen A, Gameiro SR, Palena C, et al. Efficient tumor clearance and diversified immunity through neoepitope vaccines and combinatorial immunotherapy. *Cancer Immunol Res.* 2019;7(8):1359–1370.
72. Aste-Amezaga M, D'Andrea A, Kubin M, Trinchieri G. Cooperation of natural killer cell stimulatory factor/interleukin-12 with other stimuli in the induction of cytokines and cytotoxic cell-associated molecules in human T and NK cells. *Cell Immunol.* 1994;156(2):480–492.
73. Del Vecchio M, Bajetta E, Canova S, Lotze MT, Wesa A, Parmiani G, et al. Interleukin-12: biological properties and clinical application. *Clin Cancer Res.* 2007;13(16):4677–4685.
74. Lacy MQ, Jacobus S, Blood EA, Kay NE, Rajkumar SV, Greipp PR. Phase II study of interleukin-12 for treatment of plateau phase multiple myeloma (E1A96): a trial of the Eastern Cooperative Oncology Group. *Leuk Res.* 2009;33(11):1485–1489.
75. Fallon J, Tighe R, Kradjian G, Guzman W, Bernhardt A, Neuteboom B, et al. The immunocytokine NHS-IL-12 as a potential cancer therapeutic. *Oncotarget.* 2014;5(7):1869–1884.
76. Strauss J, Heery CR, Kim JW, Jochems C, Donahue RN, Montgomery AS, et al. First-in-human phase I trial of a tumor-targeted cytokine (NHS-IL-12) in subjects with metastatic solid tumors. *Clin Cancer Res.* 2019;25(1):99–109.
77. Smalley Rumfield C, Pellom ST, Morillon II YM, Schlom J, Jochems C. Immunomodulation to enhance the efficacy of an HPV therapeutic vaccine. *J Immunother Cancer.* 2020;8(1).
78. Zitvogel L, Kepp O, Senovilla L, Menger L, Chaput N, Kroemer G. Immunogenic tumor cell death for optimal anticancer therapy: the calreticulin exposure pathway. *Clin Cancer Res.* 2010;16(12):3100–3104.
79. Tesniere A, Schlemmer F, Boige V, Kepp O, Martins I, Ghiringhelli F, et al. Immunogenic death of colon cancer cells treated with oxaliplatin. *Oncogene.* 2010;29(4):482–491.
80. Locher C, Conforti R, Aymeric L, Ma Y, Yamazaki T, Rusakiewicz S, et al. Desirable cell death during anticancer chemotherapy. *Ann N Y Acad Sci.* 2010;1209:99–108.
81. Kepp O, Galluzzi L, Martins I, Schlemmer F, Adjemian S, Michaud M, et al. Molecular determinants of immunogenic cell death elicited by anticancer chemotherapy. *Cancer Metastasis Rev.* 2011;30(1):61–69.
82. Chakraborty M, Abrams SI, Coleman CN, Camphausen K, Schlom J, Hodge JW. External beam radiation of tumors alters phenotype of tumor cells to render them susceptible to vaccine-mediated T-cell killing. *Cancer Res.* 2004;64(12):4328–4337.

83. Gameiro SR, Malamas AS, Bernstein MB, Tsang KY, Vassantachart A, Sahoo N, et al. Tumor cells surviving exposure to proton or photon radiation share a common immunogenic modulation signature, rendering them more sensitive to T cell-mediated killing. *Int J Radiat Oncol Biol Phys*. 2016;95(1):120–130.
84. Garnett CT, Palena C, Chakraborty M, Tsang KY, Schlom J, Hodge JW. Sublethal irradiation of human tumor cells modulates phenotype resulting in enhanced killing by cytotoxic T lymphocytes. *Cancer Res*. 2004;64(21):7985–7994.
85. Gelbard A, Garnett CT, Abrams SI, Patel V, Gutkind JS, Palena C, et al. Combination chemotherapy and radiation of human squamous cell carcinoma of the head and neck augments CTL-mediated lysis. *Clin Cancer Res*. 2006;12(6):1897–1905.
86. Hodge JW, Garnett CT, Farsaci B, Palena C, Tsang KY, Ferrone S, et al. Chemotherapy-induced immunogenic modulation of tumor cells enhances killing by cytotoxic T lymphocytes and is distinct from immunogenic cell death. *Int J Cancer*. 2013;133(3):624–636.
87. Kodumudi KN, Woan K, Gilvary DL, Sahakian E, Wei S, Djeu JY. A novel chemoimmunomodulating property of docetaxel: suppression of myeloid-derived suppressor cells in tumor bearers. *Clin Cancer Res*. 2010;16(18):4583–4594.
88. Garnett CT, Schlom J, Hodge JW. Combination of docetaxel and recombinant vaccine enhances T-cell responses and antitumor activity: effects of docetaxel on immune enhancement. *Clin Cancer Res*. 2008;14(11):3536–3544.
89. Kudo-Saito C, Schlom J, Camphausen K, Coleman CN, Hodge JW. The requirement of multimodal therapy (vaccine, local tumor radiation, and reduction of suppressor cells) to eliminate established tumors. *Clin Cancer Res*. 2005;11(12):4533–4544.
90. Roselli M, Cereda V, di Bari MG, Formica V, Spila A, Jochems C, et al. Effects of conventional therapeutic interventions on the number and function of regulatory T cells. *Oncoimmunology*. 2013;2(10):e27025.
91. Melief CJM, Welters MJP, Vergote I, Kroep JR, Kenter GG, Ottevanger PB, et al. Strong vaccine responses during chemotherapy are associated with prolonged cancer survival. *Sci Transl Med*. 2020;12(535):eaaz8235.
92. Welters MJ, van der Sluis TC, van Meir H, Loof NM, van Ham VJ, van Duikeren S, et al. Vaccination during myeloid cell depletion by cancer chemotherapy fosters robust T cell responses. *Sci Transl Med*. 2016;8(334):334ra52.
93. McGranahan N, Furness AJ, Rosenthal R, Ramskov S, Lyngaa R, Saini SK, et al. Clonal neoantigens elicit T cell immunoreactivity and sensitivity to immune checkpoint blockade. *Science*. 2016;351(6280):1463–1469.
94. Pardoll DM. The blockade of immune checkpoints in cancer immunotherapy. *Nat Rev Cancer*. 2012;12(4):252–264.
95. van der Burg SH, Arens R, Ossendorp F, van Hall T, Melief CJ. Vaccines for established cancer: overcoming the challenges posed by immune evasion. *Nat Rev Cancer*. 2016;16(4):219–233.
96. Massarelli E, William W, Johnson F, Kies M, Ferrarotto R, Guo M, et al. Combining immune checkpoint blockade and tumor-specific vaccine for patients with incurable human papillomavirus 16-related cancer: a phase 2 clinical trial. *JAMA Oncol*. 2019;5(1):67–73.
97. Bauml J, Seiwert TY, Pfister DG, Worden F, Liu SV, Gilbert J, et al. Pembrolizumab for platinum- and cetuximab-refractory head and neck cancer: results from a single-arm, phase II study. *J Clin Oncol*. 2017;35(14):1542–1549.

98. Ferris RL, Licitra L, Fayette J, Even C, Blumenschein Jr. G, Harrington KJ, et al. Nivolumab in patients with recurrent or metastatic squamous cell carcinoma of the head and neck: efficacy and safety in checkmate 141 by prior cetuximab use. *Clin Cancer Res.* 2019;25(17):5221–5230.
99. Mehra R, Seiwert TY, Gupta S, Weiss J, Gluck I, Eder JP, et al. Efficacy and safety of pembrolizumab in recurrent/metastatic head and neck squamous cell carcinoma: pooled analyses after long-term follow-up in KEYNOTE-012. *Br J Cancer.* 2018;119(2):153–159.
100. Kalluri R, Weinberg RA. The basics of epithelial-mesenchymal transition. *J Clin Invest.* 2009;119(6):1420–1428.
101. Thiery JP, Sleeman JP. Complex networks orchestrate epithelial-mesenchymal transitions. *Nat Rev Mol Cell Biol.* 2006;7(2):131–142.
102. Polyak K, Weinberg RA. Transitions between epithelial and mesenchymal states: acquisition of malignant and stem cell traits. *Nat Rev Cancer.* 2009;9(4):265–273.
103. Thiery JP. Epithelial-mesenchymal transitions in tumour progression. *Nat Rev Cancer.* 2002;2(6):442–454.
104. Anestakis D, Petanidis S, Kalyvas S, Nday CM, Tsave O, Kioseoglou E, et al. Mechanisms and applications of interleukins in cancer immunotherapy. *Int J Mol Sci.* 2015;16(1):1691–1710.
105. David JM, Dominguez C, Hamilton DH, Palena C. The IL-8/IL-8R axis: a double agent in tumor immune resistance. *Vaccines (Basel).* 2016;4(3).
106. David JM, Dominguez C, Palena C. Pharmacological and immunological targeting of tumor mesenchymalization. *Pharmacol Ther.* 2017;170:212–225.
107. Fernando RI, Hamilton DH, Dominguez C, David JM, McCampbell KK, Palena C. IL-8 signaling is involved in resistance of lung carcinoma cells to erlotinib. *Oncotarget.* 2016;7(27):42031–42044.
108. Fousek K, Horn LA, Palena C. Interleukin-8: a chemokine at the intersection of cancer plasticity, angiogenesis, and immune suppression. *Pharmacol Ther.* 2021;219:107692.
109. Hamilton DH, David JM, Dominguez C, Palena C. Development of cancer vaccines targeting brachyury, a transcription factor associated with tumor epithelial-mesenchymal transition. *Cells Tissues Organs.* 2017;203(2):128–138.
110. Hamilton DH, Griner LM, Keller JM, Hu X, Southall N, Marugan J, et al. Targeting estrogen receptor signaling with fulvestrant enhances immune and chemotherapy-mediated cytotoxicity of human lung cancer. *Clin Cancer Res.* 2016;22(24):6204–6216.
111. Horn LA, Fousek K, Palena C. Tumor plasticity and resistance to immunotherapy. *Trends Cancer.* 2020;6(5):432–441.
112. Fabian KP, Malamas AS, Padget MR, Solocinski K, Wolfson B, Fujii R, et al. Therapy of established tumors with rationally designed multiple agents targeting diverse immune-tumor interactions: engage, expand, enable. *Cancer Immunol Res.* 2020;9:239–252.
113. Redman JM, Steinberg SM, Gulley JL. Quick efficacy seeking trial (QuEST1): a novel combination immunotherapy study designed for rapid clinical signal assessment metastatic castration-resistant prostate cancer. *J Immunother Cancer.* 2018;6(1):91.

Chapter 10

Personalized peptide vaccines

Hidetomo Himuro and Tetsuro Sasada
Kanagawa Cancer Center Research Institute, Kanagawa, Japan

Introduction

Cancer vaccines that can directly induce/activate antitumor immunity may have a great potential as a therapeutic modality against tumors. Cancer vaccines comprise substances (peptides, proteins, mRNA, etc.) derived from tumor-associated antigens (TAAs) or tumor-specific antigens (TSAs), which are preferentially or specifically expressed in tumor cells, and are expected to induce/activate specific T-cell responses.[1–3] Considering the mechanistic basis of amplifying immune responses to tumor cells, cancer vaccines are promising and effective in theory but have not performed well so far in clinical trials.[4–7] There are several possible reasons for such failure. One of the serious issues is that most tumor antigens employed in clinical trials were not tumor-specific and could not induce strong immune responses. Therefore developing a cancer vaccine targeting TSAs might be necessary.[2,3] For example, neoantigens resulting from somatic mutations in tumor cells are tumor-specific and might be highly desirable targets for eliciting strong antitumor responses. However, since most tumor-specific somatic mutations are passenger mutations that are not shared among multiple patients, cancer vaccines targeting them need to be personalized for each patient. Recently, remarkable technological innovations, such as next-generation sequencing (NGS) and new bioinformatics tools, have enabled detailed biological and immunological analysis, facilitating the development of personalized vaccines. This article outlines and discusses the current status and future perspective of personalized peptide vaccines.

Limitation of TAA-derived peptides for cancer vaccines

As TAAs are not tumor specific but frequently shared by many patients with cancer, cancer vaccines targeting them could be commonly employed to save costs and labor. Thus various TAAs have been identified by comprehensive analysis of cancer tissues and corresponding normal tissues and applied to

Cancer Vaccines as Immunotherapy of Cancer. DOI: https://doi.org/10.1016/B978-0-12-823901-8.00007-8
© 2022 Elsevier Inc. All rights reserved.

cancer vaccine therapy.[8,9] In particular, short-chain peptides (SPs) that do not need to be processed for binding to HLA molecules are one of the most common forms of antigens for vaccination. TAA-derived SPs administered to patients with cancer are expected to bind to HLA class I (HLA-I) molecules on the cell surfaces of APCs and activate specific cytotoxic T lymphocytes (CTLs), which attack cancer cells that express the same SPs on the cell surface. So far, thousands of TAAs have been identified using several approaches, such as the systematic evolution of ligands by exponential enrichment (SELEX) and cDNA microarray analysis.[8–10] The amino acid sequences and HLA restriction of antigen epitopes derived from them are usually predicted using HLA-binding algorithms, and their immunogenicity can be confirmed using human T-cell assays in vitro or transgenic mice expressing HLA in vivo. In this way, many TAA-derived SPs that can induce/activate CTLs in an HLA-I-restricted manner have been identified and clinically applied to cancer vaccine therapy.[4–7] Alternatively, because of the uncertainty of algorithm-based prediction of antigen epitopes, a mass spectrometry (MS)-based peptidomic approach has been attempted to directly identify amino acid sequences of peptides bound to HLA molecules on the surface of cancer cells and/or APCs.[11] For example, Walter et al. have identified multiple TAA-derived SPs bound to HLA-I molecules expressed in renal cell carcinoma using an MS-based peptidomic approach and applied to cancer peptide vaccine therapy.[12]

Numerous clinical trials of cancer vaccines targeting various TAA-derived peptide epitopes, which lack complete specificity to tumors, have been extensively attempted, but with only limited success.[4–7] Indeed, although some patients with advanced cancers showed tumor shrinkage and prolonged stable disease with no serious adverse events, almost all late-phase trials, except for GP-100 peptide vaccine for melanoma,[13] failed to meet their endpoints. This might be mainly because T cells that have a high affinity to TAA-derived epitopes may have been eliminated by central immune tolerance due to negative selection in the thymus. In addition, since many previous clinical trials employed SPs in complex with incomplete Freund Adjuvant for vaccination, vaccine delivery systems might be inappropriate for inducing sufficient immune responses to TAA-derived epitopes. For example, in a phase I trial in patients with advanced melanoma, an intravenously administered liposomal RNA vaccine, which targets four nonmutated, shared TAAs, in combination with immune checkpoint-inhibitors was shown to induce antigen-specific strong $CD4^+$ and $CD8^+$ T-cell responses, resulting in durable objective responses.[14] This finding suggests the importance of vaccine delivery systems for targeting nonmutant TAAs.

Antitumor immunity may depend strongly on both the tumor cell characteristics, such as expression levels of antigens and immune-related molecules, and the immunological status of the host, such as T-cell repertoire and immunosuppressive tumor microenvironments, in patients with cancer.[15,16] It thus might differ widely among individuals since the tumor cell characteristics and the host immune cell repertoires are quite heterogeneous and diverse among patients, even among those with identical HLA types and the same pathological types of cancer. Nevertheless, although the expression of vaccine antigens in tumor cells is sometimes confirmed, the immunological statuses of the hosts are rarely evaluated before vaccination. Considering such diversity of antitumor immunity, tailored selection of vaccine antigens appropriate for individual patients, depending on their immunological statuses, could be a rational approach for developing effective cancer vaccines. For example, a novel immunotherapeutic approach called personalized peptide vaccine has been developed, in which a maximum of four HLA-matched TAA-derived SPs was selected based on the preexisting host immunity evaluated by antigen-specific CTL and/or IgG responses before vaccination.[6,8,17] A series of phase I and phase II clinical trials have shown better antigen-specific immune responses and promising clinical outcomes in patients with various types of advanced cancers, but randomized phase III trials involving patients with glioblastoma and prostatic cancer failed to meet their endpoints.[18]

Long-chain peptides (LPs) containing multiple HLA-I- and/or HLA class II (HLA-II)-restricted antigen epitopes have attracted attention, as they can stimulate both $CD4^+$ and $CD8^+$ T cells.[19] The presence of TAA-specific Th1 immune responses is important for inducing stronger antitumor immunity. In addition, TAA-derived SPs may induce nonresponsiveness to CTLs by directly binding to HLA-I molecules expressed on nonimmune cells that do not express costimulatory molecules, such as CD80/86. In contrast, as LPs cannot directly bind to HLA molecules, they need to be taken up and processed intracellularly by APCs for HLA-I-restricted cross-presentation to TAA-specific CTLs and/or HLA-II-restricted antigen presentation to TAA-specific Th cells. Therefore LPs are expected to be useful for inducing stronger antitumor effects mediated by both $CD8^+$ CTLs and $CD4^+$ Th cells without induction of nonresponsiveness, which is a problem with SP vaccines.[20] Furthermore, LPs containing multipeptides are expected to be more effective than single-peptide preparations because they are less likely to cause immune escape through loss of TAA and/or HLA-I expression.[21] Early-phase clinical trials using TAA-derived LP vaccines have been conducted with some potential results. For example, a first-in-human study of TAS0313, a novel cancer vaccine cocktail consisting of several LPs targeting multiple TAAs, showed substantial induction of specific CTL and IgG responses and tumor-infiltrating lymphocyte (TIL) accumulation in patients with advanced solid tumors.[22] In addition, another first-in-man phase I/II study of a synthetic LP vaccine targeting RhoC demonstrated induction of a potent and long-lasting T-cell immunity in the majority of patients with prostate cancer.[23] Further late-phase studies remain necessary to confirm the feasibility of LP vaccines.

Potential of neoantigen peptides

TAA-derived epitopes may induce central immune tolerance due to negative selection.[24] In contrast, it is speculated that neoantigens derived from TSAs occurring only in tumors, but not in normal cells, can efficiently target tumor cells since neoantigen-specific T cells are expected to be unsusceptible to immune tolerance. There are several classes of TSAs that can potentially generate immunogenic neoantigens: (1) genomic alterations, such as nonsynonymous point mutations, nucleotide deletion/insertions (INDELs), and gene fusions (genomic rearrangements); (2) oncogenic viruses, such as human papilloma virus, Epstein-Barr virus, and hepatitis B virus; (3) aberrant mRNA splicing; (4) aberrant RNA translation of protein-coding genes in alternative open reading frames or translation of noncoding sequences; and (5) posttranslational modifications, such as proteasomal peptide splicing and enzymatic amino acid modifications (phosphorylation, acetylation, citrullination, glycosylation, or deamidation).[9,25]

Nonsynonymous point mutations providing neoantigens are roughly divided into two types: driver and passenger mutations. Driver mutations are gene mutations that promote the transformation/progression of malignant tumor cells, whereas passenger mutations are not directly involved in such a mechanism. Targeting for neoantigens derived from driver mutations is suggested to be more advantageous for several reasons. They are less likely to be lost due to immune escape, because they are essential for malignant transformation/progression. In addition, they might be clonal, shared by all (or most) tumor cells. Furthermore, when they are shared by many patients, they could provide off-the-shelf vaccines to treat them. Recently, several driver mutations, such as KRAS[26] and p53,[27] have been reported to provide immunogenic neoantigens that might cause tumor cell destruction mediated by specific CTLs, but driver mutations are often poorly presented by HLA to prevent immune attack from the early stages of tumorigenesis.[28] Alternatively, long-term exposure to immunogenic neoantigens derived from driver mutations would make specific T cells unresponsive due to some mechanisms, such as epigenetic changes.[29] Therefore most gene mutations that provide immunogenic neoantigens are derived from passenger mutations rather than driver mutations.[30]

Preclinical studies demonstrated the importance of neoantigens derived from somatic mutations in murine tumor models.[31–35] In these murine tumor models, vaccination using neoantigen peptides or mRNAs efficiently elicited specific T-cell responses that mediated antitumor activities. Interestingly, immunotherapy-induced antitumor responses require the activity of both tumor-antigen-specific $CD8^+$ and $CD4^+$ T cells, even in tumors without MHC class II expression.[35] In humans, adoptive cell transfer of TILs or of T cells genetically engineered to express T-cell receptor (TCR) specific to TAA-derived epitopes has been reported to result in complete and durable responses with a high overall response rate in patients with advanced melanoma,[36,37] but not in those with other solid tumors. In contrast, ATC of TILs or TCR-transduced T cells specific to TSA-derived neoantigens demonstrated substantial tumor regression in those with advanced solid cancers, including cholangiocarcinoma,[38] colon cancer,[26] and breast cancer.[39] Therefore targeting TSA-derived neoantigens for immunotherapy has been expected as a novel and promising strategy.

Procedure of personalized peptide vaccines targeting neoantigens

Currently, clinical trials of cancer peptide vaccines targeting neoantigens derived from somatic mutations have been extensively planned and performed for various carcinomas. However, of note, due to the complexity and diversity of mutations and HLA molecules in each patient, neoantigens selected for vaccination are diverse and different among individuals. Studies have shown that the efficacy of cancer vaccines targeting a single tumor antigen might be limited because they cannot overcome problems, such as tumor heterogeneity and immune evasion. For example, a phase II clinical trial of an EGFRvIII-targeted peptide vaccine showed loss of EGFRvIII expression, leading to limited clinical efficacy in patients with newly diagnosed EGFRvIII-expressing glioblastoma with minimal residual disease.[40] Therefore to prevent immune escape of tumor cells due to loss of vaccine antigens, effective cancer vaccines should target multiple neoantigens identified in each individual tumor.

The following is a typical procedure for selecting neoantigens for personalized vaccine therapy in currently undergoing clinical trials[2,3] (Fig. 10.1). First, a comprehensive whole-exome sequencing analysis using DNA extracted from a patient's tumor and normal tissues is performed to identify somatic mutations in tumor cells, which result in amino acid substitutions or frameshifts caused by INDELs. Second, sequencing of RNA extracted from a patient's tumor is performed to confirm that the identified mutations are transcribed and expressed as RNAs in tumor cells because not all exons are transcribed into RNAs. Third, HLA types are determined using DNAs from normal tissues/cells in each patient, and among the amino acid changes due to somatic mutations, neoepitopes likely to be presented by the patients' HLA molecules are selected.[10,38] This step is performed by mainly predicting whether the peptides containing the mutated/substituted amino acids can bind to HLA molecules with a certain affinity or higher through an algorithm created by the machine learning method based on the results of MS using tumor cell lines and clinical tumor samples.[41–44] Fourth, if T cells (TILs or peripheral blood lymphocytes) from patients are available, the immunogenicity of the selected neoepitope candidates could be confirmed by an in vitro assessment because predicting by computer-based analysis is difficult.[45] For example, T cells from patients can be stimulated with antigen peptides

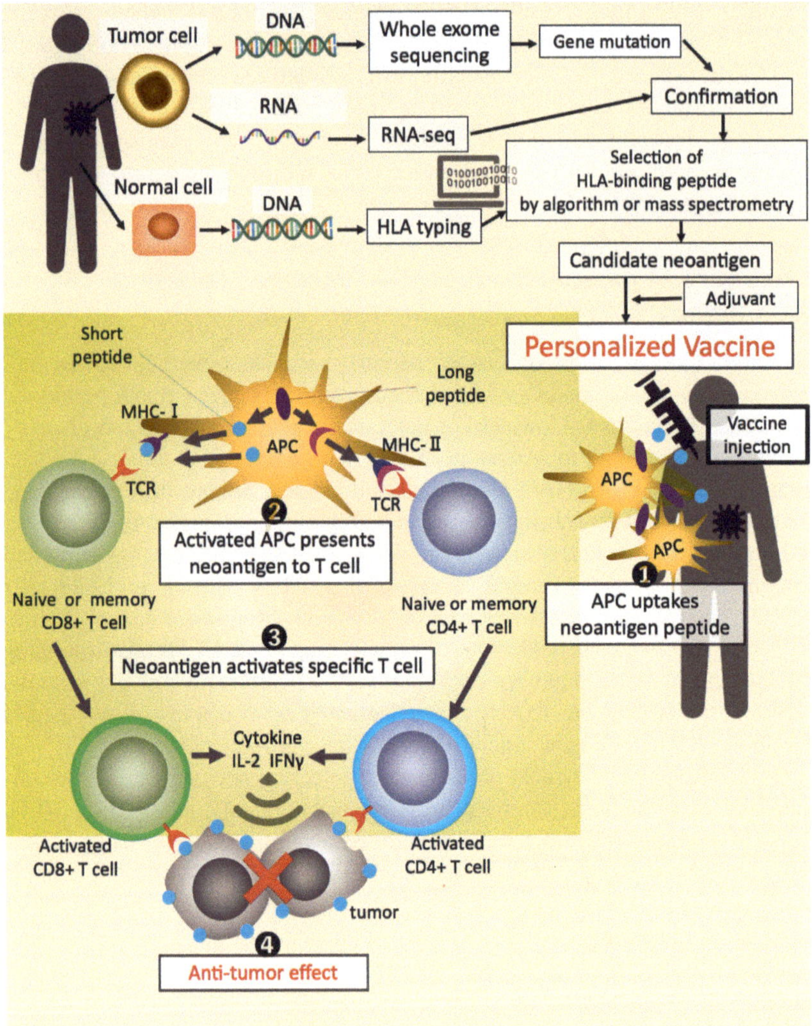

FIGURE 10.1 Work flow for personalized neoantigen-based peptide vaccines and mechanism of action.

in vitro and examined for the production of cytokines, such as IFN-γ, by using the enzyme-linked immune absorbent spot and/or the upregulation of T-cell activation markers, such as 4–1BB and CD40L, by flow cytometry. In addition, HLA multimer reagents in which a candidate peptide is incorporated into a fluorescently labeled HLA molecule can be used for biochemically evaluating antigen-specific T cells using flow cytometry.[45–47] Such techniques might provide useful information for predicting whether neoepitope candidates can induce specific immune responses, although the availability of

T cells (TILs or peripheral blood lymphocytes) from patients and low sensitivity of the assays are often problematic.

Clinical trials of personalized vaccines targeting neoantigens

Clinical trials of neoantigen-based personalized cancer vaccines have been conducted on multiple tumor types using various adjuvants and delivery approaches. Ott et al. have reported a phase I clinical trial of a personalized peptide vaccine in patients with melanoma (ClinicalTrials.gov identifier: NCT01970358).[48] Tumor-specific neoantigens were individually identified in each patient, and a maximum of 20 synthetic LPs (15–30 mer) derived from them was administered with poly ICLC in six patients. Four patients with stage III melanoma showed no signs of tumor recurrence until 32 months after vaccination. In addition, two patients with stage IV melanoma showed recurrence of the disease immediately after vaccination but subsequently experienced complete tumor regression after receiving anti-PD1 therapy. Notably, the same authors demonstrated long-term persistence of neoantigen-specific T-cell responses exhibiting a memory phenotype nearly 4 years after vaccination.[49] Furthermore, two other phase I clinical trials of a personalized cancer vaccine using other approaches were also conducted on patients with melanoma, with promising results.[50,51] Carreno et al. reported an intravenous administration of dendritic cells pulsed with 7 HLA-A2-restricted neoantigen (8–10 mer) SPs in three patients with malignant melanoma (ClinicalTrials.gov identifier: NCT00683670).[50] In this study, vaccination promoted diverse neoantigen-specific T-cell responses, confirming the high immunogenicity of neoantigens. Sahin et al. have reported an RNA-based poly-neoepitope approach in 13 patients with melanoma (ClinicalTrials.gov identifier: NCT02035956).[51] Individual mutations were comprehensively identified, and an RNA vaccine containing 10 neoepitopes designed by computational prediction in each patient was manufactured. All patients developed T-cell responses against multiple vaccine neoepitopes, with potential clinical effects. Both trials have shown substantial induction of antigen-specific T-cell responses that were not observed before treatment and suggested the feasibility and safety of neoantigen-based cancer vaccines. Furthermore, a clinical trial of a personalized peptide vaccine was conducted in patients with glioblastoma, and promising results, such as the induction of T-cell responses, have been reported.[52]

Consortiums composed of researchers from multiple countries, such as the Glioma Actively Personalized Vaccine Consortium (GAPVAC) and Cancer Vaccine Development for Hepatocellular Carcinoma, have conducted clinical trials of personalized peptide vaccines (ClinicalTrials.gov Identifiers: NCT03203005).[53,54] For example, the GAPVAC clinical trial involving 16 patients with glioblastoma used a combination of vaccines composed of a premade library of unmutated TAA-derived peptides that are often overexpressed and shared in patients' tumor tissues (APVAC1) and personalized neoantigen peptides that are variable among different patients (APVAC2).[53] Immunological assessment demonstrated that APVAC1 elicited a sustained response of central memory $CD8^+$ T cells, whereas APVAC2 induced a predominantly $CD4^+$ Th1 cell response to the predicted neoepitopes; neoepitope-specific $CD4^+$ T-cell responses were identified in 8 of the 10 patients (80%), who were evaluated for APVAC2 immunogenicity.

Collectively, the aforementioned clinical trials have suggested that neoantigens could be promising targets for cancer vaccines because immunological studies have shown that they elicit substantial and long-lasting antigen-specific T-cell responses in vaccinated patients.[48–52] Currently, more clinical trials of personalized neopeptide-based vaccines against various types of cancers, including glioblastoma, breast cancer, pancreatic cancer, pediatric brain cancer, and hepatocellular carcinoma (ClinicalTrials.gov identifiers: NCT02287428, NCT02510950, NCT02427581, NCT02600949, and NCT03068832, respectively), are ongoing, and their results are awaited.[55] In the future, to demonstrate the superiority of TSA-derived neoantigens over TAA-derived antigens as targets for cancer vaccines, the strength and breadth of specific immune responses, such as the affinity of antigen-specific TCR and phenotypes of activated antigen-specific T cells, should be precisely examined and compared between them.

Challenges to overcome the current limitations

Most neoantigens are unique and not shared among patients with cancer. Therefore preparing personalized peptides for each patient is more expensive and time-consuming than other treatment methods. Although the costs of tumor sample analysis and peptide production have been slightly reduced due to innovative technological advances, further improvements, such as faster NGS technology and automated mass production for peptides, are needed.[56] In addition, various issues should be overcome to further develop this treatment modality, as discussed below.

How to efficiently identify neoantigens

To further develop personalized cancer vaccines, appropriately identifying and selecting neoepitopes that are more likely to produce effective antitumor immunity might be the most important issue. However, there have been some problems in the accurate and efficient identification/selection of neoantigens.[57] Currently, neoantigens are usually identified by predicting their binding capabilities to HLA molecules based on the computer algorithm. For antigen processing and presentation, however, various processes are involved, including gene/protein expression and peptide editing and loading to MHC molecules.[58] Therefore various new algorithms have been developed by constantly integrating such elements for antigen presentation over time,[41–44,59] but their accuracy and efficacy in prediction have remained still unsatisfactory.[57,60] Indeed, current clinical trials of neoantigen vaccines have demonstrated that only a fraction of predicted HLA class I binding peptide candidates could induce $CD8^+$ T-cell responses and that they sometimes and unexpectedly activated $CD4^+$ T-cell responses, but not $CD8^+$ T-cell responses.[48,51,52] This might be because scientific knowledge of antigen processing and presentation is still limited and insufficient, and the underlying database has not yet matured.

Recent studies have reported new and detailed characterization on the immunogenicity of neoantigens. For example, it was suggested that the information on whether the mutation is located at an anchor residue seems to be important, although different studies led to different results. Duan et al. suggested that immunogenic neoantigens are more frequently mutated at the anchor residues,[61] whereas Fritsch et al. and Yadav et al. showed that amino acid substitutions at the nonanchor positions are more immunogenic by affecting the contact with TCR.[62,63] Capietto et al. suggested that for mutations at the anchor positions, the relative affinity to peptide-HLA complex (pHLA) between the mutant and wild-type counterpart peptide strongly affected the immunogenicity, but for those at the nonanchor positions, the absolute binding affinity of the mutant peptide to pHLA was a better predictor of T-cell responses,[64] In addition, the structural relationship between the TCR and mutated amino acid residues within the pHLA is suggested to be critical for antigen recognition and immune response mediated by T cells.[65] Especially, higher structural dissimilarity from the self (wild-type) counterpart peptide might increase the exposure and contact to specific TCR, resulting in enhanced immunogenicity.[66,67] Furthermore, cross-reactivities have been reported for $CD8^+$ CTLs that recognize both MHC class I-restricted tumor neoantigens and pathogen-derived antigens expressed by intestinal commensals, due to homology between them. For example, in a mouse model, a peptide derived from a commensal bacterium was shown to cross-react with a model neoantigen that was artificially induced into a murine tumor cell line.[68] For human pancreatic cancer, cross-reactivity between tumor neoantigens and homologous noncancer microbial antigens have also been documented.[69]

Recently, a 28-team global consortium wherein each participant predicted immunogenic epitopes from shared tumor sequencing data has developed a model of tumor neoepitope immunogenicity by integrating crucial features, such as strong pHLA binding affinity, high levels of mRNA expression in tumors, high degrees of pHLA binding stability, and recognition characteristics (agretopicity or foreignness).[60] In the future, it is expected that such consortium-based approaches could provide the opportunities to improve the bioinformatics pipelines used to predict which tumor mutations are likely to act as neoantigens for effective antitumor immunity.

As an alternative approach to the algorithm-based prediction, MS-based technology has been used along with exome sequencing to directly identify neoepitopes presented on MHC molecules from tumor tissues.[11,41–44] Recent great improvements of MS-related technology, including instrumentation speed and sensitivity, chromatographic resolution, and proteogenomic database search strategies, have facilitated the identification of HLA ligands from a variety of cell types and tumor tissues. Nevertheless, there have been a few reports regarding the MS-based identification of neoepitopes from clinical tumor tissues, possibly due to sensitivity issues, when they are expressed at minute amounts on tumor cells.[70] Further improvements are thus needed for clinical application of this technology.

Limited accuracy of HLA class II-restricted neoepitope prediction

$CD8^+$ T cells have been considered the most critical T-cell population because they directly exert antitumor effects. However, recent reports have suggested that $CD4^+$ Th cells also play an important role in antitumor immunity.[34,35,71] In addition to the indirect effects of helping other immune cells, such as the induction/activation of antitumor CTLs, $CD4^+$ T cells can directly destroy tumor cells.[72] Indeed, neoepitopes recognized by $CD4^+$ cells are reported to be abundant in melanoma tissues.[73] In addition, studies have reported that a personalized neoantigen vaccine for patients with melanoma and glioblastoma elicited a tumor-specific response primarily in $CD4^+$ T cells.[48,51–53] Of note, although most cancer peptide vaccines developed previously focused on HLA-I-restricted epitopes that can induce $CD8^+$ CTLs, recent studies have paid more attention to HLA-II-restricted epitopes that activate $CD4^+$ T cells for enhancing antitumor immunity.[51] Nevertheless, the accuracy of the currently available algorithms for predicting HLA-II-restricted epitopes is lower than that of HLA-I-restricted epitopes due to limited and insufficient database and information about HLA-II-restricted epitopes.[43,44] Therefore novel algorithms remain to be developed to efficiently predict HLA-II-restricted neoepitopes.

How to enhance the efficacy of cancer vaccines

As tumors grow, the tumor microenvironment gradually changes and suppresses antitumor immunity due to various mechanisms.[15,16] Therefore improving the immunosuppressive tumor microenvironment would be necessary to enhance the efficacy of cancer vaccines. For example, tumor cells express an immunosuppressive molecule, programmed death-ligand 1 (PD-L1), which binds to programmed death-1 (PD-1) on T cells. PD-1 is an inhibitory receptor expressed on the surface of activated T cells, and the binding of PD-L1 to PD-1 inhibits T-cell proliferation, cytokine production, and cytolytic activity, resulting in immune evasion. Therefore it might be possible that coadministration of personalized cancer vaccines and PD-1 inhibitors shows an additive or synergistic effect. Indeed, Ott et al. have conducted an open-label Phase Ib clinical trial (NEO-PV-01) of a personalized neoantigen-based peptide vaccine combined with PD-1 blockade therapy in 82 patients with advanced melanoma, nonsmall-cell lung cancer, and bladder cancer.[74] Each neoantigen vaccine consisted of up to 20 peptides (14–35 mer), and each patient received at least one vaccination. Neoantigen-specific $CD4^+$ and $CD8^+$ T-cell responses were observed after vaccination in all patients without serious treatment-related adverse events, and vaccine-induced T cells showed cytotoxic potential and trafficked to the tumor. In addition, the epitope spreading to neoantigens not included in the vaccine was detected after vaccination and was correlated with patient progression-free survival across all three carcinomas. Currently, other clinical trials of ipilimumab (ClinicalTrials.gov identifier: NCT02950766), nivolumab (ClinicalTrials.gov identifier: NCT02897765), or atezolizumab (ClinicalTrials.gov identifier: NCT03289962) in combination with a personalized neoantigen vaccine are underway, and their results are awaited.

Recently, various other mechanisms, including immunosuppressive cells and factors, have also been reported to relate to immunosuppressive tumor microenvironments.[15,16] Therefore it is speculated that immunomodulatory treatments, such as the depletion of regulatory/suppressive cells, inhibition of regulatory/suppressive molecules, and blockade of metabolic suppression, could reverse immunosuppression in the tumor microenvironment and maximize the efficacy of personalized cancer vaccines. In addition, developing and optimizing appropriate vaccine delivery systems and schedule (amount, route, timetable, etc.) is recommended to enhance the efficacy of cancer vaccines. For example, new molecular engineering technologies have been used to develop novel delivery systems, such as liposomes and nanoparticles, with various advantages: they can carry multiple hydrophilic and/or lipophilic antigens, protect antigens from degradation, control the release of antigens, enhance cellular uptake, and activate APCs.[14,75–77] Further studies remain necessary.

Immune escape of tumors

The dual role of the immune system, which not only protects the host against tumor formation but also shapes tumor immunogenicity, is termed "cancer immunoediting."[31,32] Since tumor cells need to escape from the attack of $CD8^+$ T cells for their survival, highly immunogenic mutations are often deleted during tumor development/progression, and only less immunogenic mutations are preferentially selected, leading to tumor heterogeneity. Then, tumor cell clones without immunogenic neoantigens are selected and proliferated to form nonimmunogenic (cold) tumors, leading to worse clinical effects of cancer immunotherapy.

Note that among gene mutations with amino acid substitutions, only a limited percentage of them can bind to HLA molecules, possibly due to cancer immunoediting. For example, in patients with hepatocellular carcinoma, among an average of 66 nonsynonymous gene mutations (per patient), 244 neoepitopes were predicted to be capable of binding to HLA molecules.[78] However, only 23 neoepitopes had confirmed protein expression, and none of them were shown to be naturally presented as HLA ligands on tumor cells. A similar analysis was performed for melanoma, suggesting that only 3 neoepitopes of 531 nonsynonymous gene mutations (per patient) were detected as HLA ligands.[78] Therefore due to such low frequency of neoantigens derived from nonsynonymous point mutations, other types of genomic abnormalities, such as INDELs, frameshift, and gene fusions, could be considered as the source of neoantigens.[79,80] In addition, aberrant mRNA or peptide splicing, aberrant RNA translation from long noncoding RNAs, and posttranslational modifications might be potentially a rich source of neoantigens for cancer vaccines.[81–85]

Some tumor cells might retain potentially immunogenic mutations, which are not deleted in the equilibrium phase of cancer immunoediting. In such cases, other immune-related changes might exist, such as defects in the antigen presentation machinery and the receptors of effector molecules, such as IFN-γ. For example, loss of heterozygosity in HLA is frequently detected in patients with a large number of neoantigen peptides, resulting in immune escape from $CD8^+$ CTLs.[86] In such a situation, since tumor cells are unresponsive to the immune attack by CTLs, other approaches need to be considered.

Conclusion

Over the years various efforts have been made in developing cancer peptide vaccines, but their effectiveness has been limited. This might be because less immunogenic peptides derived from TAAs had been often employed since cancer vaccines targeting more immunogenic TSAs, such as mutation-derived neoantigens, were technically difficult to realize. However, with recent development of technologies, such as NGS and MS, clinical application of TSA-derived neoantigens has become feasible. For example, early-phase clinical trials of personalized vaccine therapy with mutation-derived neoantigen peptides have shown potentially beneficial therapeutic effects. However, various issues remain to be overcome. Especially, further studies are necessary to clarify how to efficiently identify neoepitopes and how to effectively vaccinate them in a personalized manner to enhance their clinical responses.

References

1. Coulie PG, Van den Eynde BJ, van der Bruggen P, Boon T. Tumour antigens recognized by T lymphocytes: at the core of cancer immunotherapy. *Nat Rev Cancer*. 2014;14 (2):135–146.

2. Gubin MM, Artyomov MN, Mardis ER, Schreiber RD. Tumor neoantigens: building a framework for personalized cancer immunotherapy. *J Clin Invest.* 2015;125(9):3413–3421.
3. Hu Z, Ott PA, Wu CJ. Towards personalized, tumour-specific, therapeutic vaccines for cancer. *Nat Rev Immunol.* 2018;18(3):168–182.
4. Hu X, Chakraborty NG, Sporn JR, Kurtzman SH, Ergin MT, Mukherji B. Enhancement of cytolytic T lymphocyte precursor frequency in melanoma patients following immunization with the MAGE-1 peptide loaded antigen presenting cell-based vaccine. *Cancer Res.* 1996;56(11):2479–2483.
5. Rosenberg SA, Yang JC, Restifo NP. Cancer immunotherapy: moving beyond current vaccines. *Nat Med.* 2004;10(9):909–915.
6. Sasada T, Yamada A, Noguchi M, Itoh K. Personalized peptide vaccine for treatment of advanced cancer. *Curr Med Chem.* 2014;21(21):2332–2345.
7. Melero I, Gaudernack G, Gerritsen W, Huber C, Parmiani G, Scholl S, et al. Therapeutic vaccines for cancer: an overview of clinical trials. *Nat Rev Clin Oncol.* 2014;11(9):509–524.
8. Itoh K, Yamada A. Personalized peptide vaccines: a new therapeutic modality for cancer. *Cancer Sci.* 2006;97(10):970–976.
9. Leko V, Rosenberg SA. Identifying and targeting human tumor antigens for T cell-based immunotherapy of solid tumors. *Cancer Cell.* 2020;38(4):454–472.
10. Türeci O, Sahin U, Schobert I, Koslowski M, Scmitt H, Schild HJ, et al. The SSX-2 gene, which is involved in the t(X;18) translocation of synovial sarcomas, codes for the human tumor antigen HOM-MEL-40. *Cancer Res.* 1996;56(20):4766–4772.
11. Creech AL, Ting YS, Goulding SP, Sauld JFK, Barthelme D, Rooney MS, et al. The role of mass spectrometry and proteogenomics in the advancement of HLA epitope prediction. *Proteomics.* 2018;18(12):e1700259.
12. Walter S, Weinschenk T, Stenzl A, Zdrojowy R, Pluzanska A, Szczylik C, et al. Multipeptide immune response to cancer vaccine IMA901 after single-dose cyclophosphamide associates with longer patient survival. *Nat Med.* 2012;18(8):1254–1261.
13. Schwartzentruber DJ, Lawson DH, Richards JM, Conry RM, Miller DM, et al. gp100 peptide vaccine and interleukin-2 in patients with advanced melanoma. *N Engl J Med.* 2011;364(22):2119–2127.
14. Sahin U, Oehm P, Derhovanessian E, Jabulowsky RA, Vormehr M, Gold M, et al. An RNA vaccine drives immunity in checkpoint-inhibitor-treated melanoma. *Nature.* 2020;585(7823):107–112.
15. van der Burg SH, Arens R, Ossendorp F, van Hall T, Melief CJ. Vaccines for established cancer: overcoming the challenges posed by immune evasion. *Nat Rev Cancer.* 2016;16(4):219–233.
16. Spranger S, Gajewski TF. Mechanisms of tumor cell–intrinsic immune evasion. *Ann Rev Cancer Biol.* 2018;2:213–228.
17. Kimura T, Egawa S, Uemura H. Personalized peptide vaccines and their relation to other therapies in urological cancer. *Nat Rev Urol.* 2017;14(8):501–510.
18. Narita Y, Arakawa Y, Yamasaki F, Nishikawa R, Aoki T, et al. A randomized, double-blind, phase III trial of personalized peptide vaccination for recurrent glioblastoma. *Neuro Oncol.* 2019;21(3):348–359.
19. Melief CJ, van der Burg SH. Immunotherapy of established (pre)malignant disease by synthetic long peptide vaccines. *Nat Rev Cancer.* 2008;8(5):351–360.
20. Smith CM, Wilson NS, Waithman J, Villadangos JA, Carbone FR, Heath WR, et al. Cognate CD4(+) T cell licensing of dendritic cells in CD8(+) T cell immunity. *Nat Immunol.* 2004;5(11):1143–1148.

21. Campoli M, Ferrone S. HLA antigen changes in malignant cells: epigenetic mechanisms and biologic significance. *Oncogene*. 2008;27(45):5869–5885.
22. Kondo S, Shimizu T, Koyama T, Sato J, Iwasa S, Yonemori K, et al. First-in-human study of the cancer peptide vaccine TAS0313 in patients with advanced solid tumors. *Cancer Sci*. 2020;. Available from: https://doi.org/10.1111/cas.14765. Epub ahead of print.
23. Schuhmacher J, Heidu S, Balchen T, Richardson JR, Schmeltz C, Sonne J, et al. Vaccination against RhoC induces long-lasting immune responses in patients with prostate cancer: results from a phase I/II clinical trial. *J Immunother Cancer*. 2020;8(2):e001157.
24. Thaxton JE, Li Z. To affinity and beyond: harnessing the T cell receptor for cancer immunotherapy. *Hum Vaccin Immunother*. 2014;10(11):3313–3321.
25. Haen SP, Löffler MW, Rammensee HG, Brossart P. Towards new horizons: characterization, classification and implications of the tumour antigenic repertoire. *Nat Rev Clin Oncol*. 2020;17(10):595–610.
26. Tran E, Robbins PF, Lu YC, Prickett TD, Gartner JJ, Jia L, et al. T-cell transfer therapy targeting mutant KRAS in cancer. *N Engl J Med*. 2016;375(23):2255–2262.
27. Lo W, Parkhurst M, Robbins PF, Tran E, Lu YC, Jia L, et al. Immunologic recognition of a shared p53 mutated neoantigen in a patient with metastatic colorectal cancer. *Cancer Immunol Res*. 2019;7(4):534–543.
28. Marty R, Kaabinejadian S, Rossell D, Slifker MJ, van de Haar J, Engin HB, et al. MHC-I genotype restricts the oncogenic mutational landscape. *Cell*. 2017;171(6):1272–1283. e15.
29. Schietinger A, Philip M, Krisnawan VE, Chiu EY, Delrow JJ, Basom RS, et al. Tumor-specific T cell dysfunction is a dynamic antigen-driven differentiation program initiated early during tumorigenesis. *Immunity*. 2016;45(2):389–401.
30. Schumacher TN, Schreiber RD. Neoantigens in cancer immunotherapy. *Science*. 2015;348 (6230):69–74.
31. Matsushita H, et al. Cancer exome analysis reveals a T-cell-dependent mechanism of cancer immunoediting. *Nature*. 2012;482(7385):400–404.
32. DuPage M, Mazumdar C, Schmidt LM, Cheung AF, Jacks T. Expression of tumour-specific antigens underlies cancer immunoediting. *Nature*. 2012;482(7385):405–409.
33. Gubin MM, et al. Checkpoint blockade cancer immunotherapy targets tumour-specific mutant antigens. *Nature*. 2014;515(7528):577–581.
34. Kreiter S, Vormehr M, van de Roemer N, Diken M, Löwer M, Diekmann J, et al. Mutant MHC class II epitopes drive therapeutic immune responses to cancer. *Nature*. 2015;520 (7549):692–696.
35. Alspach E, Lussier DM, Miceli AP, Kizhvatov I, DuPage M, Luoma AM, et al. MHC-II neoantigens shape tumour immunity and response to immunotherapy. *Nature*. 2019;574(7780):696–701.
36. Rosenberg SA, Restifo NP. Adoptive cell transfer as personalized immunotherapy for human cancer. *Science*. 2015;348(6230):62–68.
37. Goff SL, Dudley ME, Citrin DE, Somerville RP, Wunderlich JR, Danforth DN, et al. Randomized, prospective evaluation comparing intensity of lymphodepletion before adoptive transfer of tumor-infiltrating lymphocytes for patients with metastatic melanoma. *J Clin Oncol*. 2016;34(20):2389–2397.
38. Tran E, Turcotte S, Gros A, Robbins PF, Lu YC, Dudley ME, et al. Cancer immunotherapy based on mutation-specific CD4 + T cells in a patient with epithelial cancer. *Science*. 2014;344(6184):641–645.
39. Zacharakis N, Chinnasamy H, Black M, Xu H, Lu YC, Zheng Z, et al. Immune recognition of somatic mutations leading to complete durable regression in metastatic breast cancer. *Nat Med*. 2018;24(6):724–730.

40. Sampson JH, Heimberger AB, Archer GE, Aldape KD, Friedman AH, Friedman HS, et al. Immunologic escape after prolonged progression-free survival with epidermal growth factor receptor variant III peptide vaccination in patients with newly diagnosed glioblastoma. *J Clin Oncol*. 2010;28(31):4722–4729.
41. Abelin JG, Keskin DB, Sarkizova S, Hartigan CR, Zhang W, et al. Mass spectrometry profiling of HLA-associated peptidomes in mono-allelic cells enables more accurate epitope prediction. *Immunity*. 2017;46(2):315–326.
42. Sarkizova S, Klaeger S, Le PM, Li LW, Oliveira G, et al. A large peptidome dataset improves HLA class I epitope prediction across most of the human population. *Nat Biotechnol*. 2020;38(2):199–209.
43. Abelin JG, Harjanto D, Malloy M, Suri P, Colson T, Goulding SP, et al. Defining HLA-II ligand processing and binding rules with mass spectrometry enhances cancer epitope prediction. *Immunity*. 2019;51(4):766–779.e17.
44. Racle J, Michaux J, Rockinger GA, Arnaud M, Bobisse S, Chong C, et al. Robust prediction of HLA class II epitopes by deep motif deconvolution of immunopeptidomes. *Nat Biotechnol*. 2019;37(11):1283–1286.
45. Vanhaver C, Gordon-Alonso M, Bayard A, Catanese MT, Colau D, van der Bruggen P, et al. Reverse immunology: from peptide sequence to tumor-killing human T-cell clones. *Methods Enzymol*. 2020;631:159–194.
46. Moritz A, Anjanappa R, Wagner C, Bunk S, Hofmann M, Pszolla G, et al. High-throughput peptide-MHC complex generation and kinetic screenings of TCRs with peptide-receptive HLA-A*02:01 molecules. *Sci Immunol*. 2019;4(37). eaav0860.
47. Saini SK, Tamhane T, Anjanappa R, Saikia A, Ramskov S, Donia M, et al. Empty peptide-receptive MHC class I molecules for efficient detection of antigen-specific T cells. *Sci Immunol*. 2019;4(37). eaau9039.
48. Ott PA, et al. An immunogenic personal neoantigen vaccine for patients with melanoma. *Nature*. 2017;547(7662):217–221.
49. Hu Z, Leet DE, Allesøe RL, Oliveira G, Li S, Luoma AM, et al. Personal neoantigen vaccines induce persistent memory T cell responses and epitope spreading in patients with melanoma. *Nat Med*. 2021;. Available from: https://doi.org/10.1038/s41591-020-01206-4.
50. Carreno BM, et al. Cancer immunotherapy. A dendritic cell vaccine increases the breadth and diversity of melanoma neoantigen-specific T cells. *Science*. 2015;348(6236):803–808.
51. Sahin U, et al. Personalized RNA mutanome vaccines mobilize poly-specific therapeutic immunity against cancer. *Nature*. 2017;547(7662):222–226.
52. Keskin DB, Anandappa AJ, Sun J, Tirosh I, Mathewson ND, Li S, et al. Neoantigen vaccine generates intratumoral T cell responses in phase Ib glioblastoma trial. *Nature*. 2019;565:234–239.
53. Hilf N, Kuttruff-Coqui S, Frenzel K, Bukur V, Stevanović S, Gouttefangeas C, et al. Actively personalized vaccination trial for newly diagnosed glioblastoma. *Nature*. 2019;565:240–245.
54. Buonaguro L, Mayer A, Loeffler M, et al. Hepavac-101 first-in-man clinical trial of a multi-peptide-based vaccine for hepatocellular carcinoma. In: *Proceedings of the Annual Meeting of the American Association for Cancer Research 2020*, 2020 April 27–28 and June 22–24, Philadelphia, PA: AACR; *Cancer Res*. 2020;80(16 suppl): Abstract LB-094.
55. Johanns TM, Miller CA, Liu CJ, Perrin RJ, Bender D, Kobayashi DK, et al. Detection of neoantigen-specific T cells following a personalized vaccine in a patient with glioblastoma. *Oncoimmunology*. 2019;8:e1561106.

56. Mijalis AJ, et al. A fully automated flow-based approach for accelerated peptide synthesis. *Nat Chem Biol.* 2017;13:464–466.
57. Vitiello A, Zanetti M. Neoantigen prediction and the need for validation. *Nat Biotechnol.* 2017;35(9):815–817.
58. Lee MY, Jeon JW, Sievers C, Allen CT. Antigen processing and presentation in cancer immunotherapy. *J Immunother Cancer.* 2020;8(2):e001111.
59. O'Donnell TJ, Rubinsteyn A, Laserson U. MHCflurry 2.0: improved pan-allele prediction of MHC class I-presented peptides by incorporating antigen processing. *Cell Syst.* 2020;11(1):42–48. e7.
60. Wells DK, van Buuren MM, Dang KK, Hubbard-Lucey VM, Sheehan KCF, Campbell KM, et al. Key parameters of tumor epitope immunogenicity revealed through a consortium approach improve neoantigen prediction. *Cell.* 2020;183(3):818–834. e13.
61. Duan F, Duitama J, Al Seesi S, Ayres CM, Corcelli SA, Pawashe AP, et al. Genomic and bioinformatic profiling of mutational neoepitopes reveals new rules to predict anticancer immunogenicity. *J Exp Med.* 2014;211(11):2231–2248.
62. Fritsch EF, Rajasagi M, Ott PA, Brusic V, Hacohen N, Wu CJ. HLA-binding properties of tumor neoepitopes in humans. *Cancer Immunol Res.* 2014;2(6):522–529.
63. Yadav M, Jhunjhunwala S, Phung QT, Lupardus P, Tanguay J, Bumbaca S, et al. Predicting immunogenic tumour mutations by combining mass spectrometry and exome sequencing. *Nature.* 2014;515(7528):572–576.
64. Capietto AH, Jhunjhunwala S, Pollock SB, Lupardus P, Wong J, Hänsch L, et al. Mutation position is an important determinant for predicting cancer neoantigens. *J Exp Med.* 2020;217:e20190179.
65. Sibener LV, Fernandes RA, Kolawole EM, Carbone CB, Liu F, McAffee D, et al. Isolation of a structural mechanism for uncoupling T cell receptor signaling from peptide-MHC binding. *Cell.* 2018;174:672–687.
66. Devlin JR, Alonso JA, Ayres CM, Keller GLJ, Bobisse S, Vander Kooi CW, et al. Structural dissimilarity from self drives neoepitope escape from immune tolerance. *Nat Chem Biol.* 2020;16(11):1269–1276.
67. Brennick CA, George MM, Moussa MM, Hagymasi AT, Seesi SA, Shcheglova TV, et al. An unbiased approach to defining bona fide cancer neoepitopes that elicit immune-mediated cancer rejection. *J Clin Invest.* 2021;131(3):e142823.
68. Bessell CA, Isser A, Havel JJ, Lee S, Bell DR, Hickey JW, et al. Commensal bacteria stimulate antitumor responses via T cell cross-reactivity. *JCI Insight.* 2020;5(8):e135597.
69. Balachandran VP, Łuksza M, Zhao JN, Makarov V, Moral JA, Remark R, et al. Identification of unique neoantigen qualities in long-term survivors of pancreatic cancer. *Nature.* 2017;551(7681):512–516.
70. Bassani-Sternberg M, Bräunlein E, Klar R, Engleitner T, Sinitcyn P, Audehm S, et al. Direct identification of clinically relevant neoepitopes presented on native human melanoma tissue by mass spectrometry. *Nat Commun.* 2016;7:13404.
71. Brightman SE, Naradikian MS, Miller AM, Schoenberger SP. Harnessing neoantigen specific CD4 T cells for cancer immunotherapy. *J Leukoc Biol.* 2020;107(4):625–633.
72. Oh DY, Kwek SS, Raju SS, Li T, McCarthy E, Chow E, et al. Intratumoral CD4 + T cells mediate anti-tumor cytotoxicity in human bladder cancer. *Cell.* 2020;181(7):1612–1625. e13.
73. Linnemann C, van Buuren MM, Bies L, Verdegaal EM, Schotte R, Calis JJ, et al. High-throughput epitope discovery reveals frequent recognition of neo-antigens by CD4 + T cells in human melanoma. *Nat Med.* 2015;21:81–85.

74. Ott PA, Hu-Lieskovan S, Chmielowski B, et al. A phase Ib trial of personalized neoantigen therapy plus anti-PD-1 in patients with advanced melanoma, non-small cell lung cancer, or bladder cancer. *Cell.* 2020;183(2):347–362.e24.
75. Ali OA, et al. Identification of immune factors regulating antitumor immunity using polymeric vaccines with multiple adjuvants. *Cancer Res.* 2014;74(6):1670–1681.
76. Kim J, et al. Injectable, spontaneously assembling, inorganic scaffolds modulate immune cells in vivo and increase vaccine efficacy. *Nat Biotechnol.* 2015;33(1):64–72.
77. Kuai R, Ochyl LJ, Bahjat KS, Schwendeman A, Moon JJ. Designer vaccine nanodiscs for personalized cancer immunotherapy. *Nat Mater.* 2017;16(4):489–496.
78. Löffler MW, Mohr C, Bichmann L, et al. Multi-omics discovery of exome-derived neoantigens in hepatocellular carcinoma. *Genome Med.* 2019;11(1):28.
79. Roudko V, Bozkus CC, Orfanelli T, McClain CB, Carr C, O'Donnell T, et al. Shared immunogenic poly-epitope frameshift mutations in microsatellite unstable tumors. *Cell.* 2020;183(6):1634–1649. e17.
80. Yang W, Lee KW, Srivastava RM, Kuo F, Krishna C, Chowell D, et al. Immunogenic neoantigens derived from gene fusions stimulate T cell responses. *Nat Med.* 2019;25 (5):767–775.
81. Faridi P, Li C, Ramarathinam SH, Vivian JP, Illing PT, Mifsud NA, et al. A subset of HLA-I peptides are not genomically templated: evidence for cis- and trans-spliced peptide ligands. *Sci Immunol.* 2018;3(28). eaar3947.
82. Laumont CM, Vincent K, Hesnard L, Audemard É, Bonneil É, Laverdure JP, et al. Noncoding regions are the main source of targetable tumor-specific antigens. *Sci Transl Med.* 2018;10(470):eaau5516.
83. Smart AC, Margolis CA, Pimentel H, He MX, Miao D, Adeegbe D, et al. Intron retention is a source of neoepitopes in cancer. *Nat Biotechnol.* 2018;36(11):1056–1058.
84. Bartok O, Pataskar A, Nagel R, Laos M, Goldfarb E, Hayoun D, et al. Anti-tumour immunity induces aberrant peptide presentation in melanoma. *Nature.* 2021;590(7845):332–337.
85. Engelhard VH, Obeng RC, Cummings KL, Petroni GR, Ambakhutwala AL, Chianese-Bullock KA, et al. MHC-restricted phosphopeptide antigens: preclinical validation and first-in-humans clinical trial in participants with high-risk melanoma. *J Immunother Cancer.* 2020;8(1):e000262.
86. McGranahan N, et al. Allele-specific HLA loss and immune escape in lung cancer evolution. *Cell.* 2017;171(6):1259–1271. e11.

Chapter 11

Antigen-specific T cells

Mohammad Kadivar and Sine Reker Hadrup
Section for Experimental and Translational Immunology, Department of Health Technology, The Technical University of Denmark, Kgs Lyngby, Denmark

Tools to detect antigen-specific T cells and identify vaccine targets

Why we need it?

Adaptive immune response plays a crucial role in immunity against pathogens and the surveillance of pathological cells. T and B cells are the key immune cell types of the adaptive immune system. They both are antigen specific expressing antigen receptors whose repertoire is constructed by the somatic recombination of different germline-encoded gene segments. T and B cells have the ability to tolerate self-antigens and respond toward nonself-antigens via their antigen receptors. Most human T cells express $\alpha\beta$ T-cell receptors (TCRs); however, an unconventional subset of T cells express $\gamma\delta$ TCRs. In this chapter, we are focusing on peptide antigen-specific T cells because most vaccines for the immunotherapy of human cancers target such tumor antigens. Thus we will address $\alpha\beta$ T cells since $\gamma\delta$ T cells can recognize nonpeptide antigens (such as lipid antigens) in a major histocompatibility complex (MHC)-unrestricted way.[1]

The $\alpha\beta$ T-cell TCRs constituted of TCRα and TCRβ chains, in which each chain contain a variable (V) and constant (C) regions. Vα chain is encoded by V and joining (J) gene segments (TRAV and TRAJ), whereas the Vβ chain is consisted of V, diversity (D), and J gene segments (TRBV, TRBD, and TRBJ). Gene rearrangement in V–J or V–D–J order generates high diversity in TCRs.[2] Addition of nontemplated nucleotide and deletions at gene junctional boundaries increases the diversity of TCRs. V domains contain three complementarity-determining regions, providing the main interaction site between the TCR main and the peptide presented by the MHC. It has been suggested that 10^{11}–10^{12} unique TCRs can be generated in human body to recognize different antigens.[3,4] Upon successful TCR:peptide-MHC (pMHC) interaction, the immunoreceptor tyrosine-based activation motifs in

Cancer Vaccines as Immunotherapy of Cancer. DOI: https://doi.org/10.1016/B978-0-12-823901-8.00011-X
© 2022 Elsevier Inc. All rights reserved.

CD3 chains are immediately phosphorylated and trigger T-cell activation that induces T-cell function.[5] The binding of coreceptors CD4 and CD8 to the MHC molecules enhances the avidity of TCR:pMHC interaction.[6] Interestingly, cross-reactivity can occur in TCR:pMHC interaction, meaning that a TCR can recognize various number of different peptides, and a peptide can be detected by numerous different TCRs.[7–9] However, due to the huge diversity and structural complexity of TCR-driven recognition, it is very challenging to fully map the landscape of TCR:pMHC interactions in health and diseases.

In the recent years it has been established that autologous T cells can recognize tumor antigens in patients with cancer. Such antigens can be categorized into tumor-associated antigens and tumor-specific antigens. These tumor antigens can be shared among patients, based on certain shared characteristics of their tumor, termed "shared" tumor antigens, or "patient-unique" antigens. Antigens may derive either from cancer-specific somatic mutations or from other genetic alterations, called neoantigens, as the peptide sequences are novel to the immune system, and not part of the antigen repertoire in healthy cells.[10] Neoantigens are tumor specific. The vast majority of neoantigens are "personalized," but a minority can be shared representing for example, cancer driver mutations. Most of the other shared tumor antigens are tumor-associated antigens, derived from proteins dysregulated in cancer cells, but are also present in subpopulation of healthy cells. These include, for example, cancer-testis antigens that are often highly expressed in tumor cells due to epigenetic modulation, while their normal expression is restricted to immune privileged organs, such as the testis.[11] Neoantigens are fully personalized, as the mutational landscape of tumors shares very few commonalities. Cancer vaccines are most often designed to raise a T-cell response to either shared or neoantigens. Thus a first step in vaccine development is to identify the antigens of relevance for the given cancer, and with the capacity to mediate tumor-cell killing. Despite extensive work, however, the number of tumor antigens that is suitable targets for generic vaccination of patients with cancer is still limited. Over the past few years, extensive progress has been made in the field of T-cell antigen discovery, and here, we review recent advances in TCR ligand discovery, discuss their strengths and limitations, and highlight the outstanding challenges in this field.

T-cell detection and characterization is furthermore critical to monitor if the vaccine successfully induces the desired immune response. Thus such tools are of key importance both in the design and evaluation phase of vaccine development

Screening of antigen-specific T-cell tools

Several tools have been developed to discover and characterize antigen-specific T cells. Here, we categorized the screening tools into antigen-directed methods that can be used to interrogate antigen-specific T cells in heterogeneous samples, such as blood, and TCR-directed methods, wherein antigen discovery for certain TCR(s) is intended.

Antigen-directed methods

Functional assays by coculturing T cells and target cells

One of the commonly used approaches in detecting pMHC:TCR interaction is evaluating the functional response of T cells after exposure to given antigens. The readout for these methods include enzyme-linked immune absorbent spot assay, assessment of T-cell proliferation, detection of antigen-specific cytolytic activity of T cells, and intracellular cytokine staining.[5,12,13] Several classic melanoma antigens, including MAGE-A1, MAGE-A2, MAGE-A3, and MART-1, were discovered using T-cell functional screening.[14,15] In two pioneering studies, selected T cells or tumor-infiltrating lymphocytes (TILs) were cocultured with melanoma antigen transfected cells. Cytokine release or cytolytic activity of T cells (measured by chromium release) was evaluated in the various coculture combinations of T cells and cell lines transfected with genomic DNA or cDNA libraries derived from original melanoma cells.[14,15] Also, in an animal model, interrogation of T cells using the same approach led to identification of the autoantigens recognized by highly pathogenic $CD8^+$ T cells in the nonobese diabetic mouse.[16]

Although the functional screening approaches might be robust to detect antigen-specific T cells, they might not be sensitive enough to detect low frequency T cells. Thus antigen-specific stimulation and in vitro culturing, for example, following peptide pulsing has been broadly used and developed to increase the frequency of antigen-specific T cells in a certain cell samples.[5,13,17] Often, an antigen-specific prestimulation is needed to reach the detection limit of the applied method when evaluating antigen-specific T-cell responses, either vaccine-induced or spontaneous responses. This pose a challenge to directly infer the in vivo activity of such T cells and their functional profile and frequency can be altered by the prestimulation strategy.

The synthetic minigenes are also widely used in many antigen-directed methods to expand or screen for both $CD8^+$ and $CD4^+$ antigen-specific T cells. For instance, synthetic minigenes have been used to discover autoreactive $CD8^+$ T cells toward novel epitopes from pancreatic islet in patients with type 1 diabetes.[18]

Quantitative nonfunctional T-cell detection

A widely used antigen-directed approach involves the use of pMHC multimers. In these affinity-based methods, pMHC:TCR interactions can be directly assessed by conjugating the soluble pMHC multimers to detectable tags, such as fluorochromes. The pMHC multimer-labeling methods circumvent physiological antigen processing, presentation, TCR signaling, and T-cell activation and rely mainly on the affinity between pMHC molecules and TCRs.[19] Thus the pMHC multimer labeling methods may in some cases discover epitopes that are physiologically irrelevant, while missing other low affinity but physiologically important epitopes.[20] Despite this limitation and by using flow cytometry, fluorescent-labeled pMHC multimers have been broadly used to identify, isolate, or characterize antigen-specific T cells ex vivo.[5,12] The number of different pMHC complexes that can be generated and used at once can limit the scale of screening antigen-specific T cells; however, peptide exchange methods like UV exchange have been developed to overcome this challenge.[21–23] There are few fluorochrome dyes available to use for antigen-specific T-cell detection by flow cytometry, which is one of the major drawbacks of fluorescent pMHC multimer labeling methods. To increase the specificities, combinatorial encoding of fluorescently labeled multimers approaches was developed to simultaneously interrogate a multitude of different T-cell specificities in a single sample.[24,25] We developed a set of dual-color encoded pMHC multimers to analyze up to 36 peptide specificities (depends on the available different fluorescent streptavidins and flow cytometer) in a single sample with very high sensitivity (0.002% of $CD8^+$ T cells).[24] This high level of sensitivity can be reached by eliminating the background from single-color events. A similar approach uses multivalent code exploring all possible combinations of fluorophores to identify up to 63 possible antigen specificities.[25] Using these methods, clonal and subclonal neoantigen-specific TILs were identified and monitored upon checkpoint inhibitor therapies.[26,27] It is worth mentioning that both approaches require that peptide sequences with minimal variation are avoided (or analyzed separately) to prevent cross-reactivity of T cells to such peptides within a given sample.

In mass cytometry, a higher number of labels are available to compare the fluorescent flow cytometry in the format of heavy-metal ion tags. By combining mass cytometry and pMHC multimer staining, Newell et al. developed a method using ten metal labels applied in unique combinations of three labels per MHC multimer to simultaneously identify up to 10^9 pMHC specificities in a single sample.[28] In parallel to the heavy-metal-labeled pMHC multimers, surface and intracellular markers can be analyzed to characterize antigen-specific T cells as well as other cells in the given sample.[28] This method was later adapted to assess the human lung and colorectal cancer $CD8^+$ TILs.[29] It was reported that bystander TILs that could recognize a wide range of epitopes unrelated to cancer, such as virus-derived antigens, lacked CD39 expression and were abundant in tumor samples, whereas tumor-reactive neoantigen specific T cells expressed CD39,[29] thus suggesting CD39 expression among TILs, as a key marker to differentiate tumor reactive from bystander T cells. Although these advances have enhanced our ability to define antigen-specific T cells, there are drawbacks associated with mass cytometry including low cell recovery of the instrument, time-consuming, relatively low resolution, no sorting possibility, and high cost.[30]

One of the restrictive factors in using fluorescence or metal labels is the limited number of available labels for high-throughput detection of antigen-specific T cells. To address this limitation, a method was developed using DNA barcodes to label pMHC multimers.[31] This strategy is not limited by the number of available DNA tags since up to 10^{10} different sequences of DNA barcode labels can be designed.[32] We have shown in several studies that this approach can be utilized to screen T cells for more than 1000 specificities in a single sample.[31,33–36] In this method, unique pMHCs and DNA barcodes are assembled on a PE-conjugated polysaccharide dextran. T cells isolated from blood or tissues then can be stained using a large library of DNA barcode-labeled pMHC multimers and sorted by their shared PE fluorochrome. The DNA sequence is amplified and recovered by high-throughput sequencing.

The DNA barcode-labeled pMHC multimers approach allows for the characterization of antigen-specific T cells at broad scale.[31,33–37] Recently, our laboratory showed that a significant frequency of $CD8^+$ T cells in patients with COVID-19 was SARS-CoV-2 specific; nonetheless, $CD8^+$ T-cell recognition of lower affinity was also observed in healthy donors toward SARS-CoV-2-derived epitopes.[37] However, such cross-reactive preexisting SARS-CoV-2-specific T cells showed very different immunophenotype characterization compared to SARS-CoV-2-specific T cells in patients with COVID-19. The SARS-CoV-2-specific $CD8^+$ T cells showed a strong T-cell activation in patients with COVID-19, while minimal T-cell activation was seen in cross-reactive preexisting SARS-CoV-2-specific T cells healthy individuals.[37] Along the same line, our laboratory has been shown in different studies that the detectable antigen-specific T cells are primarily derived from the activated subsets of T cells.[31,35,36] In this setting, the expression of both surface and intracellular markers can be investigated.

In both fluorescent and heavy metal pMHC-labeling systems, we need a good separation between antigen-specific and irrelevant T cells. Thus we might miss detecting low-avidity pMHC:TCR interactions, such as shared tumor-associated antigens. However, the DNA barcode-labeled pMHC multimer approach offers more sensitivity when it comes to low-affinity TCRs because even events with low fluorescently separation will still bind enough DNA barcode-labeled pMHC multimers to allow the detection of their specific DNA barcode, thus enabling the identification of the pMHC:TCR interaction.[12,31] In autoimmune diseases and cancer, a TCR might have low affinity to interact with self-antigens or neoantigens. Thus the DNA barcode-labeled pMHC multimer approach can be a better tool to use for T-cell interrogation in these diseases. For example in a study on Narcolepsy Type 1 (NT1) autoimmune disease, Pedersen et al. showed that the frequency of circulating autoreactive $CD8^+$ T cells was higher in NT1 patients compared to genetically nonsusceptible healthy individuals.[34] In cancer, DNA barcode-labeled pMHC multimer approach can help researchers to interrogate $CD8^+$ T cells with large-scale libraries of predicted neoepitopes to detect, characterize, or even manipulate the neoantigen-specific $CD8^+$ T cells. Our laboratory has reported that neoantigen-specific $CD8^+$ T cells can be frequently observed in the blood of patients across different cancers, but these are personalized antigens that require analyses and generate patient-specific neoantigen libraries.[31,35,36]

Moreover, a T cell with a low-avidity TCR toward the given pMHC, or low surface TCR expression may not interact sufficiently with pMHC multimers to pass the threshold of detection in fluorescent- or heavy metal-label approach but will still provide a detectable signal from the DNA barcode tag.[31] In fluorescent and heavy metal labeling approach, TCR downregulation or internalization following T-cell activation has previously conflicted with effective assessment of pMHC:TCR interaction in such T cells, specifically affecting those T cells with low-affinity TCRs.[31]

Furthermore, tetramer-associated TCR sequencing was developed to interrogate T-cell recognition, in which DNA barcode-labeled pMHC tetramers were constructed in a relatively rapid in vitro transcription and translation manner.[38] In this method, DNA barcodes and pMHC molecules are assembled on fluorescent-coupled streptavidins. The TCR sequence of each antigen-specific T cells can be decoded by the next-generation sequencing of single sorted tetramer-positive T cells.[38] T-cell cross-reactivity can be investigated using this method when researchers can analyze pMHC T-cell recognition and TCR sequence simultaneously. Later, similar approach has been also used for high-throughput mapping of B-cell receptor sequences to antigen specificity (LIBRA-seq).[39]

High-throughput screening of antigen-specific T cells using microfluidics

Lately, the advancement of microfluidics techniques has made new tools available to study antigen-specific T cells by analyzing pMHC:TCR interaction at higher throughput. For instance, a method has been established to monitor the kinetics of pMHC:TCR recognition in live cells within droplets using a nuclear factor of activated T cells (NFAT)−eGFP reporter.[40] This platform relies on screening functional activity of antigen-specific T cells after recognizing their target. This method introduced a tracking system for each T-cell clone in addition to a sorting procedure with single-cell TCR sequencing.[40] This approach, however, has few drawbacks. First, the coencapsulation efficiency of target and T cells in one droplet is low (approximately 20%). Second, the throughput of this system is very limited due to the few total number of trapping wells and the low frequency of the laser-based sorting (about 15 minutes to sort one droplet). Moreover, cell movement in the droplets makes it time-consuming for real-time monitoring and evaporation can be very challenging.[40]

Using microfluidics, Peng et al. described a sensitive method using nanoparticle-barcoded nucleic acid cell sorting (NACS) to isolate neoantigen-specific T cells with high capture efficiency from the tumor and blood of patients with melanoma.[41] This method is based on magnetic nanoparticles that present pMHC tetramers at high avidity by barcoded DNA linkers. The authors claimed that this method exhibits better recovery of antigen-specific T-cell populations relative to flow cytometry approaches for the detection of these low-abundance cells, mainly because of the increased avidity of the nanoparticle-presentation format for the pMHC tetramers. They showed that they could screen for 27 different specificities in only 10^4 $CD8^+$ T cells with the nanoparticle-barcoded NACS approach.[41] This approach might therefore be optimal to enumerate and isolate neoantigen-specific $CD8^+$ T cells from small samples of patient tumor or blood. The captured cells in this method could be used for further functional or genetic investigations, however, not at the single cell level. They applied nanoparticle-barcoded NACS method to identify and characterize neoantigen-specific T cells in the tumor and blood of patients with metastatic melanoma over the course of anti-PD1 checkpoint inhibitor therapy.[41]

The establishment and development of a microfluidic antigen–TCR engagement sequencing technique facilitate high-throughput neoantigen-specific T cells' identification and isolation for subsequent single-cell TCR sequencing.[42] Varadarajan et al. have used cells from blood and mucosal tissues of chronically infected with HIV-1 patients to demonstrate the considerable advantage of this method, in which only 10^4–10^5 cells are needed to identify an antigen-specific responses.[42] However, this sensitivity has not been tested for less frequent T cells, such as neoantigen-specific T cells. With this approach, functional activity of each T cell monitors using on-chip imaging cytometry. The antigen-specific T cells can be recovered readily by micromanipulation for further characterization in vitro.[42]

Altogether, these antigen-directed approaches offer great ability for understanding T-cell recognition as well as developing and optimizing novel cancer vaccines. However, the antigen-directed methods require foreknowledge of the antigenic targets. Moreover, other elements, such as instability of pMHC multimers (particularly MHC II), limited throughput of peptide synthesis, aggregation of pMHC multimers in large libraries, and inefficiency of peptide loading of various HLAs, need to be tackled to study antigen-specific T cells even more comprehensively. When such detection strategies do not meet the requirements or cannot fully cover the antigen landscape of interest, one can instead apply a multiple of functional methods. Furthermore, in cases where antigen discovery is in focus—strategies that start with a selected T-cell pool or T-cell clone, that is, the TCR-directed methods, can be applied to determine their antigen specificity. These methods will be addressed further in this chapter.

TCR-directed methods

The common strategy of TCR-directed approaches is to find the cognate antigen(s) for a TCR or a T-cell clone by screening a library of pMHC complexes. In early TCR-directed approaches, a T-cell clone of interest was screened against combinatorial peptide libraries that randomize all but one position in a peptide.[43–47]

Altered peptide ligands

Wilson et al. identified altered peptide ligands (APLs) for two different T-cell clones using 200 mixtures of oligopeptide combinatorial libraries. Proliferative responses were assessed to identify APLs (also called a mimotopes).[43] The same approach was also used to identify APLs of the MART-1 epitope. This analysis interestingly characterized peptide analogs with increased antigenic activity besides agonist peptides containing several amino acid changes.[47] Moreover, epitopes of an autoreactive $CD4^{+}$ T-cell clone have been identified in a patient with type-1 diabetes using a similar strategy.[48] This approach can be used to identify targets for cancer vaccines, especially when cross-reactivity can also be assessed for the cognate TCR repertoires.[49–51]

Screening systems using yeast or baculoviral display libraries

For antigen discovery, epitope sequences can alternatively be genetically transfected into baculoviral, yeast, or mammalian cells to be presented on the cell surface. To identify mimotopes, the pMHC complexes were anchored to the membrane of an insect cell line (Sf9 cell) by fusing them to the transmembrane and cytoplasmic domains of the baculovirus protein p64. Libraries of baculoviruses as the display platforms were constructed, "probed" with fluorescent conjugated TCRs, and sorted using flow cytometry. Several rounds of enrichment are needed to identify the recognized peptide mimotopes by sequencing the enriched baculoviral cells.[52–55] Even though inefficient transfection was a limiting factor in making large libraries at the beginning ($<10^5$ variants[53]), later study showed that a library of up to 10^8 library size could be generated.[55] Thus this strategy can support the presentation of a broad antigen landscape.

Yeast cells can also been used to display large libraries of pMHC complexes.[56,57] In this method, the protein of interest is fused to the α-agglutinin components Aga2p and Aga1 of yeast to be displayed on the surface of the cells.[58] Like baculoviral cells, yeast display libraries were stained with fluorescent conjugated TCRs and sorted by flow cytometer. Through several rounds of sorting, TCR-recognized cells would be enriched and then mapped using positional weight matrices to discover naturally presenting antigens.[58–60] Using the yeast epitope display platform, Birnbaum et al. showed the high level of TCR cross-reactivity to epitope variants in both human and mice.[61] More recently, Gee et al. used yeast display libraries to identify patient-derived antigens based on "orphan" TCRs from tumor of patients with colorectal adenocarcinoma.[62] They also reported TCR cross-reactivity in patients with cancer.[62] Using yeast display approach for antigen identification may be especially important in selecting targets for cancer vaccines in cancers with low mutational burden.

Yeast display has the advantage of being an unbiased method for interrogation of TCR specificities that relies on interaction between the TCR and pMHC and does not require any a prior knowledge of the epitope. However, despite the advantages of the yeast and baculoviral display, it has the limitation that HLA alleles need to be altered to improve proper folding, stability, and pMHC presentation.[55,56] A common limitation of both baculoviral and yeast display approaches is the need for generating soluble TCRs, that is, a nonrobust process.[63]

Screening systems using mammalian display libraries

In contrast to passive screening systems using yeast or baculoviral display libraries, mammalian cell surface display approaches offer screens in which pMHC:TCR interactions can functionally generate detecting signals to differentiate them from irrelevant interactions. Several methods with mammalian display approach have been developed recently for antigen discovery.[64–68] Joglekar et al. have developed a method by using chimeric receptors called signaling and antigen-presenting bifunctional receptors (SABRs).[64] These receptors are made of to an intracellular CD28–CD3ζ signaling domain attached to an extracellular pMHC complex. They used NFAT–GFP reporter Jurkat cells as the pMHC display platform. Upon successful pMHC:TCR interaction between pMHC on the antigen-presenting cell (APC) and TCR on the clone, cognate APC would express GFP. GFP-expressing cells could then be sorted and sequenced to identify the recognized epitope.[64] In this work, a library of approximately 12,000 SABRs presenting HLA-A0201-restricted epitopes has been successfully tested to identify the targets recognized by public TCRs of known specificities. Furthermore, this method has been used for neoantigen discovery.[64] Not relying on the synthesis of peptides, MHCs, or soluble TCRs is considerable advantage of this approach. However, there are some technical limitations of using SABR libraries. This approach is able to offer up to 10^6 epitopes for antigen discovery, less than yeast display method. Moreover, for screening primary T cells, antigen-unspecific cytotoxicity might be introduced by the T cells toward SABR-expressing cells, and a substantial frequency of antigen-specific T cells is required to obtain the desired signal.

In murine models, Kisielow et al. developed a similar method of antigen display on MHC II molecules, by fusing MHC II to the intracellular signaling domain of TCR, which they called MCRs. Through such MCRs they display antigens in MHC class II and can monitor TCR engagement through NFAT-GFP reporters, and successful MRC:TCR interaction hence can be identified in the same way as SABRs.[65] Using this approach, they discovered the cognate peptides of virus- and tumor-specific T cells in murine models and presented proof of concept for human T cells. Interestingly, they applied the MCR approach to identify immunogenic tumor neoantigens and demonstrated that vaccination with a peptide naturally recognized by TILs efficiently protected mice from tumor challenge.[65]

Two interacting cells can transfer membrane proteins between them by a natural phenomenon called trogocytosis.[69] Based on trogocytosis, Li et al. have established a method in which biotinylated membrane proteins were transferred from T-cell clones specifically to target cells that present cognate pMHC complexes. By streptavidin staining, they could identify and isolate cognate target cells, enabling sequencing the epitopes they presented.[67] Using the trogocytosis-based antigen discovery method, they identified the cognate epitopes of two public and one neoantigen-reactive TCRs.[67]

Kula et al. described a similar method, called T-Scan, a high-throughput platform for the discovery of antigens productively recognized by T cells. In this method, they used lentivirus to deliver antigen libraries into granzyme B sensitive reporter into cells for endogenous processing and presentation on MHC molecules.[66] Upon recognition of epitope-displaying cells, granzyme B secreted by T cells led to the reconstitution of an active fluorescent protein in the APCs, and the antigens mediating recognition were identified by sequencing of the sorted fluorescent-positive APCs.[66] In this work, the authors used T-Scan to identify viral antigens and mapped the reactivity of a tumor-derived TCR.[66]

In an alternative strategy, Sharma et al. used a fluorescence resonance energy transfer (FRET) reporter protein in their target cell. The FRET reporter protein had a granzyme B recognition site, which upon T-cell recognition cleaves the FRET fusion protein into fluorescent GFP protein. The loss of cyan FRET signal and increased GFP signal then could be used to detect and isolate the cognate APCs. By sequencing the encoded minigenes in APCs, T-cell epitopes would be identified.[68]

In all these TCR-directed methods, the posttranslational modifications would be missed, since cognate antigens can be identified by their genetic sequences. Although posttranslational modifications can be critical for T-cell recognition,[70] it is still unknown how much such modification affects the T-cell recognition in certain disease contexts. This limitation also prevents us to study nonpeptide antigens. Despite the fact that priori knowledge of the antigen is not needed, these mammalian cell-based approaches can screen epitope libraries on intermediate scale of 10^4-10^5, which is lower than yeast display method. For most T cell epitope screening initiatives this scale is insufficient to cover the full antigen space of interest, and consequently some preselection of antigens is still needed, e.g. within the mutagenome of a patients tumor. Further studies are needed to determine the minimum T-cell abundance required to identify antigen-specific T cells in these methods, as the required frequency of antigen-specific T cells needed for a positive readout may provide a limitation for the use of such techniques to broadly screen for T-cell recognition. The majority of tumor antigen specific T-cell responses detected in peripheral blood mononuclear cells of patients with cancer range from 0.001% to 1% of $CD8^+$ T cells.

Future perspectives

The challenge of understanding the pMHC:TCR interaction is complicated by different factors. For instance, antigen-specific T cells are at a very low frequency in the blood, making it difficult to identify them. Moreover, both TCRs and pMHCs are highly polyspecific, allowing cross-reactivity among peptides, even if these share minimal sequence similarity. Researchers also need to develop very sensitive biochemical techniques for the detection of low affinity pMHC:TCR interactions. Besides, making large libraries of epitopes with proper antigen presentation is very complicated and demanding. However, all the resent improvement in technologies to interrogate pMHC-TCR interactions is gradually improving our understanding of T-cell-mediated antigen recognition, and the factors contributing to immunogenicity. Recently, substantial improvements has been made in computational technologies that can support the antigen selection through prediction of MHC binding and presentations, that is, in silico tools, such as NetMHC[71] and MHC flurry[72] (predicting MHC binding). Furthermore, novel strategies are emerging, aiming to predict antigen recognition based on TCR characteristics. Due to the huge diversity in TCRs, global prediction of antigen specificity solely based on TCR sequence is yet in the horizon, but in restricted antigen compartments such tools has recently emerged. These includes grouping lymphocyte interactions by paratope hotspots[73] (GLIPH-clustering TCRs according to their likely shared targets), TCR repertoire utilities for solid tumors[74] (TRUST), and immuno-similarity measurement using alignments of receptors of T cells[74] (iSMART-identifying tumor-associated TCRs), but these still cannot accurately predict pMHC:TCR interactions. As computational development and large high-quality data sets go hand in hand, it is plausible that within the coming years we can have improvements of computational tools based on the large set of data being generated and the novel insight to the pMHC-TCR recognition space provided (Tables 11.1 and 11.2).

TABLE 11.1 Antigen-directed methods for high-throughput detection of antigen-specific T cells.

Strategy for high-throughput detection of Ag-specific T cells	Reference	Specificity (library size)	Sensitivity (minimum abundance of reactive TCR in test population)	Sample size ($\times 10^6$)	Detection
DNA barcode labeling of pMHC multimers	Bentzen et al.[31]	1031	0.005% of CD8 T cells	0.20–10 PBMCs	DNA barcodes sequencing
Three-dimensional combinatorial encoding of metal labeled pMHC multimers	Newell et al.[28]	109	0.001% of CD8 T cells (after MHC tetramer-based enrichment)	16–302 PBMCs	Cytometry by time of flight
Two-dimensional combinatorial encoding of fluorescent-labeled pMHC multimers	Hadrup et al.[24]	36	0.002% of CD8 T cells	2 PBMCs	Flow cytometry
Fluorescent-labeled pMHC multimers	Newell et al.[25]	16	0.01% of CD8 T cells	2 PBMCs	Flow cytometry

MHC, Major histocompatibility complex; PBMCs, peripheral blood mononuclear cells; pMHC, peptide-MHC; TCR, T-cell receptor.

TABLE 11.2 TCR-directed high-throughput methods for antigen discovery.

Strategy for high-throughput antigen discovery	Reference	Specificity (library size)	Number of presenting cells/ specificity	Detection
Baculoviruses	Crawford et al.[55]	10^5–108	3×10^7 SF9 cells	Sequencing of encoded mimitopes
Yeast display	Birnbaum et al.[57]	2×10^8	5×10^7 yeast cells	Sequencing of encoded mimitopes
Mammalian cell display	Kula et al.[66]	10^4–2.5×10^5	1×10^6 antigen-presenting cells	Sequencing of encoded mimitopes

TCR, T-cell receptor.

References

1. Godfrey DI, Le Nours J, Andrews DM, Uldrich AP, Rossjohn J. Unconventional T cell targets for cancer immunotherapy. *Immunity*. 2018;48.
2. Davis MM, Bjorkman PJ. T-cell antigen receptor genes and T-cell recognition. *Nature*. 1988;334.
3. La Gruta NL, Gras S, Daley SR, Thomas PG, Rossjohn J. Understanding the drivers of MHC restriction of T cell receptors. *Nat Rev Immunol*. 2018;18.
4. Robins HS, et al. Overlap and effective size of the human CD8 + T cell receptor repertoire. *Sci Transl Med*. 2010;2.
5. Joglekar AV, Li G. T cell antigen discovery. *Nat Methods*. 2020;. Available from: https://doi.org/10.1038/s41592-020-0867-z.
6. Reinherz EL, Wang J h. Codification of bidentate pMHC interaction with TCR and its co-receptor. *Trends Immunol*. 2015;36.
7. Mason D. A very high level of crossreactivity is an essential feature of the T-cell receptor. *Immunol Today*. 1998;19.
8. Nikolich-Žugich J, Slifka MK, Messaoudi I. The many important facets of T-cell repertoire diversity. *Nat Rev Immunol*. 2004;4.
9. Sewell AK. Why must T cells be cross-reactive? *Nat Rev Immunol*. 2012;12.
10. Buonaguro L, Tagliamonte M. Selecting target antigens for cancer vaccine development. *Vaccines*. 2020;8.
11. Haen SP, Löffler MW, Rammensee HG, Brossart P. Towards new horizons: characterization, classification and implications of the tumour antigenic repertoire. *Nat Rev Clin Oncol*. 2020;17.
12. Bentzen AK, Hadrup SR. Evolution of MHC-based technologies used for detection of antigen-responsive T cells. *Cancer Immunol Immunother*. 2017;66.

13. McCutcheon M, et al. A sensitive ELISPOT assay to detect low-frequency human T lymphocytes. *J Immunol Methods*. 1997;210.
14. Van Der Bruggen P, et al. A gene encoding an antigen recognized by cytolytic T lymphocytes on a human melanoma. *Science (80-)*. 1991;254.
15. Kawakami Y, et al. Cloning of the gene coding for a shared human melanoma antigen recognized by autologous T cells infiltrating into tumor. *Proc Natl Acad Sci U S A*. 1994;91.
16. Wong FS, et al. Identification of an MHC class I-restricted autoantigen in type 1 diabetes by screening an organ-specific cDNA library. *Nat Med*. 1999;5.
17. Bethune MT, et al. Isolation and characterization of NY-ESO-1-specific T cell receptors restricted on various MHC molecules. *Proc Natl Acad Sci U S A*. 2018;115.
18. Hondowicz BD, et al. Discovery of T cell antigens by high-throughput screening of synthetic minigene libraries. *PLoS One*. 2012;7.
19. Altman JD, et al. Phenotypic analysis of antigen-specific T lymphocytes. *Science (80-)*. 1996;274.
20. Martinez RJ, Evavold BD. Lower affinity T cells are critical components and active participants of the immune response. *Front Immunol*. 2015;6.
21. Saini SK, et al. Empty peptide-receptive MHC class I molecules for efficient detection of antigen-specific T cells. *Sci Immunol*. 2019;4.
22. Saini SK, et al. Dipeptides catalyze rapid peptide exchange on MHC class I molecules. *Proc Natl Acad Sci U S A*. 2015;112.
23. Luimstra JJ, et al. A flexible MHC class I multimer loading system for large-scale detection of antigen-specific T cells. *J Exp Med*. 2018;215.
24. Hadrup SR, et al. Parallel detection of antigen-specific T-cell responses by multidimensional encoding of MHC multimers. *Nat Methods*. 2009;6.
25. Newell EW, Klein LO, Yu W, Davis MM. Simultaneous detection of many T-cell specificities using combinatorial tetramer staining. *Nat Methods*. 2009;6.
26. McGranahan N, et al. Clonal neoantigens elicit T cell immunoreactivity and sensitivity to immune checkpoint blockade. *Science (80-)*. 2016;351.
27. Van Rooij N, et al. Tumor exome analysis reveals neoantigen-specific T-cell reactivity in an ipilimumab-responsive melanoma. *J Clin Oncol*. 2013;31.
28. Newell EW, et al. Combinatorial tetramer staining and mass cytometry analysis facilitate T-cell epitope mapping and characterization. *Nat Biotechnol*. 2013;31.
29. Simoni Y, et al. Bystander CD8 + T cells are abundant and phenotypically distinct in human tumour infiltrates. *Nature*. 2018;557.
30. Spitzer MH, Nolan GP. Mass cytometry: single cells, many features. *Cell*. 2016;165.
31. Bentzen AK, et al. Large-scale detection of antigen-specific T cells using peptide-MHC-I multimers labeled with DNA barcodes. *Nat Biotechnol*. 2016;34.
32. Xu Q, Schlabach MR, Hannon GJ, Elledge SJ. Design of 240,000 orthogonal 25mer DNA barcode probes. *Proc Natl Acad Sci U S A*. 2009;106.
33. Saini SK, et al. Human endogenous retroviruses form a reservoir of T cell targets in hematological cancers. *Nat Commun*. 2020;11.
34. Pedersen NW, et al. CD8 + T cells from patients with narcolepsy and healthy controls recognize hypocretin neuron-specific antigens. *Nat Commun*. 2019;10.
35. Viborg N, et al. T cell recognition of novel shared breast cancer antigens is frequently observed in peripheral blood of breast cancer patients. *Oncoimmunology*. 2019;8.
36. Hansen UK, et al. Tumor-infiltrating T cells from clear cell renal cell carcinoma patients recognize neoepitopes derived from point and frameshift mutations. *Front Immunol*. 2020;11.

37. Saini SK, et al. SARS-CoV-2 genome-wide T cell epitope mapping reveals immunodominance and substantial CD8 + T cell activation in COVID-19 patients. *Sci Immunol.* 2021;6.
38. Zhang SQ, et al. High-throughput determination of the antigen specificities of T cell receptors in single cells. *Nat Biotechnol.* 2018;36.
39. Setliff I, et al. High-throughput mapping of B cell receptor sequences to antigen specificity. *Cell.* 2019;179.
40. Segaliny AI, et al. Functional TCR T cell screening using single-cell droplet microfluidics. *Lab Chip.* 2018;18.
41. Peng S, et al. Sensitive detection and analysis of neoantigen-specific T cell populations from tumors and blood. *Cell Rep.* 2019;28.
42. Varadarajan N, et al. Rapid, efficient functional characterization and recovery of HIV-specific human CD8 + T cells using microengraving. *Proc Natl Acad Sci U S A.* 2012;109.
43. Wilson DB, et al. Immunogenicity. I. Use of peptide libraries to identify epitopes that activate clonotypic CD4 + T cells and induce T cell responses to native peptide ligands. *J Immunol.* 1999;163.
44. Hemmer B, et al. Identification of high potency microbial and self ligands for a human autoreactive class II-restricted T cell clone. *J Exp Med.* 1997;185.
45. Gavin MA, Dere B, Grandea AG, Hogquist KA, Bevan MJ. Major histocompatibility complex class I allele-specific peptide libraries: identification of peptides that mimic an H-Y T cell epitope. *Eur J Immunol.* 1994;24.
46. Gavin MA, Bevan MJ. Major histocompatibility complex allele-specific peptide libraries and identification of T-cell mimotopes. *Methods Mol Biol.* 1998;87.
47. Pinilla C, et al. Combinatorial peptide libraries as an alternative approach to the identification of ligands for tumor-reactive cytolytic T lymphocytes. *Cancer Res.* 2001;61.
48. Hiemstra HS, et al. The identification of CD4 + T cell epitopes with dedicated synthetic peptide libraries. *Proc Natl Acad Sci U S A.* 1997;94.
49. Sherev T, Wiesmüller KH, Walden P. Mimotopes of tumor-associated T-cell epitopes for cancer vaccines determined with combinatorial peptide libraries. *Appl Biochem Biotechnol B: Mol Biotechnol.* 2003;25.
50. Linnemann T, et al. Mimotopes for tumor-specific T lymphocytes in human cancer determined with combinatorial peptide libraries. *Eur J Immunol.* 2001;31.
51. Nino-Vasquez JJ, et al. A powerful combination: the use of positional scanning libraries and biometrical analysis to identify cross-reactive T cell epitopes. *Mol Immunol.* 2004;40.
52. Szardenings M. Phage display of random peptide libraries: applications, limits, and potential. *J Recept Signal Transduct.* 2003;23.
53. Crawford F, Huseby E, White J, Marrack P, Kappler JW. Mimotopes for alloreactive and conventional T cells in a peptide-MHC display library. *PLoS Biol.* 2004;2.
54. Wang Y, et al. Using a baculovirus display library to identify MHC class I mimotopes. *Proc Natl Acad Sci U S A.* 2005;102.
55. Crawford F, et al. Use of baculovirus MHC/peptide display libraries to characterize T-cell receptor ligands. *Immunol Rev.* 2006;210.
56. Wen F, Sethi DK, Wucherpfennig KW, Zhao H. Cell surface display of functional human MHC class II proteins: yeast display vs insect cell display. *Protein Eng Des Sel.* 2011;24.
57. Birnbaum ME, Dong S, Garcia KC. Diversity-oriented approaches for interrogating T-cell receptor repertoire, ligand recognition, and function. *Immunol Rev.* 2012;250.

58. Brophy SE, Holler PD, Kranz DM. A yeast display system for engineering functional peptide-MHC complexes. *J Immunol Methods*. 2003;272.
59. Adams JJ, et al. T cell receptor signaling is limited by docking geometry to peptide-major histocompatibility complex. *Immunity*. 2011;35.
60. Boder ET, Wittrup KD. Yeast surface display for screening combinatorial polypeptide libraries. *Nat Biotechnol*. 1997;15.
61. Birnbaum ME, et al. Deconstructing the peptide-MHC specificity of t cell recognition. *Cell*. 2014;157.
62. Gee MH, et al. Antigen identification for orphan T cell receptors expressed on tumor-infiltrating lymphocytes. *Cell*. 2018;172.
63. Davis MM, Boyd SD. Recent progress in the analysis of αβ T cell and B cell receptor repertoires. *Curr Opin Immunol*. 2019;59.
64. Joglekar AV, et al. T cell antigen discovery via signaling and antigen-presenting bifunctional receptors. *Nat Methods*. 2019;16.
65. Kisielow J, Obermair FJ, Kopf M. Deciphering CD4+ T cell specificity using novel MHC−TCR chimeric receptors. *Nat Immunol*. 2019;20.
66. Kula T, et al. T-Scan: a genome-wide method for the systematic discovery of T cell epitopes. *Cell*. 2019;178.
67. Li G, et al. T cell antigen discovery via trogocytosis. *Nat Methods*. 2019;16.
68. Sharma G, Rive CM, Holt RA. Rapid selection and identification of functional CD8+ T cell epitopes from large peptide-coding libraries. *Nat. Commun*. 2019;10.
69. Joly E, Hudrisier D. What is trogocytosis and what is its purpose? *Nat Immunol*. 2003;4.
70. Engelhard VH, Altrich-Vanlith M, Ostankovitch M, Zarling AL. Post-translational modifications of naturally processed MHC-binding epitopes. *Curr Opin Immunol*. 2006;18.
71. Jurtz V, et al. NetMHCpan-4.0: improved peptide−MHC class I interaction predictions integrating eluted ligand and peptide binding affinity data. *J Immunol*. 2017;199.
72. O'Donnell TJ, Rubinsteyn A, Laserson U. MHCflurry 2.0: improved pan-allele prediction of MHC class I-presented peptides by incorporating antigen processing. *Cell Syst*. 2020;11.
73. Glanville J, et al. Identifying specificity groups in the T cell receptor repertoire. *Nature*. 2017;547.
74. Zhang H, et al. Investigation of antigen-specific T-cell receptor clusters in human cancers. *Clin Cancer Res*. 2020;26.

Chapter 12

Adjuvants for improving cancer vaccines

Jenci Hawthorne[1], Haval Shirwan[2] and Esma Yolcu[2]
[1]Department of Obstetrics, Gynecology, and Women's Health, University of Louisville School of Medicine, Louisville, KY, United States, [2]Department of Child Health, Molecular Microbiology and Immunology, School of Medicine, University of Missouri, Columbia, MO, United States

Introduction

Cancer continues to be a leading cause of death in the United States. Standard treatments, such as radiation, surgery, and chemotherapy, show limited efficacy in effectively eradicating tumors and are associated with off-target severe toxicities. Thus there is an acute need for the development of safe and more effective alternatives to standard cancer treatments. Progress in the field of immunology combined with mounting evidence that cancerous cells serve as an internal danger triggering an immune response led to the notion of cancer immunotherapy. Adoptive T-cell therapies and immune checkpoint inhibitors have shown impressive clinical success for selected cancer types,[1] thus validating the immune system as an effective vehicle to battle cancer. Mounting evidence also indicates that the immune system can serve as a double-edged source; a robust acute response eliminating cancer and a subdued chronic response facilitating progression. Cancer evolves a complex set of immune evasion mechanisms that favor chronic inflammation. Immunotherapeutic approaches designed to account for the complex interactions between cancer and the immune system and exploit such interactions for the generation of robust, acute responses may prevail in efficacy and safety.

Therapeutic vaccines are preferred component of conventional treatments for cancer because of their safety profile and generation of long-term immunological memory critical for the control of recurrence, which is a major cause of cancer morbidity. However, the success of cancer vaccines hinges on their ability to generate robust, tumor-restricted immune effector response that translates into clinical efficacy. Accumulating knowledge of cancer vaccines over few decades indicates that an effective vaccine will require three

Cancer Vaccines as Immunotherapy of Cancer. DOI: https://doi.org/10.1016/B978-0-12-823901-8.00005-4
© 2022 Elsevier Inc. All rights reserved.

major elements: (1) an antigen that primes T-cell responses restricted to the tumor; (2) an immune adjuvant that drives the primed T cells for robust expansion and sustains a balanced $CD4^+$ and $CD8^+$ T-cell response of a significant magnitude and duration; and (3) and that the adjuvant down-regulates various immune evasion mechanisms employed by the progressing tumor.

Subunit vaccines based on tumor-associated antigens (TAA) or tumor-specific antigens (TSA) represent one of the first cancer immunotherapeutic approaches pursued since mid-1900. However, clinical translation of this approach was only realized in 2010 when US Food and Drug Administration (FDA) approved a dendritic cell-based vaccine (Provenge) against advanced prostate cancer.[2] Although the clinical efficacy of Provenge was minimal, it was a major milestone in the field that spearheaded the development of various present-day cancer immunotherapies with demonstrated remarkable clinical efficacy. Although mechanisms responsible for the inefficacy of TAA or TSA vaccines are yet to be fully delineated, the immune status of cancer patients, extent of tumor burden, and design of vaccine formulation may provide an explanation.

The perceived key elements of a vaccine formulation include single or multiple TAA/TSA and an adjuvant to enhance the breadth and magnitude of the immune response. The nature of tumor antigen is critical to the vaccine efficacy as it distinctly marks the tumor for destruction by immune cells.[3] A tumor antigen that is both mutated and overexpressed by the tumor and actively involved in tumor survival, progression, and immune evasion may serve as an ideal choice. Such a tumor antigen is expected to show persistent expression by the tumor and be refractive to antigenic drift and escape caused by vaccine-induced immune pressure. Traditional tumor antigens range from common cancer mutations to immune-privileged germline cells that include cancer-testis antigens, fetal antigens, overexpressed self-antigens, differentiation antigens, mutated proteins, aberrantly expressed proteins, and antigens of oncoviruses.[4] Cancer-testis antigens are generally temporally expressed during development on germline cells, such as melanoma-associated antigen (MAGE)-A1, MAGE-A3, MAGE-A4, and NY-ESO-1, and can arise either oncogenically or stochastically during tumor development.[4–9] Other well-defined tumor antigens include differentiation antigens, such as tyrosinase, gp100, MART-1,[10–14] or antigens, that are overexpressed in cancer cells, such as hTERT, HER2, mesothelin, and MUC1.[15–18] Inasmuch as all these classic tumor antigens, except oncoviral antigens, are self-proteins, they may not generate effective immune responses due to being subject to varying degrees of preexisting immunological tolerance.[19,20] This notion is consistent with the clinical failure of vaccines based on such antigens.[19,20]

Recent clinical studies have shown that as the tumor progresses, it expresses nonautologous antigens, classified as neoantigens, generated by tumor-specific DNA alterations.[21] Neoantigens are attractive targets for cancer vaccines as they are tumor-specific and are not subject to tolerance mechanisms. Thus they have the potential to generate strong, tumor-restricted immune responses with potential clinical efficacy and safety. However, there are various challenges for designing vaccines based on neoantigens. First, these antigens are patient-specific and require a personalized vaccine approach that is time and resource intensive. Second, neoantigens are also subject to immunoediting and escape, requiring a collection of neoantigens in vaccine formulation for efficacy. Third, the neoantigen repertoire is quite dynamic and may evolve in the course of tumor progression, thus representing a moving target for the rational design of effective vaccines. Fourth, various immune evasion mechanisms targeting innate, adaptive, and regulatory immunity are operational in a progressing tumor.[22,23] Therefore immune adjuvants as single agents or in combination that exploit tumor neoantigens in situ without their identification for the generation of robust effector responses has the potential to realize the promise of neoantigens as targets for immunotherapy.

Therapeutic immune responses against cancer involve a complex array of soluble and cellular mechanisms that involves all three arms of the immune system—innate, adaptive, and regulatory. However, T cells are the potent driver of the immunological cascade and major effectors of tumor immunosurveillance and immunotherapy. The generation of an effective T-cell response is contingent on the transduction of three signals (Fig. 12.1). Signals 1 is delivered by the recognition of tumor antigens as peptides processed and presented in the context of major histocompatibility complex on the surface of antigen-presenting cells (APCs), such as dendritic cells (DCs). Signal 1 needs to be qualified by Signal 2 delivered through the interaction of costimulatory receptors, such as CD28, 4−1BB, OX-40, and CD27, on T cells with their respective ligands on DCs.[24] Activated T cells then require various cytokines in the form of Signal 3 to expand, persist, and establish a memory response.[25] The lack of signal 2 or 3 is associated with abortive T-cell responses.[26] Immune adjuvants with pleiotropic functions that target the 3 signals for enhancement, and simultaneously down-regulate tumor immune evasion mechanisms have the potential to transform the field of therapeutic cancer vaccines. Adjuvants can be used either as a component of defined tumor antigen-based vaccine formulations or standalone exploiting the *in situ* repertoire of neoantigens in a progressing tumor for the generation of a therapeutic response. This chapter will focus on immune adjuvants that target the 3 signals required for the generation of productive T-cell responses and designer adjuvants that target immune evasion mechanisms for downregulation in addition to boosting T-cell signals.

Adjuvants

Immune adjuvants are a diverse class of molecules with distinct functions. Some manifest their function through well-elucidated pathways, such as

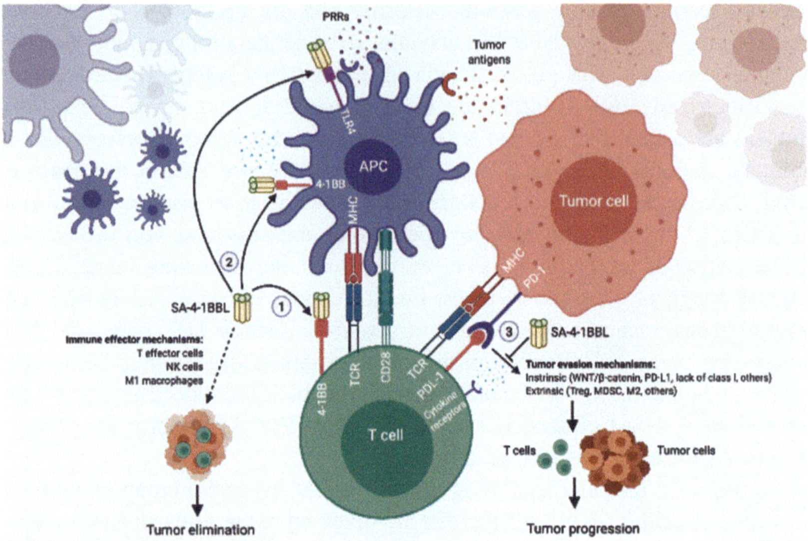

FIGURE 12.1 Immune modulators with pleiotropic functions as the next-generation designer adjuvants for therapeutic cancer vaccines. T-cell recognition of tumor-associated antigens or neoantigens accompanied by costimulation via CD28/CD80 or 86 results in activation and proliferation that is sustained by cytokines elaborated by T effector cells or dendritic cells in the tumor-draining lymph node cells. T effector cells then home to the tumor for eradication. However, tumor has evolved various intrinsic and extrinsic mechanisms that evade the immune system. Immune adjuvants are critical for initiating, expanding, and maintaining T effector responses. The existing, US Food and Drug Administration-approved adjuvants may not overcome various immune evasion mechanisms to generate a clinical response as component of therapeutic vaccines. Designer adjuvants that directly impact both immune effectors and regulators have the potential to confer the clinical efficacy of vaccines. SA-4-1BBL has the potential as such an adjuvant as it delivers costimulation through its receptor on T cells, leading to T-cell expansion, acquisition of Th1 effector function, and long-term survival. SA-4-1BBL also interacts with 4-1BB and TLR4 on dendritic cells, leading to their activation, antigen uptake, and cross-presentation. SA-4-1BBL also overcomes Treg-mediated immune suppression as well as tumor-mediated conversion of T effector cells into Treg cells.

pattern recognition receptors, and others through yet to be defined processes. Adjuvants are critical for the generation of effective and long-lasting immune responses to pathogens as external dangers by enhancing the magnitude, breadth, longevity, and quality of immune response to a specific antigen. Adjuvants primarily target innate immunity for activation and generation of signals required for productive acquired immune responses. As such, they generate rapid immune responses to a target antigen and reduce the need for high dose of antigen as well as frequency of immunization.[27,28] Adjuvants were developed as component of recombinant subunit vaccine formulations against infectious diseases.[29] The danger of using live attenuated virus for vaccination and realization that vaccines based on inactivated viruses do not generate a protective, long-lasting immune response culminated in intense effort to discover and develop immune adjuvants for more than half-a-century. In spite of this intense effort, alum was the only FDA-approved adjuvant up until 2009 when monophosphoryl lipid A, agonist of toll-like receptor 4 (TLR4), as a part of adjuvant system was approved for Cervarix vaccine for human papilloma virus.[30,31] Accumulating knowledge of the immune system and need for the development of effective adjuvant system led to the discovery and approved use of several more adjuvants. The list of adjuvants in use today is impressive and include mineral salts, emulsions, liposomes, virosomes (phospholipid vesicles incorporating viral proteins), immune-stimulating complexes, cytokines, saponins, and bacterial exotoxins.[29] Irrespective of their mode of actions, these adjuvants primarily affect innate immunity for the generation of protective cellular and humoral responses against a target pathogen in a healthy individual with measurable clinical success for the most part.

Adjuvants are even more important for the efficacy of vaccines for cancer as an internal danger that employs sophisticated immune evasion mechanisms for progression. Vaccine adjuvants are mainly designed and classified based on their functional ability to target the three signals required for the generation of a productive T-cell response. However, such classifications may be ambiguous, as many adjuvants, such as TLR4 agonists, may provide more than one signal by inducing upregulation of costimulatory molecules and production of cytokines, which drive Signals 2 and 3, respectively.[29,32] Given the critical role T-cell immunity plays in the eradication of cancer, we will classify adjuvants based on their impact on the three signals required for a therapeutic immune response.

Signal 1: adjuvants targeting innate immunity

Adjuvants facilitating Signal 1 mainly improve the delivery of antigens to DCs as well as their uptake, processing, and presentation to T cells. In some cases, these adjuvants also protect the antigen from degradation, thus improving stability and persistence. This class of adjuvants includes alum, nanoparticles, gold particles, lipid vesicles, oil- and water-based formulations, and bacterial ghosts, to name some.[33–35]

Alum (aluminum phosphate and aluminum hydroxide) is a widely used adjuvant for over a century that targets the innate immune system for activation through various mechanisms, including serving as antigen depot, targeting the antigen to DCs, and induction of inflammation, altogether culminating in the generation of acquired cellular, Th1 and Th2, and humoral immune responses.[31,36,37] The aluminum salts are administered in a basic solution of trisodium phosphate or in a phosphate solution with sodium hydroxide often referred to as a depot or mineral salt adjuvant.[25] Alum has been used in several current infectious agent vaccine formulations, including

Anthrax, DT, DTaP (Daptacel), DTaP (Infanrix), DTaP-IPV (Kinrix), DTaP-IPV (Quadracel), DTaP-HepB-IPV (Pediarix), DTaP-IPV/Hib (Pentacel), Hep A (Havrix), Hep A (Vaqta), Hep B (Engerix-B), Hep B (Recombivax), Hep A/Hep B (Twinrix), HIB (PedvaxHIB), HPV (Gardasil 9), Japanese encephalitis (Ixiaro), MenB (Bexsero, Trumenba), Pneumococcal (Prevnar 13), Td (Tenivac), Td (Mass Biologics), Tdap (Adacel), and Tdap (Boostrix). In the realm of cancer vaccines, alum has had limited use plausibly due to its inability to generate robust Th1 and $CD8^+$ killing responses required for the eradication of cancer.[38,39] In a randomized, multicenter clinical trial, alum was used as an adjuvant for an antiidiotypic vaccine mimicking TAA NeuGcGM3 ganglioside (racotumomab)-based vaccine as a switch maintenance therapy postchemo in 176 patients with advanced nonsmall-cell lung cancer (NSCLC) with reported safety and significant improvement in overall survival and progression-free survival as compared with placebo.[40]

Oil-in-water emulsions, including MF59 (Novartis) and AS03 (GlaxoSmithKline), were shown to enhance tissue inflammation, leading to the generation of humoral and mixed Th1 and Th2 immune responses.[41] Enhanced recruitment of granulocytes and DCs and their uptake of antigen have been identified as primary mechanisms for adjuvant response.[42,43] In the infectious disease realm, MF59 is commonly used as an adjuvant in Fluad, an influenza vaccine. The use of MF59 as an adjuvant for cancer vaccines has been limited due to its ability to primarily induce a Th2 response. As such, MF59 has been used in combination with other adjuvants for cancer vaccines. In preclinical cancer studies, MF59 in combination with CpG ODN as adjuvant component of a recombinant HSP65-MUC1 vaccine was shown to induce robust tumor immunity in mice.[44,45] This adjuvant combination generated a potent Th1 response with enhanced immunogenicity of HSP65-MUC1 resulting in significantly higher titers of IgG2c, increased interferon gamma (IFNγ) mRNA expression, and the generation of antigen-specific cytotoxic T lymphocytes. When prophylactically applied, HSP65-MUC1 vaccine adjuvanted with MF59 and CpG ODN inhibited the growth of $MUC1^+$ B16 melanoma, leading to the prolonged survival of tumor-bearing mice. AS04 adjuvant system, a combination of monophosphoryl lipid (MPL) and alum, that manifests its adjuvant activity through TLR4 and inflammasome pathways generates efficient humoral and Th1 responses. Although widely used in infectious disease settings and is the adjuvant component of the approved Cervarix HPV vaccine, AS04 has not been extensively studied in cancer models.[36]

Bacterial ghosts, empty cell envelopes derived from gram negative bacteria, represent another class of adjuvants used for protein as well as DNA vaccines. Bacterial ghosts have several appealing features as vaccine adjuvates. First, they contain various innate immune agonists, such as lipopolysaccharides, flagella, and peptidoglycan, that target immune and nonimmune cells for the production of proinflammatory cytokines and chemokines that culminates in the generation of adaptive cellular and humoral immune responses. Second, bacterial ghosts can be used as a vehicle to target DCs for the delivery and uptake of DNA coding for antigens of interest as vaccine. Third, bacterial ghosts can be engineered to express antigens of interest on the surface of outer membrane that allows the generation of antibodies to conformational epitopes.[46] Bacterial ghosts as adjuvants have been primarily used for vaccines against infections in veterinary medicine settings. Although has significant potential for efficacy, this class of adjuvants has not been tested in humans and their use for cancer vaccines in preclinical models is limited. In a lung cancer preclinical model, vaccination with bacterial ghosts adjuvanted tumor cell lysate as vaccine resulted in the generation of a strong $CD8^+$ T-cell response and increased overall survival.[47]

Mineral oil-based emulsions, such as Montanides,[48] are another class of adjuvants used for vaccination. This adjuvant platform targets the delivery of antigens to DCs for increased uptake, processing, and presentation to T cells.[36] Montanide was used as the adjuvant component of a subunit vaccine containing a synthetic peptide derived from the gp100 melanoma-associated antigen in combination with high dose interleukin (IL)-2 to treat advanced melanoma with a fraction of patients showing objective cancer responses.[49] A select class of heat shock proteins (Hsp), such as Hsp70, was also used as immune adjuvants and delivery vehicles for tumor antigens to DCs with demonstrated efficacy in inducing innate as well as adaptive immune responses.[50,51] Saponin adjuvants are also commonly used, specifically QS-21.[52] Its mechanism of action is suggested to be from its triterpene aldehyde group, which is capable of triggering the ASC/NALP3 inflammasome responses.[52] It has also demonstrated a robust Th1 response in preclinical studies.[53,54] Saponin adjuvants have also been harnessed in combinatorial adjuvant settings, such as ISCOMATRIX, which includes the saponin adjuvant with antigens in a micellar structure.[55] Adjuvants that target innate immunity although proven effective for preventive vaccines against infections, they have shown limited efficacy in settings of therapeutic vaccines against cancer. The lack of therapeutic efficacy of these adjuvants may be due to their inability to directly enhance adaptive immune cells, such as $CD4^+$ and $CD8^+$ T cells, and/or overcome immune evasion mechanisms employed by the progressing cancer.

Signal 2: adjuvants harnessing costimulatory mechanisms

Adjuvants that target Signal 2 for T-cell activation include a diverse group of physically and chemically unrelated molecules, such as saponins, muramyl dipeptide, and highly immunogenic immune-stimulating complex.[56–58] These adjuvants are responsible for the activation of DCs, leading to the upregulation of various costimulatory ligands, such as CD80, CD86, and 4-1BBL, that engage their counterreceptors on T effector (Teff) cells for the transduction of costimulatory signals.[24] Costimulation is critical for the activation of Teff cells and their expansion and acquisition of effector functions. Costimulation also plays a critical role in the activation and sustained function of innate immune cells, including DCs and NK cells. The lack of Signal 2 is associated with immune nonresponsiveness or tolerance. Irrespective of their significant potential, the use of costimulatory ligands as adjuvants has been limited due to the inability of natural ligands maintaining their function in soluble forms. Although agonistic antibodies to the costimulatory receptors have been investigated as adjuvants, they do not recapitulate the function of natural ligands and also their use is associated with toxicity.[59,60] These limitations led to efforts to design costimulatory ligands that have robust immune stimulatory functions without adverse effects.

Designer immune adjuvants represent a new and novel class of adjuvants that can directly target APCs, innate, adaptive, and regulatory immune cells for modulation. These combined features are highly desirable for cancer immunotherapy and particularly vaccine design, given that cancer evolves mechanisms that evade a wide range of immune cells for progression. We have generated such a designer adjuvant, SA-4-1BBL, an oligomeric form of the ligand to the CD137 (4-1BB) costimulatory receptor. CD137 is a member of the tumor necrosis factor receptor (TNFR) superfamily, which also includes OX-40, CD30, CD40, CD70, herpes virus entry mediator, glucocorticoid-induced TNFR-related protein, and others.[61] Signaling through these receptors serves as a key communication among various cell types, particularly cells of the immune system. CD137 is constitutively expressed on neutrophils, mast cells, monocytes, macrophages, a subpopulation of DCs, and regulatory T (Treg) cells.[62,63] NK and NKT cells, T cells, and B cells express CD137 following activation. Signaling through CD137 drives Teff cell proliferation, survival, acquisition of effector function, and long-term memory.[64] Signaling via CD137 in NK and NKT cells promotes survival, expansion, and production of IFNγ, which is likely a major player in the immune response to cancer.[65] CD137 signaling selectively promotes type 1 cytokines, such as IL-2, IFNγ, IL-6, tumor necrosis factor alpha, and IL-12, implicating this receptor in Th1 effector responses that are critical for the eradication of chronic infections and malignant cells.[66–68] CD137 signaling is shown to restore CD80/86-mediated downregulation of CD28 on Teff cells, resulting in T-cell proliferative capacity.[69] This aspect of signaling is important as it serves as a positive feedback loop to maintain CD28 signaling for sustained T-cell responses.

SA-4-1BBL as a designer adjuvant has several unique features preferred for cancer immunotherapy.[28] This molecule serves as a robust adjuvant for both $CD4^+$ and $CD8^+$ T-cell activation, expansion, and establishment of long-term memory.[70–72] SA-4-1BBL also blocks tumor-mediated conversion of Teff cells into Treg cells through stimulating the heightened secretion of IFNγ in effectors.[73] Treg cells consume IL-2 as a mechanism to suppress

Teff cells. SA-4-1BBL as a designer adjuvant overcomes Treg suppression by stimulating Teff cells to secrete copious amount of IL-2.[74] SA-4-1BBL as an adjuvant for TAA-based subunit vaccines showed robust therapeutic efficacy against various preclinical tumor models.[72,73] A subunit vaccine containing SA-4-1BBL and a synthetic peptide serving as dominant $CD8^+$ T-cell epitope for human papillomavirus (HPV) E7 or complete E7 protein had >70% efficacy in eradicating E7-expressing TC-1 tumors.[72,75,76] The therapeutic efficacy of SA-4-1BBL as a component of subunit vaccine containing survivin as a self TAA was shown in the 3LL lung cancer model.[70,72] SA-4-1BBL was used to generate conjugate vaccines and is shown to target the delivery of such antigens to DCs in vivo for enhanced uptake, processing, and cross-presentation to $CD8^+$ T cells for improved therapeutic efficacy in preclinical tumor models.[70]

A variation in designer adjuvants based on costimulatory ligands includes the TRICOM platform, which is a viral vector-based expression system for B7.1, LFA-3, and ICAM-1 along with the tumor-associated antigen of interest.[77,78] The therapeutic efficacy of TRICOM has been assessed in numerous clinical trials, including metastatic breast and ovarian carcinoma[79,80] and prostate cancer.[81–83] A phase I/II study vaccinating patients with prostate cancer with rFowlpox-PSA (L155)-TRICOM (PROSTVAC-F/TRICOM) demonstrated enhanced immunogenicity and clinical response in patients that allowed testing in phase III clinical trials.[13,83,84] Phase II clinical trials demonstrated similar outcomes to the DC-based vaccine, Provenge, suggesting no change in time progression of advanced disease but improved overall survival compared to placebo.[78] Thus designer adjuvants based on costimulatory ligands have the potential to target cells of innate, adaptive, and regulatory immunity for a therapeutic outcome in cancer setting with negligible toxicity.

Signal 3: overcoming tumor regulatory mechanisms

Signal 3 in the form of cytokines plays a critical role in the expansion of activated T cells and long-term persistence. The use of cytokines as a part of vaccine formulations is becoming more prevalent and has been shown to influence both cellular and humoral immune responses, resulting in the efficacy of poorly immunogenic vaccines in both preclinical and clinical settings.[85] In particular, $CD8^+$ T cells, as dominant effectors of tumor clearance, undergo clonal expansion and develop effector function require not only Signals 1 and 2, but also Signal 3 provided by IL-12 or IFNγ.[86,87] Adjuvants, such as alum, TriMix, and chitosan, facilitate Signal 1 and modulate DC activity for enhanced vaccine-specific Th2 responses and production of associated cytokines, such as IL-4, IL-5, and IL-13. Given the controversy on the role of Th2 responses in tumor clearance, the focus has been on type I cytokines, such as IFNα, IFNγ, IL-2, IL-12, IL-15, IL-18, and IL-21, as adjuvants with demonstrated impact on the enhancement of vaccine efficacy against tumors.[88,89] However, the major challenges for successful clinical use of these cytokine-based vaccine adjuvants include dose optimization to retain efficacy in the absence of toxicity and/or immunosuppression, as well as the addition of substantial cost to traditional vaccine formulations.

Combinatorial adjuvants

Immune response to pathogens is highly effective due to the elaboration of various danger signals delivered by pathogens that are sensed by the innate immunity that in turn provides required cues for the generation of a more sophisticated and long-lasting adaptive immune response. Vaccines against cancer need to be formulated to simulate danger signals delivered by pathogens for efficacy. This will require the incorporation of more than one adjuvant with a distinct mode of action. Although enormous efforts have been made in understanding the role of combination therapy on tumor-immune interaction at both molecular and cellular levels, the clinical transition remains in its infancy. However, the development of cancer vaccines received a great boost after a better understanding of DCs' biology, the requisite role these cells play in the initiation and regulation of ensuing adaptive immune responses, and how these interactions can be channeled for vaccine efficacy through adjuvant designs. Accumulating knowledge of cancer immune evasion mechanisms further emphasized the importance of adjuvants and laid the ground for designer adjuvants with pleiotropic functions directly targeting not only innate but also adaptive and regulatory immunity. It is a common consensus that rather than quantity, it is important to generate high-quality and high avidity polyclonal and polyfunctional effector $CD8^+$ T cells that eliminate primary tumor while establishing long-lasting memory able to control recurrences. Designer adjuvants have the potential to achieve this objective.

Almost all approved vaccines are established for infectious agents and are of preventive nature. Inasmuch these vaccines are administered to healthy individuals and their efficacy depends on the generation of humoral responses; the majority is formulated to contain a single adjuvant. Many of these promising formulations have been sidelined owing to a number of limitations that include the induction of immune responses of low potency or of inappropriate nature, that is, Th1versus Th2 response. Vaccines formulated with adjuvants targeting immune pathways with complementary or synergistic actions are expected to generate immune responses with a greater tumor eradication efficacy. The advantage of combined adjuvants has been demonstrated for preventive vaccines against the HPV that require the generation of high titers antibodies for efficacy. A vaccine formulated with alum as a single adjuvant showed modest efficacy in preventing HPV infection. In marked contrast, alum combined with MPL, a derivative of lipopolysaccharide, induced potent and long-lasting humoral responses with improved efficacy.[31] Utility of combined adjuvants for generating robust immune responses with improved clinical efficacy was further demonstrated for a second subunit HPV vaccine adjuvanted with a novel TLR4 agonist, CIA06, and aluminum salts.[90] The efficacy of combined use of adjuvants was also shown in the B16-F10 melanoma preclinical model, where Poly IC and CpG showed better therapeutic efficacy as compared with vaccines with single adjuvant.[91] Similarly, the combination of MF59 with the TLR9 agonist CpG demonstrated enhanced prophylactic efficacy by inducing a potent Th1 antitumor response in a preclinical melanoma tumor model.[45]

Srivastava et al. demonstrated the use of a multiadjuvanted approach by combining two different classes of adjuvants, SA-4-1BBL and MPL, with their distinct immune cell targets, signaling pathways, and significant roles as individual agents for the activation and maintenance of innate and cellular immune responses.[92] This study demonstrated that the combination of these adjuvants targeted the innate, adaptive, and regulatory arms of the immune system, resulting in the eradication of tumors in various preclinical models without detectable toxicity. Most importantly, the therapeutic efficacy of the SA-4-1BBL/MPL combined adjuvants was associated with significant infiltration of $CD8^+$ T cells into the tumor and marked reduction of $CD4^+CD25^+FoxP3^+$ Treg cells with a favorable intratumoral Teff:Treg cell ratio, an indicator of potential clinical relevance. Similarly, the combination of MF59 with the TLR9 agonist CpG demonstrated enhanced prophylactic efficacy by inducing a potent Th1 antitumor response in a preclinical melanoma tumor model.[45]

The mechanisms of synergy in vivo between components of adjuvant combinations can be difficult to tease out. Signaling by TLR4 and TLR7/8 adjuvants through different pathways in the same or disparate innate immune cell types may have synergistic or antagonistic effects, either through signal amplification[93,94] or through paracrine signaling.[95,96] Alternatively, simultaneously targeting adjuvant receptors on innate immune cells and costimulatory and inhibitory receptors on adaptive immune cells may enhance or suppress effector and memory T-cell responses. Therefore the choice of the appropriate adjuvant combination will require comprehensive knowledge of both the signaling pathways and cell expression patterns of targeted receptors.

The accumulating knowledge in cancer immunobiology and the complex nature of immune system-cancer interaction led to better vaccine design and the need to include adjuvants with the distinct course of immune actions. PROSTVAC, a combination vaccine comprising CD80 costimulatory molecule, intracellular adhesion molecule 1, and lymphocyte function-associated antigen 3 has shown potential in clinical studies. PROSTVAC-VF immunotherapy showed 44% reduction in the death rate and an 8.5-month improvement in median overall survival in men with metastatic prostate cancer.[2] In addition, an improved clinical response was observed in a phase III study with a formulation containing gp100:209–217 peptide plus incomplete Freund's adjuvant (Montanide ISA-51) in combination with IL-2 in patients with locally advanced stage III and stage IV melanoma.[48] Additional adjuvant combinations have been employed in the preclinical and clinical settings. These adjuvant systems have included combinations of several common vaccine adjuvants ranging from alum and MPL to CpG ODN.[97,98] The benefit of these adjuvant systems includes their current clinical use in human vaccines, such as AS03 for influenza and AS04 for hepatitis B virus and HPV.[99] Current studies for combinatorial adjuvant systems in cancer include AS02, which is comprised of MPL, QS-21, and oil-in-water emulsions, use in patients with melanoma and NSCLC, which express MAGE-A3^{+}.[100] These clinical trials demonstrated the efficacy of AS02 adjuvant system to generate MAGE-A3-specific antibody and T-cell responses.[101] In addition, AS15, composed of CpG ODN, MPL, QS-21, and liposomes, was studied in patients with melanoma and NSCLC in phase II and III clinical trials with reported immune-stimulatory efficacy.[97,102]

Almost all clinically approved adjuvants target APCs for activation and antigen presentation, which translate into efficacy for preventive vaccines against infections. However, this vaccine scheme is inefficient in generating immune responses of sufficient depth, magnitude, and complexity required for the eradication of cancer. Thus it is essential to continue to discover and develop adjuvants with well-defined pleiotropic immune functions and used them as single agent or in combination to directly target not only innate but also adaptive and regulatory immune responses for the generation of therapeutic immune responses. Importantly, adjuvant combination can also be exploited for cancer immunotherapy without the use of exogenous antigens by taking advantage of in situ tumor neoantigens. For example, injection of a mixture of TLR9 and TLR7/8 agonists directly into established solid tumors in mice led to increased recruitment of tumor-infiltrating CD8^{+} T cells and NK cells and reduced numbers of immunosuppressive MDSC that translated into significantly reduced tumor burden and establishment of long-term protection against rechallange.[103]

Conclusions and future prospects

The two major obstacles that confront the field of therapeutic cancer vaccines are: (1) the discovery of neoantigens that represent recurrent mutations of driver genes that are not patient and tumor specific with the potential of developing universal cancer vaccines and (2) immune adjuvants with pleiotropic direct effects on innate, adaptive, and regulatory immune cells that culminate in robust effector responses with clinical efficacy. Technological advances in recombinant DNA combined with bioinformatics and in silico prediction of neoantigens followed by the confirmation of immunogenicity using screening assays set the ground for the identification and use of neoantigens for therapeutic cancer vaccines. Renewed interest in adjuvant discovery and establishment of adjuvant systems tailored to generate desired effector immune responses against the cancer type will allow optimization of vaccine formulations for clinical efficacy. Combining vaccines with other immunotherapeutic approaches may serve as an effective strategy for cancers refractory to standard treatments. Finally, immune adjuvants that have pleiotropic functions on the cells of innate, adaptive, and regulatory immunity as single agents or in combination may prove efficacious in exploiting tumor neoantigens in situ for the generation of therapeutic immune responses, thus bypassing the need for the discovery of neoantigens and their use as personalized vaccines.

Acknowledgment

This study was funded in parts by DOD grant W81XWH-20-1-0459, LC190524 to ESY.

References

1. Yang Y. Cancer immunotherapy: harnessing the immune system to battle cancer. *J Clin Invest.* 2015;125:3335–3337.
2. Kantoff PW, Higano CS, Shore ND, et al. Sipuleucel-T immunotherapy for castration-resistant prostate cancer. *N Engl J Med.* 2010;363:411–422.
3. Xie Y, Akpinarli A, Maris C, et al. Naive tumor-specific CD4(+) T cells differentiated in vivo eradicate established melanoma. *J Exp Med.* 2010;207:651–667.
4. Butterfield LH. Cancer vaccines. *Bmj.* 2015;350:h988.
5. De Smet C, Lurquin C, van der Bruggen P, De Plaen E, Brasseur F, Boon T. Sequence and expression pattern of the human MAGE2 gene. *Immunogenetics.* 1994;39:121–129.
6. Gnjatic S, Ritter E, Büchler MW, et al. Seromic profiling of ovarian and pancreatic cancer. *Proc Natl Acad Sci U S A.* 2010;107:5088–5093.
7. Hofmann O, Caballero OL, Stevenson BJ, et al. Genome-wide analysis of cancer/testis gene expression. *Proc Natl Acad Sci U S A.* 2008;105:20422–20427.
8. Simpson AJ, Caballero OL, Jungbluth A, Chen YT, Old LJ. Cancer/testis antigens, gametogenesis and cancer. *Nat Rev Cancer.* 2005;5:615–625.
9. Karbach J, Neumann A, Atmaca A, et al. Efficient in vivo priming by vaccination with recombinant NY-ESO-1 protein and CpG in antigen naive prostate cancer patients. *Clin Cancer Res.* 2011;17:861–870.
10. Bakker AB, Schreurs MW, de Boer AJ, et al. Melanocyte lineage-specific antigen gp100 is recognized by melanoma-derived tumor-infiltrating lymphocytes. *J Exp Med.* 1994;179:1005–1009.
11. Kawakami Y, Eliyahu S, Delgado CH, et al. Identification of a human melanoma antigen recognized by tumor-infiltrating lymphocytes associated with in vivo tumor rejection. *Proc Natl Acad Sci U S A.* 1994;91:6458–6462.
12. Parkhurst MR, Fitzgerald EB, Southwood S, Sette A, Rosenberg SA, Kawakami Y. Identification of a shared HLA-A*0201-restricted T-cell epitope from the melanoma antigen tyrosinase-related protein 2 (TRP2). *Cancer Res.* 1998;58:4895–4901.

13. Correale P, Nieroda C, Zaremba S, et al. In vitro generation of human cytotoxic T lymphocytes specific for peptides derived from prostate-specific antigen. *JNCI J Natl Cancer Inst.* 1997;89:293–300.
14. Lam K-W, Li C-Y, Yam LT, Sun T, Lee G, Ziesmer S. Improved immunohistochemical detection of prostatic acid phosphatase by a monoclonal antibody. *Prostate.* 1989;15:13–21.
15. Vonderheide RH, Hahn WC, Schultze JL, Nadler LM. The telomerase catalytic subunit is a widely expressed tumor-associated antigen recognized by cytotoxic T lymphocytes. *Immunity.* 1999;10:673–679.
16. Disis ML, Wallace DR, Gooley TA, et al. Concurrent trastuzumab and HER2/neu-specific vaccination in patients with metastatic breast cancer. *J Clin Oncol.* 2009;27:4685–4692.
17. Chang K, Pastan I. Molecular cloning of mesothelin, a differentiation antigen present on mesothelium, mesotheliomas, and ovarian cancers. *Proc Natl Acad Sci U S A.* 1996;93:136–140.
18. Finn OJ, Gantt KR, Lepisto AJ, Pejawar-Gaddy S, Xue J, Beatty PL. Importance of MUC1 and spontaneous mouse tumor models for understanding the immunobiology of human adenocarcinomas. *Immunol Res.* 2011;50:261–268.
19. Calvo Tardón M, Allard M, Dutoit V, Dietrich PY, Walker PR. Peptides as cancer vaccines. *Curr Opin Pharmacol.* 2019;47:20–26.
20. Hollingsworth RE, Jansen K. Turning the corner on therapeutic cancer vaccines. *NPJ Vaccines.* 2019;4:7.
21. Srivastava PK. Neoepitopes of cancers: looking back, looking ahead. *Cancer Immunol Res.* 2015;3:969–977.
22. Rosenthal R, Cadieux EL, Salgado R, et al. Neoantigen-directed immune escape in lung cancer evolution. *Nature.* 2019;567:479–485.
23. Garrido F, Ruiz-Cabello F, Aptsiauri N. Rejection vs escape: the tumor MHC dilemma. *Cancer Immunol Immunother.* 2017;66:259–271.
24. Kober J, Leitner J, Klauser C, et al. The capacity of the TNF family members 4–1BBL, OX40L, CD70, GITRL, CD30L and LIGHT to costimulate human T cells. *Eur J Immunol.* 2008;38:2678–2688.
25. HogenEsch H, O'Hagan DT, Fox CB. Optimizing the utilization of aluminum adjuvants in vaccines: you might just get what you want. *NPJ Vaccines.* 2018;3:51.
26. Chappert P, Schwartz RH. Induction of T cell anergy: integration of environmental cues and infectious tolerance. *Curr Opin Immunol.* 2010;22:552–559.
27. Bowen WS, Svrivastava AK, Batra L, Barsoumian H, Shirwan H. Current challenges for cancer vaccine adjuvant development. *Expert Rev Vaccines.* 2018;17:207–215.
28. Sharma RK, Yolcu ES, Shirwan H. SA-4–1BBL as a novel adjuvant for the development of therapeutic cancer vaccines. *Expert Rev Vaccines.* 2014;13:387–398.
29. Dubensky Jr. TW, Reed SG. Adjuvants for cancer vaccines. *Semin Immunol.* 2010;22:155–161.
30. Brewer JM. (How) do aluminium adjuvants work? *Immunol Lett.* 2006;102:10–15.
31. Didierlaurent AM, Morel S, Lockman L, et al. AS04, an aluminum salt- and TLR4 agonist-based adjuvant system, induces a transient localized innate immune response leading to enhanced adaptive immunity. *J Immunol.* 2009;183:6186–6197.
32. Li JK, Balic JJ, Yu L, Jenkins B. TLR agonists as adjuvants for cancer vaccines. *Adv Exp Med Biol.* 2017;1024:195–212.
33. Petrovsky N. Novel human polysaccharide adjuvants with dual Th1 and Th2 potentiating activity. *Vaccine.* 2006;24(suppl 2). S2–26–29.

34. Tissot AC, Renhofa R, Schmitz N, et al. Versatile virus-like particle carrier for epitope based vaccines. *PLoS One*. 2010;5:e9809.
35. Muhammad A, Champeimont J, Mayr UB, Lubitz W, Kudela P. Bacterial ghosts as carriers of protein subunit and DNA-encoded antigens for vaccine applications. *Expert Rev Vaccines*. 2012;11:97–116.
36. Coffman RL, Sher A, Seder RA. Vaccine adjuvants: putting innate immunity to work. *Immunity*. 2010;33:492–503.
37. Gavin AL, Hoebe K, Duong B, et al. Adjuvant-enhanced antibody responses in the absence of toll-like receptor signaling. *Science*. 2006;314:1936–1938.
38. Cuzzubbo S, Mangsbo S, Nagarajan D, Habra K, Pockley AG, McArdle SEB. Cancer vaccines: adjuvant potency, importance of age, lifestyle, and treatments. *Front Immunology*. 2021;11.
39. Hogenesch H. Mechanism of immunopotentiation and safety of aluminum adjuvants. *Front Immunol*. 2012;3:406.
40. Alfonso S, Valdés-Zayas A, Santiesteban ER, et al. A randomized, multicenter, placebo-controlled clinical trial of racotumomab-alum vaccine as switch maintenance therapy in advanced non-small cell lung cancer patients. *Clin Cancer Res*. 2014;20:3660–3671.
41. Ott G, Barchfeld GL, Chernoff D, Radhakrishnan R, van Hoogevest P, Van Nest G. MF59. Design and evaluation of a safe and potent adjuvant for human vaccines. *Pharm Biotechnol*. 1995;6:277–296.
42. Dupuis M, Murphy TJ, Higgins D, et al. Dendritic cells internalize vaccine adjuvant after intramuscular injection. *Cell Immunol*. 1998;186:18–27.
43. Mosca F, Tritto E, Muzzi A, et al. Molecular and cellular signatures of human vaccine adjuvants. *Proc Natl Acad Sci U S A*. 2008;105:10501–10506.
44. Maynard SK, Marshall JD, MacGill RS, et al. Vaccination with synthetic long peptide formulated with CpG in an oil-in-water emulsion induces robust E7-specific CD8 T cell responses and TC-1 tumor eradication. *BMC Cancer*. 2019;19:540.
45. Yang M, Yan Y, Fang M, et al. MF59 formulated with CpG ODN as a potent adjuvant of recombinant HSP65-MUC1 for inducing anti-MUC1 + tumor immunity in mice. *Int Immunopharmacol*. 2012;13:408–416.
46. Hajam IA, Dar PA, Won G, Lee JH. Bacterial ghosts as adjuvants: mechanisms and potential. *Veterinary Res*. 2017;48:37.
47. Kraśko JA, Žilionytė K, Darinskas A, et al. Bacterial ghosts as adjuvants in syngeneic tumour cell lysate-based anticancer vaccination in a murine lung carcinoma model. *Oncol Rep*. 2017;37:171–178.
48. Schwartzentruber DJ, Lawson DH, Richards JM, et al. gp100 peptide vaccine and interleukin-2 in patients with advanced melanoma. *N Engl J Med*. 2011;364:2119–2127.
49. Rosenberg SA, Yang JC, Schwartzentruber DJ, et al. Immunologic and therapeutic evaluation of a synthetic peptide vaccine for the treatment of patients with metastatic melanoma. *Nat Med*. 1998;4:321–327.
50. Nishikawa M, Takemoto S, Takakura Y. Heat shock protein derivatives for delivery of antigens to antigen presenting cells. *Int J Pharm*. 2008;354:23–27.
51. Walker KB, Keeble J, Colaco C. Mycobacterial heat shock proteins as vaccines—a model of facilitated antigen presentation. *Curr Mol Med*. 2007;7:339–350.
52. Ragupathi G, Gardner JR, Livingston PO, Gin DY. Natural and synthetic saponin adjuvant QS-21 for vaccines against cancer. *Expert Rev Vaccines*. 2011;10:463–470.
53. Marty-Roix R, Vladimer GI, Pouliot K, et al. Identification of QS-21 as an inflammasome-activating molecular component of saponin adjuvants. *J Biol Chem*. 2016;291:1123–1136.

54. Gin DY, Slovin SF. Enhancing immunogenicity of cancer vaccines: QS-21 as an immune adjuvant. *Curr drug Ther*. 2011;6:207–212.
55. Davis ID, Chen W, Jackson H, et al. Recombinant NY-ESO-1 protein with ISCOMATRIX adjuvant induces broad integrated antibody and CD4(+) and CD8(+) T cell responses in humans. *Proc Natl Acad Sci U S A*. 2004;101:10697–10702.
56. Mowat AM, Maloy KJ, Donachie AM. Immune-stimulating complexes as adjuvants for inducing local and systemic immunity after oral immunization with protein antigens. *Immunology*. 1993;80:527–534.
57. den Brok MH, Büll C, Wassink M, et al. Saponin-based adjuvants induce cross-presentation in dendritic cells by intracellular lipid body formation. *Nat Commun*. 2016;7. 13324.
58. Ogawa C, Liu Y-J, Kobayashi KS. Muramyl dipeptide and its derivatives: peptide adjuvant in immunological disorders and cancer therapy. *Curr Bioact Compd*. 2011;7:180–197.
59. Carlring J, Barr TA, Buckle AM, Heath AW. Anti-CD28 has a potent adjuvant effect on the antibody response to soluble antigens mediated through CTLA-4 by-pass. *Eur J Immunol*. 2003;33:135–142.
60. Attarwala H. TGN1412: from discovery to disaster. *J Young Pharm*. 2010;2:332–336.
61. Croft M. The role of TNF superfamily members in T-cell function and diseases. *Nat Rev Immunol*. 2009;9:271–285.
62. Futagawa T, Akiba H, Kodama T, et al. Expression and function of 4–1BB and 4–1BB ligand on murine dendritic cells. *Int Immunol*. 2002;14:275–286.
63. Melero I, Murillo O, Dubrot J, Hervas-Stubbs S, Perez-Gracia JL. Multi-layered action mechanisms of CD137 (4–1BB)-targeted immunotherapies. *Trends Pharmacol Sci*. 2008;29:383–390.
64. Watts TH. TNF/TNFR family members in costimulation of T cell responses. *Annu Rev Immunol*. 2005;23:23–68.
65. Melero I, Johnston JV, Shufford WW, Mittler RS, Chen L. NK1.1 cells express 4–1BB (CDw137) costimulatory molecule and are required for tumor immunity elicited by anti-4–1BB monoclonal antibodies. *Cell Immunol*. 1998;190:167–172.
66. Halstead ES, Mueller YM, Altman JD, Katsikis PD. In vivo stimulation of CD137 broadens primary antiviral CD8 + T cell responses. *Nat Immunol*. 2002;3:536–541.
67. Bukczynski J, Wen T, Ellefsen K, Gauldie J, Watts TH. Costimulatory ligand 4–1BBL (CD137L) as an efficient adjuvant for human antiviral cytotoxic T cell responses. *Proc Natl Acad Sci U S A*. 2004;101:1291–1296.
68. Long AH, Haso WM, Shern JF, et al. 4–1BB costimulation ameliorates T cell exhaustion induced by tonic signaling of chimeric antigen receptors. *Nat Med*. 2015;21:581–590.
69. Habib-Agahi M, Jaberipour M, Searle PF. 4–1BBL costimulation retrieves CD28 expression in activated T cells. *Cell Immunol*. 2009;256:39–46.
70. Sharma RK, Schabowsky RH, Srivastava A, et al. 4–1BB ligand as an effective multifunctional immunomodulator and antigen delivery vehicle for the development of therapeutic cancer vaccines. *Cancer Res*. 2010;70:3945–3954.
71. Kwon BS, Weissman SM. cDNA sequences of two inducible T-cell genes. *Proc Natl Acad Sci U S A*. 1989;86:1963–1967.
72. Sharma RK, Elpek KG, Yolcu ES, et al. Costimulation as a platform for the development of vaccines: a peptide-based vaccine containing a novel from of 4–1BBL eradicates established tumors. *Cancer Res*. 2009;69:4319–4326.
73. Madireddi S, Schabowsky RH, Srivastava AK, Sharma RK, Yolcu ES, Shirwan H. SA-4–1BBL costimulation inhibits conversion of conventional CD4 + T cells into CD4 + FoxP3 + T regulatory cells by production of IFN-gamma. *PLoS One*. 2012;7:e42459.

74. Barsoumian HB, Yolcu ES, Shirwan H. 4−1BB signaling in conventional T cells drives IL-2 production that overcomes CD4 + CD25 + FoxP3 + T regulatory cell suppression. *PLoS One*. 2016;11:e0153088.
75. Sharma RK, Elpek KG, Yolcu ES, et al. Costimulation as a platform for the development of vaccines: a peptide-based vaccine containing a novel form of 4−1BB ligand eradicates established tumors. *Cancer Res*. 2009;69:4319−4326.
76. Sharma RK, Srivastava AK, Yolcu ES, et al. SA-4−1BBL as the immunomodulatory component of a HPV-16 E7 protein based vaccine shows robust therapeutic efficacy in a mouse cervical cancer model. *Vaccine*. 2010;28:5794−5802.
77. Heery CR, Palena C, McMahon S, et al. Phase I study of a poxviral TRICOM-based vaccine directed against the transcription factor brachyury. *Clin Cancer Res*. 2017;23:6833−6845.
78. Madan RA, Bilusic M, Heery C, Schlom J, Gulley JL. Clinical evaluation of TRICOM vector therapeutic cancer vaccines. *SemOncol*. 2012;39:296−304.
79. Gulley JL, Arlen PM, Tsang KY, et al. Pilot study of vaccination with recombinant CEA-MUC-1-TRICOM poxviral-based vaccines in patients with metastatic carcinoma. *Clin Cancer Res*. 2008;14:3060−3069.
80. Mohebtash M, Tsang K-Y, Madan RA, et al. A pilot study of MUC-1/CEA/TRICOM poxviral-based vaccine in patients with metastatic breast and ovarian cancer. *Clin Cancer Res*. 2011;17:7164−7173.
81. Madan RA, Mohebtash M, Arlen PM, et al. Ipilimumab and a poxviral vaccine targeting prostate-specific antigen in metastatic castration-resistant prostate cancer: a phase 1 dose-escalation trial. *Lancet Oncol*. 2012;13:501−508.
82. Heery CR, Madan RA, Stein MN, et al. Samarium-153-EDTMP (Quadramet®) with or without vaccine in metastatic castration-resistant prostate cancer: A randomized Phase 2 trial. *Oncotarget*. 2016;7:69014−69023.
83. Arlen PM, Skarupa L, Pazdur M, et al. Clinical safety of a viral vector based prostate cancer vaccine strategy. *J Urol*. 2007;178:1515−1520.
84. Madan RA, Heery CR, Gulley JL. Combination of vaccine and immune checkpoint inhibitor is safe with encouraging clinical activity. *Oncoimmunology*. 2012;1:1167−1168.
85. Schijns VE, Degen WG. Vaccine immunopotentiators of the future. *Clin Pharmacol Ther*. 2007;82:750−755.
86. Schmidt CS, Mescher MF. Peptide antigen priming of naive, but not memory, CD8 T cells requires a third signal that can be provided by IL-12. *J Immunol*. 2002;168:5521−5529.
87. Curtsinger JM, Valenzuela JO, Agarwal P, Lins D, Mescher MF. Type I IFNs provide a third signal to CD8 T cells to stimulate clonal expansion and differentiation. *J Immunol*. 2005;174:4465−4469.
88. Morrow MP, Yan J, Pankhong P, et al. Unique Th1/Th2 phenotypes induced during priming and memory phases by use of interleukin-12 (IL-12) or IL-28B vaccine adjuvants in rhesus macaques. *Clin Vaccine Immunol*. 2010;17:1493−1499.
89. Johansen P, Mohanan D, Martínez-Gómez JM, Kündig TM, Gander B. Lymphogeographical concepts in vaccine delivery. *J Control Rel*. 2010;148:56−62.
90. Han JE, Wui SR, Park SA, et al. Comparison of the immune responses to the CIA06-adjuvanted human papillomavirus L1 VLP vaccine with those against the licensed HPV vaccine Cervarix™ in mice. *Vaccine*. 2012;30:4127−4134.
91. Whitmore MM, DeVeer MJ, Edling A, et al. Synergistic activation of innate immunity by double-stranded RNA and CpG DNA promotes enhanced antitumor activity. *Cancer Res*. 2004;64:5850−5860.

92. Srivastava AK, Dinc G, Sharma RK, Yolcu ES, Zhao H, Shirwan H. SA-4−1BBL and monophosphoryl lipid A constitute an efficacious combination adjuvant for cancer vaccines. *Cancer Res*. 2014;74:6441−6451.
93. Bohnenkamp HR, Papazisis KT, Burchell JM, Taylor-Papadimitriou J. Synergism of Toll-like receptor-induced interleukin-12p70 secretion by monocyte-derived dendritic cells is mediated through p38 MAPK and lowers the threshold of T-helper cell type 1 responses. *Cell Immunol*. 2007;247:72−84.
94. Fischetti L, Zhong Z, Pinder CL, Tregoning JS, Shattock RJ. The synergistic effects of combining TLR ligand based adjuvants on the cytokine response are dependent upon p38/JNK signalling. *Cytokine*. 2017;99:287−296.
95. Coccia M, Collignon C, Hervé C, et al. Cellular and molecular synergy in AS01-adjuvanted vaccines results in an early IFNγ response promoting vaccine immunogenicity. *NPJ Vaccines*. 2017;2:25.
96. Sköld AE, Mathan TSM, van Beek JJP, et al. Naturally produced type I IFNs enhance human myeloid dendritic cell maturation and IL-12p70 production and mediate elevated effector functions in innate and adaptive immune cells. *Cancer Immunol Immunother*. 2018;67:1425−1436.
97. Garcon N, Van Mechelen M. Recent clinical experience with vaccines using MPL- and QS-21-containing adjuvant systems. *Expert Rev Vaccines*. 2011;10:471−486.
98. Garçon N, Chomez P, Van Mechelen M. GlaxoSmithKline adjuvant systems in vaccines: concepts, achievements and perspectives. *Expert Rev Vaccines*. 2007;6:723−739.
99. Temizoz B, Kuroda E, Ishii KJ. Vaccine adjuvants as potential cancer immunotherapeutics. *Int Immunol*. 2016;28:329−338.
100. Vantomme V, Dantinne C, Amrani N, et al. Immunologic analysis of a phase I/II study of vaccination with MAGE-3 protein combined with the AS02B adjuvant in patients with MAGE-3-positive tumors. *J Immunother*. 2004;27:124−135.
101. Atanackovic D, Altorki NK, Stockert E, et al. Vaccine-induced CD4+ T cell responses to MAGE-3 protein in lung cancer patients. *J Immunol*. 2004;172:3289−3296.
102. Slingluff Jr. CL, Petroni GR, Olson WC, et al. A randomized pilot trial testing the safety and immunologic effects of a MAGE-A3 protein plus AS15 immunostimulant administered into muscle or into dermal/subcutaneous sites. *Cancer Immunol Immunother*. 2016;65:25−36.
103. Zhao BG, Vasilakos JP, Tross D, Smirnov D, Klinman DM. Combination therapy targeting toll like receptors 7, 8 and 9 eliminates large established tumors. *J Immunother Cancer*. 2014;2:12.

Chapter 13

Nucleic acid-based vaccines

Antonella Conforti[1,2], Fabio Palombo[2,3] and Luigi Aurisicchio[1,2,3]
[1]*Evvivax s.r.l., Rome, Italy,* [2]*Takis s.r.l., Rome, Italy,* [3]*NeoMatrix s.r.l., Rome, Italy*

Overview

For over 100 years vaccination has been based on two classical approaches, either protein-based vaccines or live-attenuated infectious agents synthetizing the immunizing antigen. Then, the last decades of 19th century witnessed the development of nucleic acid-based vaccines, the so-called "third-generation vaccines." To stimulate an immune response against a self or nonself antigen, this relatively new generation uses genetic material, either DNA or RNA, in both cases providing the instructions for making a specific protein which the immune system may recognize. Once inserted into host cells, this genetic material is translated into a protein antigen, which then may trigger an immune response. Early gene therapy experiments investigated whether the direct injection of DNA or RNA expression vectors could offer an alternative to viral vectors, then it was observed[1] that intramuscular injection of nonreplicative DNA or RNA expression vectors in cationic lipid vesicles resulted in the expression of gene products in muscle cells and, surprisingly, they later found that for plasmid DNA vectors this occurred even without the lipid delivery system. In the oncology field, genetic vaccines can elicit immune responses against a wide variety of tumor antigens, mainly tumor-specific antigens, found only on cancer cells but not on healthy cells, and tumor-associated antigens (TAAs), which have elevated levels on tumor cells but are also expressed at lower levels on healthy cells.[2] After delivering in vivo an expression cassette carrying (in part or entirely, and usually controlled by a strong constitutive promoter) the coding region of one or more antigens of choice in the subject to be vaccinated, the genetic material is up taken by resident cells [e.g., muscle cells, fibroblasts, monocytes, and dendritic cells (DCs)] and this results in the endogenous production of the selected antigen(s). It is well known that the extent of gene expression is proportional to the amount of transduced cells and of intact plasmid copies entering the nucleus.[3] Once produced and secreted by transduced cells (usually myocytes or dermal fibroblasts, depending on the injection route), the antigen is endocytosed by host professional antigen-presenting cells

Cancer Vaccines as Immunotherapy of Cancer. DOI: https://doi.org/10.1016/B978-0-12-823901-8.00002-9
© 2022 Elsevier Inc. All rights reserved.

(APCs), processed and cross-presented to the immune system in the draining lymph nodes.[4]

Since early gene therapy experiments, genetic vaccines have been extensively investigated, thus increasing the therapeutic landscape for treating both infectious diseases and cancer. Indeed, the concept of genetic vaccines is an appealing immunotherapeutic approach in many diverse fields, potentially translatable to the clinics, thanks to several beneficial features. First, nucleic acid-based vaccines have shown a highly satisfactory safety profile, without the potential risk of integration or pathogenicity,[5] and for this reason, they are considered the ideal therapeutic strategy in cancer immunotherapy or for vaccinating immunocompromised people. Second, genetic vaccination can elicit both T-cell activation and antibody production in response to even small amounts of expressed protein and, unlike whole virus vectors, can be more easily administered in multidose regimens without generating preexisting immunity (Fig. 13.1). In addition, manufacturing process of genetic vaccines implies some advantages: both DNA and RNA are inexpensively and easily constructed, directly from the genetic sequence of the desired antigen and, once fully established, this process can be applied also to other systems, so that it can be easily adjusted according to the diversity in HLA in various field settings, to include the most immunogenic antigens and modulators for a specific population. Hence, as they recently demonstrated their full capacity during the COVID-19 pandemic, in which two RNA-based vaccines outperformed classical virus vector-based vaccines in terms of preventing disease efficacy,[7] the use of nucleic acids in vaccine development programs is constantly growing in a wide range of traditional pharmaceutical markets, such as cancers and allergies, as well as infectious diseases and it is increasingly demonstrating its safety and efficacy in early- and mid-stage human clinical trials.[8,9] Importantly, these vaccines can be quickly designed based on genetic sequencing alone and this technical advantage may be essential not only to respond to the emergency outbreak of an epidemic[10] but also to generate personalized vaccines for treating tumors.[11]

Nevertheless, these vaccination technologies still present some drawbacks to consider. As for DNA vaccines immunogenicity, a number of diverse approaches have been set up to increase DNA potency, such as immunostimulants (cytokines and immunostimulatory molecules), delivery routes and devices (with intramuscular injection followed by electroporation having been found to be the most effective in inducing strong immune responses), and different combination strategies (e.g., DNA prime followed by viral vector/peptide/recombinant protein heterologous boosts). In addition, to avoid any animal contamination in the manufacturing process, researchers are exploring alternative manufacturing strategies, such as PCR-generated linear DNA fragments.[12]

Conversely, as for RNA vaccines, there still remain some challenges to overcome, such as stability, potential toxicity, and manufacture limitations.

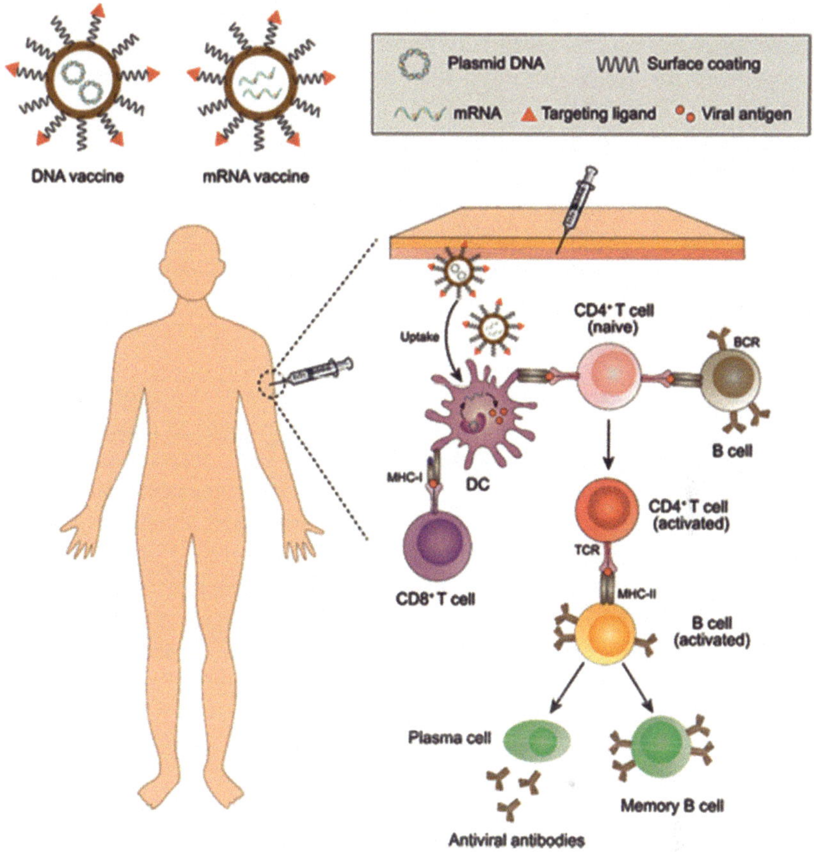

FIGURE 13.1 Schematic representation of the immune system activation following nucleic acid-based vaccination.[6]

Actually, vaccine developers have spent many efforts to increase RNA stability and cellular delivery through encapsulation into nanoparticles and to reduce its intrinsic reactogenicity, by means of modified nucleosides over the past decade.[13] To develop an efficient RNA vaccine for recent COVID-19 pandemic, RBD antigen expressed by BNT162b1 vaccine has been fused to a T4 fibritin-derived "foldon" trimerization domain to increase its immunogenicity by multivalent display. Moreover, 1-methylpseudouridine has been incorporated instead of uridine to reduce innate immune sensing and to increase RNA translation in vivo.[7]

Besides several remarkable breakthroughs in immunotherapy, such as the use of antibodies to inhibitory T-cell checkpoint molecules and the infusion of ex vivo engineered tumor-specific T cells, leading to a dramatic control of previously refractory tumors,[14] genetic cancer vaccines still have to show notable success in the field of cancer immunotherapy. Indeed, immunotherapeutic approach may be potentially translatable to the clinics, where it could complement traditional therapies, such as surgery, chemotherapy, and radiation therapy. There are several reasons justifying this approach, particularly the ease of administration and the possibility to target pathways involved in tumor development. Besides this, genetic vaccines show other practical advantages, such as being more cost effective compared to other vaccination platforms (i.e., recombinant protein, tumor cells, or viral vectors) and allowing for multiple administrations while keeping safe and immunogenic.[15]

DNA-based vaccines

DNA-based vaccines offer a highly versatile and powerful platform technology, with several advantages to consider. First, they induce elevated levels of proinflammatory cytokines [interleukin (IL)-1, IL-6, tumor necrosis factor-α, interferon (IFN)-γ, IL-17] at the site of injection, thus promoting homing and activation of DCs and allowing for a prolonged antigen expression and consequently higher epitope display on MHC class I molecules and induction of $CD8^+$ responses. Second, by means of easy genetic engineering practices, it is possible to manipulate the antigen of interest to generate more immunogenic versions or fusion genes between the TAAs and epitopes known to facilitate the generation of a T-helper memory response.[16] Moreover, DNA-based vaccines have the potential to be produced in large scale as a generic manufacturing platform, with easy and economic GMP production and high level of stability and safety, due to the absence of viral elements. Finally, DNA vaccines can be administered repeatedly because they do not elicit specific antivector immunity, and there are not preexisting antibodies against DNA.[1,2]

Although presenting these advantages, initially DNA vaccines presented some drawbacks to efficient immunization, with naked DNA injection failing to show strong vaccine immunogenicity against human papilloma virus (HPV)-derived cancer antigen,[18] possibly due to the combination of inefficient cellular delivery of DNA plasmids, low levels of antigen production, and lack of stimulation of the innate immune system. New delivery technologies have put acceleration to the rapidly growing sector of genetic cancer vaccines. One of these vaccination tools is electroporation also known as Electro-Gene-Transfer (EGT). EGT is a powerful method of DNA delivery currently employed for several medical applications, such as DNA vaccination and gene therapy for cancer treatment. Due to its safety, efficacy, and ease of application, EGT is the ideal candidate vaccine platform as a substitute to viral gene delivery methods. This procedure consists of exposing a cell or a tissue to an external electric field that increases cell membrane permeability to molecules that otherwise would cross the plasma membrane with low efficiency, and for this reason, it has been widely used in different biomedical applications.[19,20] Due to the rapid turnover of tumor cells and to nonintegration of the plasmid, a short-lived and low transgene expression may be observed when electroporating tumor cells, whereas a longer expression, in the order of a few weeks, may be expected when transferring genes to skin and, even at higher expression levels, to myocytes.[21] Comparison between DNA injection alone and DNA-EGT has demonstrated an increase in both cellular and humoral responses after electric fields were applied. It has been demonstrated that the addition of electroporation provides a 10–100-fold augmentation of immune response and defense against pathogens in humans and numerous animal models of diseases, such as HIV/SIV, malaria, hepatitis B and C, HPV, anthrax, and influenza.[22] Furthermore, EGT may enhance immune responses through increased protein expression, secretion of inflammatory chemokines and cytokines, and recruitment of monocytes and APCs at the site of electroporation.[23,24] In addition, in vivo electroporation causes transient and reversible cell damage, also resulting in local inflammation and release of cytokines, which may act as a danger signal, consequently enhancing vaccine potency. As a result, antigen-specific humoral and cellular immune responses are stronger when plasmid DNA is delivered by electroporation when compared to the levels achieved by intramuscular injection of DNA alone.[16] Moreover, combinations of heterologous modalities of immunization, comprising both DNA and adenoviral vector administrations, have proven to induce superior immune reactions as compared to single modality vaccines. Actually, an important advantage of heterologous prime–boost protocols comprising the sequential use of adenoviral vector and plasmid DNA is that it is possible to break tolerance by exploiting adenovirus strong immunogenicity as priming agent and then use DNA for repeated boosting because of the lack of anamnestic responses against the vector.

DNA electrotransfer in vivo is in many cases more efficient than other nonviral methods of gene delivery, such as gene gun in the liver,[25] liposomes in the brain or the cornea,[26,27] sonoporation in the muscle,[28] or cationic lipids in the synovial tissue.[29] Moreover, since gene expression can be transient, it is possible to repeat the electrotransfer procedure to reach same levels of transfection as following the first treatment.[30] Electrotransfer of multiple genes in parallel is also feasible and by adapting the procedure to the target tissue, electrotransfer has been successfully applied to various tissues, including skeletal muscle, skin, tumors, liver, lungs, kidneys, brain, retina, cornea, and heart with minimal tissue damage.[31]

Many preclinical studies have demonstrated the robust immunological and therapeutic effect exerted by DNA electroporation (DNA-EGT) in small and large animal models.[32,33] The BALB/NeuT murine model develops spontaneously mammary tumors driven by HER-2/neu overexpression. HER-2/neu oncoprotein is a tyrosine kinase receptor overexpressed in several human and animal tumors and associated with poor prognosis, and for this reason, it is an optimal target for cancer immunotherapy. Aurisicchio et al. showed that the combination of a genetic vaccine targeting HER-2/neu, based on the heterologous combination of adenoviral vector and DNA delivered by EGT, with an endosomal toll-like receptor(TLR)-9 agonist, known to induce potent Th1-type immune responses,[34] has a significant therapeutic effect in leading to tumor regression/stabilization.[1] Moreover, this effect was associated with immunological parameters involving innate and adaptive immunities, such as NK cell activation through TLR9 and production of HER-2/neu-specific antibodies and Th1-skewed T cells. After being vaccinated at 13 weeks of age in the presence of relatively small mammary hyperplasia and in situ carcinomas, only the group of mice that received the DNA-EGT vaccine in combination with a TLR9 agonist produced a marked and long-lasting antitumor response, disease stabilization, and tumor shrinkage.

A genetic vaccine based on DNA electroporation (DNA-EGT) and adenovirus targeting murine TERT (mTERT) has elicited significant immune responses against mouse TERT[35] and therapeutic effects in two different murine models of spontaneous carcinogenesis.[36] Besides these encouraging results in mouse models, a DNA-EGT and adenoviral vaccine targeting human TERT were also evaluated for innate humoral and cell-mediated immune responses in rhesus monkeys,[37] given that human and rhesus TERT protein sequence share 96% identity and tolerance is expected to play a major role in determining vaccination efficacy also in this animal model. As expected, results showed that DNA-EGT/Ad can break immune tolerance to TERT and achieve a durable immune response in nonhuman primates, thus also paving the way for clinical applications of DNA-EGT in humans.

Another strategy to enhance the immunogenicity of genetic vaccines is the choice of encoded neoantigens, resulting from tumor-specific somatic mutations, that can be recognized as nonself by the host immune system, resulting in increased specificity, efficacy, and safety of the antitumor therapies.[38] Different studies have shown the effectiveness of DNA-based vaccines encoding neoantigens for cancer immunity and antitumor activity, also in combination with electroporation.[39–43]

RNA-based vaccines

RNA vaccines represent a promising alternative to conventional vaccination platforms because of their high immunogenicity, capacity for rapid development, and potential for low-cost manufacture and safe administration. In the past, their use for vaccine development has been limited by the instability and inefficient in vivo delivery of RNA. Recent technological advances have largely overcome these issues, and multiple RNA vaccine platforms against infectious diseases and several types of cancer have demonstrated encouraging results in both animal models and humans. As shown by vaccines recently approved for Covid-19 pandemic, many technical advancements have been introduced in the coding and noncoding sequences of the expression cassette including the use of modified necleosides that can affect the immunostimulatory property. Synthetic RNA is single-stranded, contains a 5′ cap-UTRs embracing the coding region and a 3′ poly(A) tail, thus resembling naturally occurring processed mature RNA molecules, and is generated by in vitro transcription from a linear DNA template. In the cytoplasm, RNA can directly encode the antigen and interact with the immunological sensors as well. Its activity as a natural ligand of TLR3, 7, and 8, or of retinoic acid-inducible gene 1 (RIG-I) and melanoma differentiation-associated protein 5 (MDA5) triggers the release of type I IFN and proinflammatory cytokines.[44] IFN-stimulated genes (ISGs), such as IFN-inducible double-stranded (ds)RNA-activated protein kinase, and 2′,5′-oligoadenylate synthetase with subsequent RNase-L expression, stimulate the antiviral defense, characterized by stalled RNA translation and increased targeting of RNA for degradation.[45,46] The intrinsic RNA immunostimulatory activity is particularly important for breaking immune tolerance when nonmutated cancer-associated self-antigens are targeted; however, it can be attenuated by modified nucleosides, such as pseudouridine, methylpseudouridine, N^6-2-thiouridine, 5-methylcytidine, or 5-methyladenosine,[13] and by the removal of dsRNA.[47,48]

Efficient in vivo RNA delivery is critical to achieving therapeutic relevance. Indeed, exogenous RNA must pass through the barrier of the lipid membrane in order to reach the cytoplasm and be translated to functional protein. So far, some of the major approaches for the delivery of RNA vaccines that have been described are: ex vivo loading of DCs, followed by reinfusion of the transfected cells, lipid-based delivery, and naked RNA delivery. In all cases, the route of administration and delivery format of RNA vaccines can greatly affect outcomes. A variety of RNA cancer vaccine formats have been developed using common delivery routes (intradermal, intramuscular, subcutaneous, or intranasal) and some unconventional routes of vaccination [intranodal, intravenous (IV), intrasplenic, or intratumoural].

DCs can be loaded with mRNA both ex vivo and in situ.[49] For the ex vivo condition, immature DCs are obtained from patients' peripheral blood. After the maturation of DCs, DCs are loaded with antigen-encoding mRNA and subsequently administrated back to patients. RNA loading into DCs can be realized by several techniques and, among them, the electroporation technique is the most frequently used.[50] Though initial investigations have applied the mRNA electroporation to inflammatory diseases, most of the electroporation-related mRNA vaccine studies are about cancer. In tumor-bearing mouse models (B16/F10.9) with poor immunogenicity and massive metastases, the administration of DCs pulsed with encoding-OVA mRNA (tumor-derived) showed a consistent reduction of metastatic lung sites.[51] Other studies demonstrated the antitumoral efficacy of DC-based mRNA vaccines administered with mRNA encoding for proinflammatory cytokines[52] or other immunity-regulating proteins, such as 4−1BB ligand[53] and CD83.[54]

Intranodal administration of naked mRNA is an unconventional but efficient means of vaccine delivery. Nonformulated or naked RNA vaccines administered intradermally or into lymphnodes (intranodally) have been shown to prime T-cell responses in mice[55,56] and humans.[57,58] The rationale for intranodal injection is to deliver the vaccine directly into the area of T-cell priming for processing by resident DCs.

One of the challenges is that the systematic administration of mRNA vaccines causes the aggregation of serum proteins and mRNA degradation. To overcome this obstacle, various mRNAs formulations with molecular carriers have been investigated to enhance mRNA translational activities and protect it from degradation. Lipid nanoparticles (LNPs) have become one of the most appealing and commonly used RNA delivery tools. LNPs often consist of four components: an ionizable cationic lipid, which promotes self-assembly into virus-sized ($\sim$100 nm) particles and allows endosomal release of mRNA to the cytoplasm; lipid-linked polyethylene glycol (PEG), which increases the half-life of formulations; cholesterol, a stabilizing agent; and naturally occurring phospholipids, which support lipid bilayer structure. RNA-LNP can be modified to have a negative net charge when prepared by molar excess of negatively charged RNA, or by shielding positive surface charges with PEG. RNA-LPN with a negative net charge are subject to specific uptake by APCs in vivo without the need for targeting moieties.[59] The administration methods determine the distribution and expression kinetics of lipid-based mRNA vaccines in vivo. Local delivery is realized by subcutaneous, intramuscular, intradermal, and intranodal administration. Scientists applied these methods to APCs and other immune cells, eliciting locally strong and long-lasting immune responses.[60,61] The IV administration of RNA-LPNs is commonly used in systematic diseases, like hematological diseases.[62]

Genetic vaccines in human clinical trials

Since the first phase I dose escalation clinical trial of DNA plasmid IL-12 electroporated vaccine, conducted in 24 patients with metastatic melanoma, demonstrated safety and efficacy of this technology,[63] several clinical studies for DNA vaccination in patients with cancer have been performed or are currently ongoing (see Table 13.1). Actually, electroporation-based DNA delivery technology dramatically enhances cellular uptake of DNA vaccines. Although electroporation effectively delivers plasmids to a wide variety of tissues in preclinical animal studies, translation of EGT applications has thus far focused on muscle, skin, and directly to tumors. The skeletal muscle is the most widely used tissue for EGT in clinics[64] because of its dimensions, accessibility, and its organization in long parallel fibers, which offers an

TABLE 13.1 Nucleic acid-based human clinical trials currently ongoing in cancer treatment (last updated on May 2021 on clinicaltrials.gov).

	Trial	Status	Conditions	Interventions	Location
DNA	INO-3107 with electroporation (EGT) in participants with HPV6- and/or HPV11-associated recurrent respiratory papillomatosis	Not yet recruiting	Respiratory papillomatosis	Drug: INO-3107	United States
	HPV DNA vaccine via electroporation for HPV16-positive cervical neoplasia	Recruiting	• Human papillomavirus type 16 • Cervical intraepithelial neoplasia grade II • Cervical intraepithelial neoplasia, grade III	Drug: pNGVL4aCRTE6E7L2	United States
	INO-5401 + INO-9012 in combination with atezolizumab in locally advanced unresectable or metastatic/recurrent urothelial carcinoma	Active, not recruiting	Urothelial carcinoma	• Biological: INO-5401 • Biological: INO-9012 • Drug: atezolizumab	United States
	INO-5401 and INO-9012 delivered by electroporation (EGT) in combination with cemiplimab (REGN2810) in Newly-diagnosed glioblastoma (GBM)	Active, not recruiting	Glioblastoma	• Biological: INO-5401 • Biological: INO-9012 • Biological: cemiplimab • Radiation: radiation therapy • Drug: temozolomide	United States
	DNA plasmid-encoding interleukin-12/HPV DNA plasmidsTherapeutic vaccine INO-3112 and durvalumab in treating patients with recurrent or metastatic human papillomavirus-associated cancers	Recruiting	• Human papillomavirus-16 positive • Human papillomavirus-18 positive • Metastatic malignant neoplasm • Recurrent anal canal carcinoma	• Biological: DNA plasmid encoding IL-12/HPV DNA plasmids therapeutic vaccine MEDI0457 • Biological: durvalumab	United States

(*Continued*)

TABLE 13.1 (Continued)

	Trial	Status	Conditions	Interventions	Location
			• Recurrent cervical carcinoma • Recurrent malignant neoplasm • Recurrent penile carcinoma • Recurrent vaginal carcinoma • Recurrent vulvar carcinoma • Refractory malignant neoplasm		
	hTERT immunotherapy alone or in combination with IL-12 DNA followed by electroporation in adults with solid tumors at high risk of relapse	Completed	• Breast cancer • Lung cancer • Pancreatic cancer • Head and neck cancer • Ovarian cancer • Colorectal cancer • Gastric cancer • Esophageal cancer • Hepatocellular carcinoma	• Biological: INO-1400 • Biological: INO-9012 • Biological: INO-1401	United States
	INVAC-1 anticancer hTERT DNA immunotherapy	Completed	Solid tumors	Biological: INVAC-1	France
	Safety of GX-188E DNA therapeutic vaccine administered by electroporation to cervical intraepithelial neoplasia grade 3	Completed	Cervical intraepithelial neoplasia 3	Biological: GX-188E	South Korea
	Safety study of DNA vaccine delivered by intradermal electroporation to treat colorectal cancer	Unknown	Colorectal cancer	• Biological: tetwtCEA DNA (wt CEA with tetanus toxoid Th epitope) • Biological: GM-CSF • Drug: cyclophosphamide	Sweden
	Phase I trial of intratumoral pIL-12 electroporation in malignant melanoma	Completed	Malignant melanoma	• Biological: IL-12p DNA	United States

	A study of V934/V935 vaccine in cancer participants with selected solid tumors (V934-002)	Completed	• Nonsmall-cell lung carcinoma • Breast cancer • Melanoma • Upper GI tract carcinoma • Colon carcinoma • Renal cell carcinoma • Bladder carcinoma • Prostate cancer	• Biological: V935 • Biological: V934	United States
RNA	A clinical trial investigating the safety, tolerability, and therapeutic effects of BNT113 in combination with pembrolizumab versus pembrolizumab alone for patients with a form of head and neck cancer positive for human papilloma virus 16 and expressing the protein PD-L1	Recruiting	• Unresectable head and neck squamous cell carcinoma • Metastatic head and neck cancer • Recurrent head and neck cancer	Biological: BNT113Biological: pembrolizumab	United States
	Evaluation of the safety and tolerability of i.v. administration of a cancer vaccine in patients with advanced melanoma	Active, not recruiting	Melanoma	Biological: Lipo-MERIT	Germany
	A study of autogene cevumeran (RO7198457) as a single agent and in combination with atezolizumab in participants with locally advanced or metastatic tumors	Recruiting	• Melanoma • Nonsmall-cell lung cancer • Bladder cancer • Colorectal cancer • Triple negative breast cancer • Renal cancer • Head and neck cancer • Other solid cancers	• Drug: autogene cevumeran • Drug: atezolizumab	United States
	A study to evaluate the efficacy and safety of RO7198457 in combination with pembrolizumab versus pembrolizumab alone in participants with previously untreated advanced melanoma	Recruiting	Advanced melanoma	• Biological: RO7198457 • Drug: pembrolizumab	United States

(*Continued*)

TABLE 13.1 (Continued)

Trial	Status	Conditions	Interventions	Location
A phase II clinical trial comparing the efficacy of RO7198457 versus watchful waiting in patients with ctDNA-positive, resected stage II (high risk) and stage III colorectal cancer	Recruiting	• Colorectal cancer stage II • Colorectal cancer stage III	Drug: RO7198457	United States
Safety, tolerability, and immunogenicity of mRNA-4157 alone in participants with resected solid tumors and in combination with pembrolizumab in participants with unresectable solid tumors	Recruiting	Solid tumors	• Biological: mRNA-4157 • Biological: pembrolizumab	United States
An efficacy study of adjuvant treatment with the personalized cancer vaccine mRNA-4157 and pembrolizumab in patients with high-risk melanoma (KEYNOTE-942)	Recruiting	Melanoma	• Biological: mRNA-4157 • Biological: pembrolizumab	United States
mRNA-based, personalized cancer vaccine against neoantigens expressed by the autologous cancer	Terminated	• Melanoma • Colon cancer • Gastrointestinal cancer • Genitourinary cancer • Hepatocellular cancer	Biological: National Cancer Institute-4650	United States

EGT, Electro-Gene-Transfer; *GM-CSF*, granulocyte-macrophage colony-stimulating factor; *HPV*, human papilloma virus; *hTERT*, human TERT; *IL*, interleukin; *mRNA*, messenger RNA.

optimal orientation relative to the direction of the electric field, promoting maximum delivery across the entire length of the fibers. Since skeletal muscle cells do not divide, gene expression following electrotransfer is stable for a long period. Most importantly, skeletal muscle produces biologically active proteins and releases them into the bloodstream and consequently muscle can be used as protein delivery system for distant targets.[65] The skin is the second most broadly used tissue for EGT.[66] It is accessible for treatment over large areas, and some epidermal cells (keratinocytes) can also produce and release proteins into the bloodstream. The first human clinical trial using EGT was for the treatment of skin cancer.[22] An interesting phase I trial, which has collected promising results in terms of safety and immunogenicity, concerned the treatment of HPV16/18 delivering VGX-3100 vaccine using electroporation[67]. In this study, a significant increase in antibody titer and of T cells reactive to the oncoproteins E6 and E7 of both HPV16 and HPV18 was detected, persisting over several months. The results confirmed that DNA-EGT is safe and able to significantly elicit the immune response.

Another clinical study was designed to cotarget two well-known TAAs, namely HER2 and CEA (carcino-embryonic antigen), through the coinjection of two plasmids carrying expression cassettes for codon usage optimized versions of the respective cDNAs. The vaccine had been previously tested in mice double transgenic for both human HER2 and human CEA, where it was able to break tolerance and to induce strong antibody and CMI. The vaccine, called V930, was administered in a dose-escalation protocol to a total of 28 patients, in 5 biweekly injections by muscle electroporation. The clinical trial was only designed to assess safety, tolerability, and immunogenicity. The vaccine was, indeed, safe and well tolerated with no major adverse events. However, immunogenicity data were not encouraging: none of the 28 patients vaccinated with V930 developed a detectable immune response against CEA and HER2, probably because the codelivery of two plasmids bearing expression cassettes for the two antigens may have interfered with each other. Indeed, preclinical studies had already shown a lower level of immune responses in mice coinjected with the two plasmids versus mice injected with a single plasmid; thus it is reasonable that scale up from mice to human muscles might have magnified this phenomenon.[68]

A therapeutic cancer vaccine brought into phase I/II was V934/V935, directed against the protein component of human TERT and based on a heterologous prime–boost administration of Ad vector and plasmid DNA by EGT, in 37 patients with selected solid tumors. Results of this trial demonstrated the safety and feasibility of this vaccination strategy.[69] Another phase I study, based on a vaccine targeting human TERT and administered by intradermal route followed by electroporation or by a needle-free injection system (delivering vaccine intradermally), proved to be safe and immunogenic in patients with relapsed or refractory solid tumors.[70] In conclusion, albeit requiring complete validation in phase II randomized efficacy studies,

EGT-based DNA vaccines demonstrate to be promising vaccination platforms in cancer immunotherapy treatments.

A newly developed intradermal DNA delivery is the tattoo technology. The tattoo device has a cartridge of nine fine metal perforating needles that oscillate at a constant high frequency and puncture the skin, leading to DNA transfer to skin-associated cells. The expression of reporter genes results in robust T-cell responses.[71] Recently tattoo-immunization was applied in a phase I study to assess the toxicity and efficacy of inducing tumor-specific T-cell immunity against melanoma.[72]

On the basis of encouraging preclinical data, clinical trials have recently been initiated to evaluate the immunogenicity and safety of nucleoside-modified RNA-LNP vaccines in humans for the first time (see Table 13.1). One example is a HPV16 E6 and E7 mRNA-encoding RNA-LNP vaccine, which is being evaluated in a randomized phase II trial in combination with pembrolizumab in patients with $HPV16^{+}$-$L1^{+}$ and PD-1 head and neck squamous cell carcinoma. In a phase I clinical trial, 89 patients with melanoma were vaccinated with an RNA-LNP vaccine encoding melanoma TAAs MAGE-A3, transmembrane phosphatase with tensin homology (TPTE), NY-ESO-1, and tyrosinase.[73] Escalating doses from 7.2 to 100 μg resulted in detectable immune response at all doses with the best response rate reported at the highest dose in conjunction with anti-PD1 and a partial response in 5 out of 10 patients (50% response rate). The RNA-LNP vaccine has been used in truly personalized treatment targeting neoantigens in various tumor types and treatment lines in phase I and II clinical studies. Intramuscularly administered, LNP-formulated nucleoside-modified mRNA vaccines encoding 20 patient-specific mutated neoepitopes are currently being evaluated in combination with pembrolizumab in patients with solid cancer (phase I) and melanoma (randomized phase II). Data from a recent phase I clinical trial in four patients with gastrointestinal cancer suggest that vaccination with neoantigen-specific RNA-LNP is safe and induces mutation-specific T-cell responses[74] with no objective clinical responses observed yet in the limited number of patients tested. In another trial, patients with stage III or IV melanoma were treated with ipilimumab, a monoclonal antibody against CTL antigen 4, and DCs loaded with mRNA encoding melanoma-associated antigens plus TriMix. This intervention resulted in durable tumor reduction in a proportion of individuals with recurrent or refractory melanoma.[75]

Conclusions

To date, cancer vaccines based upon nucleic acids have reached phase I/II human clinical trials showing safety, tolerability, and immunogenicity and have proven to be a reliable anticancer strategy, in terms of safety and effectiveness, in combination with other immunotherapeutic strategies. However, the appropriate design of cancer vaccine clinical trials, defining clear therapeutic endpoints, is crucial to demonstrate their clinical efficacy. A major hurdle is to identify highly immunogenic tumor-specific antigens. Apart from the shared tumor-associated antigens and oncogenic viral antigens, tumor antigens are highly variable across different individuals, and some, less immunogenic, may escape spontaneous recognition by the host immune system. However, even if vaccines can successfully induce a tumor-reactive T-cell response, a suppressive microenvironment could prevent effective T-cell infiltration and cause T-cell exhaustion. Although identifying immunogenic TAAs and overcoming suppressive tumor microenvironment still remain major hurdles for genetic vaccines, the recent discovery and identification of neoantigens paves the way for personalized vaccine treatment applications, and in the next future DNA- and RNA-encoded neoantigens will become the frontrunner in the personalized vaccine campaign.

References

1. Wolff JA, Malone RW, Williams P, et al. Direct gene transfer into mouse muscle in vivo. *Science*. 1990;247:1465–1468.
2. Buonaguro L, Petrizzo A, Tornesello ML, Buonaguro FM. Translating tumor antigens into cancer vaccines. *Clin Vaccine Immunol*. 2010;18(1):23–34.
3. Wolff JA, Budker V. The mechanism of naked DNA uptake and expression. *Adv Genet*. 2005;54:3–20.
4. Ulmer JB, Otten GR. Priming of CTL responses by DNA vaccines: direct transfection of antigen presenting cells vs cross-priming. *Dev Biol*. 2000;104:9–14.
5. Ledwith BJ, Manam S, Troilo PJ, et al. Plasmid DNA vaccines: investigation of integration into host cellular DNA following intramuscular injection in mice. *Intervirology*. 2000;43(4–6):258–272.
6. Ho W, Gao M, Li F, et al. Next-generation vaccines: nanoparticle-mediated DNA and mRNA delivery. *Adv Healthc Mater*. 2021;10(8):e2001812.
7. Sahin U, Muik A, Vogler I, et al. BNT162b2 vaccine induces neutralizing antibodies and poly-specific T cells in humans. *Nature*. 2021;. Available from: https://doi.org/10.1038/s41586-021-03653-6. Epub ahead of print. PMID: 34044428.
8. Jihui L, Shreedevi AK, Yong YJ, Corey JB. Engineering DNA vaccines against infectious diseases. *Acta Biomater*. 2018;80:31–47.
9. Liu MA. A comparison of plasmid DNA and mRNA as vaccine technologies. *Vaccines (Basel)*. 2019;7:37.
10. Conforti A, Marra E, Roscilli G, et al. Are genetic vaccines the right weapon against COVID-19? *Mol Ther*. 2020;28(7):1555–1556.
11. Beck JD, Reidenbach D, Salomon N, et al. mRNA therapeutics in cancer immunotherapy. *Mol Cancer*. 2021;20(1):69.
12. Allen A, Wang C, Caproni LJ, et al. Linear doggybone DNA vaccine induces similar immunological responses to conventional plasmid DNA independently of immune recognition by TLR9 in a pre-clinical model. *Cancer Immunol Immunother*. 2018;67:627–638.
13. Andries O, Cafferty MC, De Smedt S, et al. N1-methylpseudouridine-incorporated mRNA outperforms pseudouridine-incorporated mRNA by providing enhanced protein expression and reduced immunogenicity in mammalian cell lines and mice. *J Control Release*. 2015;217:337–344.

14. Yang Y. Cancer immunotherapy: harnessing the immune system to battle cancer. *J Clin Invest*. 2015;125(9):3335–3337.
15. Lopes A, Vandermeulen G, Préat V. Cancer DNA vaccines: current preclinical and clinical developments and future perspectives. *J Exp Clin Cancer Res*. 2019;38(1):146.
16. Aurisicchio L, Ciliberto G. Genetic cancer vaccines: current status and perspectives. *Expert Opin Biol Ther*. 2012;12(8):1043–1058.
17. Peruzzi D, Mesiti G, Ciliberto G, et al. Telomerase and HER-2/neu as targets of genetic cancer vaccines in dogs. *Vaccine*. 2010;28:1201–1208.
18. Trimble CL, Peng S, Kos F, et al. A phase I trial of a human papillomavirus DNA vaccine for HPV16 + cervical intraepithelial neoplasia 2/3. *Clin Cancer Res*. 2009;15 (1):361–367.
19. Kotnik T, Frey W, Sack M, et al. Electroporation-based applications in biotechnology. *Trends Biotechnol*. 2015;33(8):480–488.
20. Lambricht L, Lopes A, Kos S, et al. Clinical potential of electroporation for gene therapy and DNA vaccine delivery. *Expert Opin Drug Deliv*. 2016;13(2):295–310.
21. Gehl J. Gene electrotransfer in clinical trials. *Methods Mol Biol*. 2014;1121:241–246.
22. Heller R, Heller LC. Gene electrotransfer clinical trials. *Adv Genet*. 2015;89:235–262.
23. Babiuk S, Baca-Estrada ME, Foldvari M, et al. Increased gene expression and inflammatory cell infiltration caused by electroporation are both important for improving the efficacy of DNA vaccines. *J Biotechnol*. 2004;110(1):1–10.
24. Liu J, Kjeken R, Mathiesen I, et al. Recruitment of antigen-presenting cells to the site of inoculation and augmentation of human immunodeficiency virus type 1 DNA vaccine immunogenicity by in vivo electroporation. *J Virol*. 2008;82:5643–5649.
25. Muramatsu T, Shibata O, Ryoki S, et al. Foreign gene expression in the mouse testis by localized in vivo gene transfer. *Biochem Biophys Res Commun*. 1997;233(1):45–49.
26. Blair-Parks K, Weston BC, Dean DA. High-level gene transfer to the cornea using electroporation. *J Gene Med*. 2002;4(1):92–100.
27. Tsujie M, Isaka Y, Nakamura H, et al. Electroporation-mediated gene transfer that targets glomeruli. *J Am Soc Nephrol*. 2001;12(5):949–954.
28. Sheyn D, Kimelman-Bleich N, Pelled G, et al. Ultrasound-based nonviral gene delivery induces bone formation in vivo. *Gene Ther*. 2008;15(4):257–266.
29. Ohashi S, Kubo T, Kishida T, et al. Successful genetic transduction in vivo into synovium by means of electroporation. *Biochem Biophys Res Commun*. 2002;293(5):1530–1535.
30. Hoover F, Magne Kalhovde J. A double-injection DNA electroporation protocol to enhance in vivo gene delivery in skeletal muscle. *Anal Biochem*. 2000;285(1):175–178.
31. Trezise AE. In vivo DNA electrotransfer. *DNA Cell Biol*. 2002;21(12):869–877.
32. Capone S, Zampaglione I, Vitelli A, et al. Modulation of the immune response induced by gene electrotransfer of a hepatitis C virus DNA vaccine in nonhuman primates. *J Immunol*. 2006;177:7462–7471.
33. Rizzuto G, Cappelletti M, Mennuni C, et al. Electro gene-transfer results in a high level transduction of rat skeletal muscle and corrects anaemia of renal failure. *Hum Gene Ther*. 2000;11:1891–1900.
34. Kandimalla ER, Bhagat L, Zhu FG, et al. A dinucleotide motif in oligonucleotides shows potent immunomodulatory activity and overrides species-specific recognition observed with CpG motif. *Proc Natl Acad Sci USA*. 2003;100:14303–14308.
35. Mennuni C, Ugel S, Mori F, et al. Preventive vaccination with telomerase ontrols tumor growth in genetically engineered and carcinogen-induced mouse models of cancer. *Cancer Res*. 2008;68:9865–9874.

36. Conforti A, Cipriani B, Peruzzi D, et al. A TLR9 agonist enhances therapeutic effects of telomerase genetic vaccine. *Vaccine*. 2010;28(20):3522–3530.
37. Dharmapuri S, Peruzzi D, Mennuni C, et al. Coadministration of telomerase genetic vaccine and a novel TLR9 agonist in nonhuman primates. *Mol Ther*. 2009;17(10):1804–1813.
38. Yarchoan M, Johnson BA, Lutz ER, et al. Targeting neoantigens to augment antitumour immunity. *Nat Rev Cancer*. 2017;17(9):569.
39. Aurisicchio L, Pallocca M, Ciliberto G, Palombo F. Palombo F. The perfect personalized cancer therapy: cancer vaccines against neoantigens. *J Exp Clin Cancer Res*. 2018;37(1):86.
40. Aurisicchio L, Peruzzi D, Koo G, et al. Immunogenicity and therapeutic efficacy of a dual-component genetic cancer vaccine cotargeting carcinoembryonic antigen and HER2/neu in preclinical models. *Hum Gene Ther*. 2014;25(2):121–131.
41. Duperret EK, Perales-Puchalt A, Stoltz R, et al. A synthetic DNA, multi-neoantigen vaccine drives predominately MHC class I CD8 + T-cell responses, impacting tumor challenge. *Cancer Immunol Res*. 2019;7(2):174–182.
42. Li L, Zhang X, Wang X, et al. Optimized polyepitope neoantigen DNA vaccines elicit 8neoantigen-specific immune responses in preclinical models and in clinical translation. *Genome Med*. 2021;13(1):56.
43. Lione L, Salvatori E, Petrazzuolo A, et al. Antitumor efficacy of a neoantigen cancer vaccine delivered by electroporation is influenced by microbiota composition. *Oncoimmunology*. 2021;10(1). Available from: https://doi.org/10.1080/2162402X.2021.1898832 In this issue.
44. Pastor F, Berraondo P, Etxeberria I, et al. An RNA toolbox for cancer immunotherapy. *Nat Rev Drug Discov*. 2018;17(10):751–767.
45. Kroczynska B, Mehrotra S, Arslan AD, et al. Regulation of interferon-dependent mRNA translation of target genes. *J Interf Cytokine Res*. 2014;34(4):289–296.
46. Munir M, Berg M. The multiple faces of proteinkinase R in antiviral defense. *Virulence*. 2013;4(1):85–89.
47. Baiersdörfer M, Boros G, Muramatsu H, et al. A facile method for the removal of dsRNA contaminant from in vitro-transcribed mRNA. *Mol Ther Nucleic Acids*. 2019;15:26–35.
48. Karikó K, Muramatsu H, Ludwig J, Weissman D. Generating the optimal mRNA for therapy: HPLC purification eliminates immune activation and improves translation of nucleoside-modified, protein-encoding mRNA. *Nucleic Acids Res*. 2011;39(21). e142.
49. Benteyn D, Heirman C, Bonehill A, et al. RNA-based dendritic cell vaccines. *Expert Rev Vaccines*. 2015;14:161–176.
50. Ahmed R, Sayegh N, Graciotti M, Kandalaft LE. Electroporation as a method of choice to generate genetically modified dendritic cell cancer vaccines. *Curr Opin Biotechnol*. 2020;65:142–155.
51. Boczkowski D, Nair SK, Snyder D, Gilboa E. Dendritic cells pulsed with RNA are potent antigen-presenting cells in vitro and in vivo. *J Exp Med*. 1996;184(2):465–472.
52. Minkis K, Kavanagh DG, Alter G, et al. Type 2 Bias of T cells expanded from the blood of melanoma patients switched to type 1 by IL-12p70 mRNA-transfected dendritic cells. *Cancer Res*. 2008;68(22):9441–9450.
53. De Keersmaecker B, Heirman C, Corthals J, et al. The combination of 4–1BBL and CD40L strongly enhances the capacity of dendritic cells to stimulate HIV-specific T cell responses. *J Leukoc Biol*. 2011;89:989–999.
54. Aerts-Toegaert C, Heirman C, Tuyaerts S, et al. CD83 expression on dendritic cells and T cells: correlation with effective immune responses. *Eur J Immunol*. 2007;37:686–695.

55. Bialkowski L, van Weijnen A, Van der Jeught K, et al. Intralymphatic mRNA vaccine induces CD8 T-cell responses that inhibit the growth of mucosally located tumours. *Sci Rep*. 2016;6:22509.
56. Kreiter S, Selmi A, Diken M, et al. Intranodal vaccination with naked antigen-encoding RNA elicits potent heller LC, Heller R. Electroporation gene therapy preclinical and clinical trials for melanoma. *Curr Gene Ther*. 2010;10(4):312–317.
57. Rittig SM, Haentschel M, Weimer KJ, et al. Intradermal vaccinations with RNA coding for TAA generate CD8 + and CD4 + immune responses and induce clinical benefit in vaccinated patients. *Mol Ther*. 2011;19(5):990–999.
58. Sahin U, Derhovanessian E, Miller M, et al. Personalized RNA mutanome vaccines mobilize poly-specific therapeutic immunity against cancer. *Nature*. 2017;547 (7662):222–226.
59. Guan S, Rosenecker J. Nanotechnologies in delivery of mRNA therapeutics using nonviral vector-based delivery systems. *Gene Therapy*. 2017;24(3):133–143.
60. Goswami R, Chatzikleanthous D, Lou G. Mannosylation of LNP results in improved potency for self-amplifying RNA (SAM) vaccines. *ACS Infect Dis*. 2019;5:1546–1558.
61. Le Moignic A, Malard V, Benvegnu T, et al. Preclinical evaluation of mRNA trimannosylated lipopolyplexes as therapeutic cancer vaccines targeting dendritic cells. *J Control Release*. 2018;278:110–121.
62. Fan YN, Li M, Luo YL, et al. Cationic lipid-assisted nanoparticles for delivery of mRNA cancer vaccine. *Biomater Sci*. 2018;6:3009–3018.
63. Daud AI, DeConti RC, Andrews S, et al. Phase I trial of interleukin-12 plasmid electroporation in patients with metastatic melanoma. *J Clin Oncol*. 2008;26(36):5896–5903.
64. Prud'homme GJ, Glinka Y, Khan AS, et al. Electroporation enhanced non viral gene transfer for the prevention or treatment of immunological, endocrine and neoplastic diseases. *Curr Gene Ther*. 2006;6(2):243–273.
65. Sokołowska E, Błachnio-Zabielska AU. A critical review of electroporation as a plasmid delivery system in mouse skeletal muscle. *Int J Mol Sci*. 2019;20(11):2776.
66. Gothelf A, Gehl J. Gene electrotransfer to skin; review of existing literature and clinical perspectives. *Curr Gene Ther*. 2010;10(4):287–299.
67. Bagarazzi ML, Yan J, Morrow MP, et al. Immunotherapy against HPV16/18 generates potent TH1 and cytotoxic cellular immune responses. *Sci Transl Med*. 2012;4(155). 155ra38.
68. Aurisicchio L, Mancini R, Ciliberto G. Cancer vaccination by electro-gene-transfer. *Expert Rev Vaccines*. 2013;12(10):1127–1137.
69. Aurisicchio L, Fridman A, Mauro D, et al. Safety, tolerability and immunogenicity of V934/V935 hTERT vaccination in cancer patients with selected solid tumors: a phase I study. *J Transl Med*. 2020;18(1):39.
70. Teixeira L, Medioni J, Garibal J, et al. A first-in-human phase I study of INVAC-1, an optimized human telomerase DNA vaccine in patients with advanced solid tumors. *Clin Cancer Res*. 2020;26(3):588–597.
71. Bins AD, Jorritsma A, Wolkers MC, et al. A rapid and potent DNA vaccination strategy defined by in vivo monitoring of antigen expression. *Nat Med*. 2005;8:899–904.
72. Quaak SG, van den Berg JH, Toebes M, et al. GMP production of pDERMATT for vaccination against melanoma in a phase I clinical trial. *Eur J Pharm Biopharm*. 2008;70 (2):429–438.
73. Sahin U, Oehm P, Derhovanessian E, et al. An RNA vaccine drives immunity in checkpoint-inhibitor-treated melanoma. *Nature*. 2020;585(7823):107–112.

74. Cafri G, Gartner JJ, Zaks T, et al. mRNA vaccine-induced neoantigen-specific T cell immunity in patients with gastrointestinal cancer. *J Clin Invest*. 2020;130(11):5976–5988.
75. Wilgenhof S, Corthals J, Heirman C, et al. Phase II study of autologous monocyte-derived mRNA electroporated dendritic cells (TriMixDC-MEL) plus ipilimumab in patients with pretreated advanced melanoma. *J Clin Oncol*. 2016;34(12):1330–1338.

Index

Note: Page numbers followed by "*f*" and "*t*" refer to figures and tables, respectively.

D

L

M

N

Q

R

S

T

V

W

FSC
www.fsc.org
MIX
Papier | Fördert
gute Waldnutzung
FSC® C083411

Kolibri 360 GmbH
Ferdinand-Jühlke-Straße 7
99095 Erfurt, Deutschland
produktsicherheit@kolibri360.de